Strive for a 5: Preparing for the AP® Statistics Examination

to accompany

The Practice of Statistics, Sixth Edition

Daren Starnes

Josh Tabor

Dan Yates

David Moore

Jason M. Molesky

Lakeville Area Public Schools

Michael Legacy

Greenhill School

ISBN-13: 978-1-3192-0990-2
ISBN-10: 1-3192-0990-4

Printed in the United States of America.

W. H. Freeman and Company
Bedford, Freeman & Worth
One New York Plaza
Suite 4500
New York, NY 10004-1562
highschool.bfwpub.com/catalog

TABLE OF CONTENTS

Preface

Strive for a 5: Preparing for the AP® Statistics Examination is designed for use with *The Practice of Statistics, Sixth Edition* by Daren Starnes and Josh Tabor. It is intended to help you evaluate your understanding of the material covered in the textbook, reinforce the key concepts you need to learn, develop your conceptual understanding and communication skills, and prepare you to take the AP® Statistics Exam. This book is divided into two sections: a study guide and a test preparation section.

THE STUDY GUIDE SECTION

The study guide is designed for you to use throughout your course in AP® Statistics. As each chapter is covered in your class, use the study guide to help you identify and learn the important statistics concepts and terms. For each section, this guide provides learning targets, checks for understanding, practice problems, and hints to help you master the material and verify your understanding before moving on to the next section.

This guide is designed to help you clearly identify what it is you need to learn, whether or not you have learned it, and what you need to do to close any gaps in your understanding. That is, it will help you answer the questions:

"Where am I going?".
"Where am I now?"
"How can I close the gap?"

For each chapter, the study guide is organized as follows.

- **Overview**: The chapter content is summarized to provide a quick overview of the material covered. The material is broken down into the specific learning targets to help you answer the question "Where am I going?"
- **Chapter Sections:** Each section includes a presentation of the important content to help guide your learning. The section content includes the following.
  - **Learning Targets**: These "I can…" statements provide a guide for you to keep track of your learning. Check these off as you become confident with each of the concepts.
  - **Vocabulary**: Space for you to record and study definitions of important terms.
  - **Concept Explanations**: Important concepts are explained further to aid you in your study of the key points of each section.
  - **Checks for Understanding**: Mini assessments for each learning target help you answer the question "Where am I now?"
- **Chapter Summary**: The key concepts of the chapter are briefly reviewed and a summary of all of the learning targets is provided to help you check your learning.
- **Multiple Choice Questions**: These multiple-choice questions focus on the key concepts that you should grasp after reading the chapter and are designed for quick exam preparation. After checking your answers, you should reflect on the big ideas from the chapter and identify the concepts you may need to study more.
- **FRAPPY!:** A "Free-Response AP Problem, Yay!" is provided after each chapter. Each FRAPPY is modeled after actual AP Statistics Exam questions and provide an opportunity for you to practice your communication skills to maximize your score. These problems are meant to help you determine what further work is needed to address the question "How can I close the gap?"
- **Vocabulary Crossword Puzzle**: These puzzles provide a fun way for you to check your understanding of key definitions from the chapter.

Answers to all questions and problems in the study guide are available in the back of the book. Checking your answers will help you determine whether or not you need additional work on specific learning targets.

THE TEST PREPARATION SECTION

The "Preparing for the AP Statistics Examination" section of this guide is meant to help you better understand how the exam is constructed and scored, how best to study for the exam, and how to make sure you highlight what you have learned when answering exam questions. Two full-length exams and answer modeled on the actual Advanced Placement Exam (40 multiple-choice questions, 5 free response questions, and an investigative task) are included to help you get a feel for the actual test. Try to simulate the actual exam conditions as much as possible when taking these practice tests!

I hope that your use of "Strive for a 5: Preparing for the AP® Statistics Examination" will assist you in your study of statistical concepts, help you earn as high a score as possible on the exam, and provide you with an interest and desire to further your study of statistics. I'd like to thank Daren Starnes, Josh Tabor, and Ann Heath for inviting me to update this guide for the latest edition of The Practice of Statistics. I'd also like to thank Michael Legacy for contributing his expertise, once again, in writing two excellent practice exams. Special thanks to my AP® Stats friends for their continued email, phone, and text discussions on how to better teach and learn this subject we love!

Most importantly, I'd like to thank my wife, Anne, for her patience and support through all of my stats projects, my kiddos, Addison and Aidan, for providing much needed play breaks, and my trusty boxer, Surly, for his company as I typed away in my office. Best wishes for a great year of studies and best of luck on your AP Statistics Exam!

Jason Molesky
"StatsMonkey"

About the Authors

Jason M. Molesky
Northfield, MN

Jason taught AP Statistics since 2002 and served as an AP Statistics Reader from 2007-2016, including 4 years as a Table Leader and member of a Rubric Team. Jason has enjoyed working with the Practice of Statistics Team on a variety of roles, from reviewing, to media development, to the Strive Guide. He is currently enjoying a new role in education, working with schools to transform teaching and learning through the use of digital tools. He still maintains close contact with the AP® Statistics world and continues to host the "StatsMonkey" website, a clearinghouse for AP® Statistics resources. In his spare time, Jason enjoys spending time with his wife, their two children, and his trusty boxer.

Michael Legacy
Greenhill School, Dallas, TX

Michael is a past member of the AP® Statistics Development Committee (2001-2005) and a former Table Leader at the Reading. He currently reads the Alternate Exam and is a lead teacher at many AP® Summer Institutes. Michael is the author of the 2007 College Board AP® Statistics Teacher's Guide and was named the Texas 2009-2010 AP® Math/Science Teacher of the Year by the Siemens Corporation.

CHAPTER 1: DATA ANALYSIS

"Statistical thinking will one day be as necessary for efficient citizenship as the ability to read and write." H.G. Wells

Chapter Overview

Statistics is the science of data. We begin our study of the subject by mastering the art of examining and describing sets of data. This chapter introduces you to the concept of exploratory data analysis. You will learn how to use a variety of graphical tools to display data as well as how to describe data numerically.

By the end of this chapter, you should understand the difference between categorical and quantitative data, how to display and describe these two types of data, and how to move from data analysis to inference. These skills are the basis for our entire study of Statistics. Use this guide to help ensure you have a firm grasp on these concepts!

Sections in this Chapter

Plan Your Learning

Use the following *suggested* guide to help plan your reading and assignments in "The Practice of Statistics, 6th Edition." Note: your teacher may schedule a different pacing or assign different problems. Be sure to follow his or her instructions!

Read	Ch1 Introduction	1.1: pp 9-16	1.1: pp 17-23	1.2: pp 30-34
Do	1,5,7 MC 9-10	13,15,17,19,21,23	27,29,33,35 MC 40-43	45,49,51,59,63

Read	1.2: pp 34-47	1.3: pp 54-65	1.3: pp 65-74	Chapter Summary
Do	55,65,69,77 MC 80-85	87,89,91,95,97,101, 103,105,121	109,111,113,115 MC 123-126	Review Exercises FRAPPY!

Use this "Strive Guide" to help identify learning targets for each section, summarize key concepts, practice additional problems, and check your understanding as you read the text.

Introduction - Statistics: The Science and Art of Data

Before You Read: Section Summary

In this section, you will learn the basic terms and definitions necessary for our study of statistics. You will also be introduced to the idea of moving from data analysis to inference. One of our goals in statistics is to use data from a representative sample to make an inference about the population from which it was drawn. In order to do this, we must be able to identify the type of data we are dealing with, as our choice of statistical procedures will depend upon that distinction.

Learning Targets:

____ I can identify individuals and variables in a set of data
____ I can classify variables as categorical or quantitative

While You Read: Key Vocabulary and Concepts

statistics:
individuals:
variable:
categorical variable:
quantitative variable:
distribution:
inference:

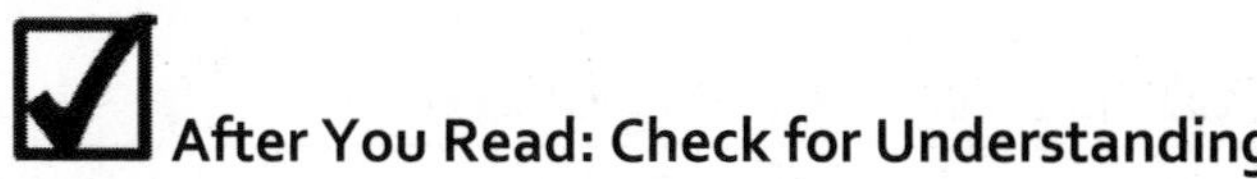

After You Read: Check for Understanding

Concept 1: Individuals and Variables

Sets of data contain information about individuals. The characteristics of the individuals that are measured are called variables. Variables can take on different values for different individuals. It is these values and the variation in them that we will be learning how to study in this course.

Concept 2: Types of Variables

Variables can fall into one of two categories: categorical or quantitative. When the characteristic we measure places an individual into one of several groups, we call it a categorical variable. When the characteristic we measure results in numerical values for which it makes sense to find an average, we call it a quantitative variable. This distinction is important, as the methods we use to describe and analyze data will depend upon the type of variable we are studying. In the rest of this chapter, we'll learn how to display and describe the distribution of a variable.

Check for Understanding: ____ *I can identify individuals and variables in a set of data.*

Mr. Buckley gathered some information on his class and organized it in a table similar to the one below:

Student	Gender	ACT Score	Favorite Subject	Grade Point Average
James	M	34	Statistics	3.89
Jen	F	35	Biology	3.75
DeAnna	F	32	History	4.00
Jonathan	M	28	Literature	3.00
Doug	M	33	Algebra	2.89
Sharon	F	30	Spanish	3.25

1) Who are the individuals in this dataset?

2) What variables were measured? Identify each as quantitative or categorical.

3) Describe the distribution of ACT scores.

4) Could we infer from this set of data that students who prefer math and science perform better on the ACT? Explain.

Section 1.1: Analyzing Categorical Data

Before You Read: Section Summary

In this section, you will learn how to describe and analyze categorical variables. You will learn how to display data with pie charts and bar graphs as well as how to describe these displays. You will also learn how to analyze the relationship between two categorical variables using marginal and conditional distributions.

Learning Targets:

____ I can make and interpret bar graphs for categorical data
____ I can identify what makes some graphs of categorical data misleading
____ I can calculate marginal and joint relative frequencies from a two-way table
____ I can calculate conditional relative frequencies from a two–way table
____ I can use bar graphs to compare distributions of categorical data
____ I can describe the nature of the association between two categorical variables

While You Read: Key Vocabulary and Concepts

frequency table:
relative frequency table:
bar graph:
pie chart:
two-way table:
marginal relative frequency:
joint relative frequency:
conditional relative frequency:
association:

After You Read: Check for Understanding

Concept 1: Displaying Categorical Data

A frequency table (or a relative frequency table) displays the counts (or percents) of individuals that take on each value of a variable. Tables are sometimes difficult to read and they don't always highlight important features of a distribution. Graphical displays of data are much easier to read and often reveal interesting patterns and departures from patterns in the distribution of data. We can use pie charts and bar graphs to display the distribution of categorical variables. Note, when constructing graphical displays, we must be careful not to distort the quantities. Beware of pictographs and watch the scales when displaying or reading graphs!

Check for Understanding: ____ *I can make and interpret bar graphs for categorical data.*

A local business owner was interested in knowing the coffee-shop preferences of her town's residents. According to her survey of 250 residents, 75 preferred "Goodbye Blue Monday", 50 liked "The Ugly Mug", 38 chose "Morning Joe's", 50 said "One Mean Bean", 25 brewed their own coffee, and 12 preferred the national chain.

1) Construct a bar graph to display the data. Describe what you see.

2) Construct a pie chart to display the data. How does this display differ from the bar chart?

Concept 2: Two-Way Tables

Bar graphs and pie charts are helpful when analyzing a single categorical variable. However, often we want to explore the relationship between two categorical variables. To do this, we organize our data in a two-way table with a row variable and a column variable. The counts of the individuals in each intersecting category make up the entries in the table.

When exploring a two-way table, you should start by describing each variable separately. This can be done by describing the marginal distribution of the row or column variable. To explore the relationship between the variables, study the conditional distributions by graphing each distribution of one variable for a fixed value of the other. If there is no association between the two variables, the graphs of the conditional distributions will be similar. If there is an association between the two variables, the graphs of the conditional distributions will be different.

Check for Understanding: ____ *I can describe the nature of the association between two categorical variables.*

A survey of 1000 randomly chosen residents of a Minnesota town asked "Where do you prefer to purchase your daily coffee?" The two-way table below shows the responses.

<u>**Coffee preference by gender**</u>

Preference	**Male**	**Female**	**Total**
National Chain	95	65	**160**
One Mean Bean	15	85	**100**
The Ugly Mug	145	25	**170**
Goodbye Blue Monday	170	90	**260**
Home-brewed	100	160	**260**
Don't drink coffee	10	40	**50**
Total	**535**	**465**	**1000**

Based on the data, can we conclude that there is an association between gender and coffee preferences? Use appropriate graphical and numerical evidence to support your conclusion.

Section 1.2: Displaying Quantitative Data with Graphs

Before You Read: Section Summary

In this section, you will learn how to display quantitative data. Like categorical data, we are interested in describing the distribution of quantitative data. Dotplots, stemplots, and histograms are helpful tools for revealing the shape, center, and variability of a distribution of quantitative data. These basic plots will come in very handy throughout the course, so be sure to master them in this section!

Learning Targets:

- ____ I can make and interpret dotplots, stemplots, and histograms of quantitative data
- ____ I can identify the shape of a distribution from a graph
- ____ I can describe the overall pattern (shape, center, and variability) of a distribution and identify any major departures from the pattern (outliers)
- ____ I can compare distributions of quantitative data using dotplots, stemplots, and histograms

While You Read: Key Vocabulary and Concepts

dotplot:
stemplot:
histogram:
shape:
center:
variability:
outliers:
symmetric:
skewed to the right:
skewed to the left:

After You Read: Check for Understanding

Concept 1: Statistical Opinions Can Vary!

The reason we construct graphs of quantitative data is so that we can get a better understanding of it. Constructing a plot helps us examine the data and identify its overall pattern. When examining the distribution of a quantitative variable, start by looking for the overall pattern and then describe any major departures from it. Note its shape. Is it symmetric? Is it skewed? How many peaks does it have? What is its center? How variable are the data? Are the values bunched up around the center, or are they spread out? Finally, are there any outliers? Do any values fall far away from the rest of the distribution? The answers to each of these questions are very important when describing a set of data. Be sure to address all of them as you explore data sets throughout this course. As you explore data, don't forget Statistical Opinions Can Vary (Shape, Outliers, Center, Variability)!

Concept 2: Dotplots and Stemplots

Dotplots and stemplots are the easiest graphs to construct, especially if you have a small set of data. These plots are helpful for describing distributions because they keep the data intact. That is, you can determine the individual data values directly from the plot. When comparing two sets of data, we can construct back-to-back stemplots or dotplots that share the same scale. Remember, when constructing plots, always label your axes and provide a key!

Check for Understanding: ____ *I can make and interpret dotplots and stemplots.*

A recent study by the Environmental Protection Agency (EPA) measured the gas mileage (miles per gallon) for 30 models of cars. The results are below:

EPA-Measured MPG for 30 Cars

36.3	32.7	40.5	36.2	38.5	36.3	41.0	37.0	37.1	39.9
41.0	37.3	36.5	37.9	39.0	36.8	31.8	37.2	40.3	36.9
36.7	33.6	34.2	35.1	39.7	39.3	35.8	34.5	39.5	36.9

1) Construct a dotplot to display this data and describe the distribution.

2) Construct a stemplot to display this data.

Concept 2: Histograms

When dealing with larger sets of data, dotplots and stemplots can be a bit cumbersome and time-consuming to construct. In these cases, it may be easier to construct a histogram. Instead of plotting each value in the data set, a histogram displays the frequency of values that fall within equal-width classes. Be sure not to confuse histograms with bar graphs. Even though they look similar, they describe different types of data!

Check for Understanding: ____ *I can make and interpret a histogram.*

The EPA expanded its study to include a total of 50 car models. The results are below:

EPA-Measured MPG for 50 Cars

36.3	32.7	40.5	36.2	38.5	36.3	41.0	37.0	37.1	39.9
41.0	37.3	36.5	37.9	39.0	36.8	31.8	37.2	40.3	36.9
36.7	33.6	34.2	35.1	39.7	39.3	35.8	34.5	39.5	36.9
36.9	41.2	37.6	36.0	35.5	32.5	37.3	40.7	36.7	32.9
42.1	37.5	40.0	35.6	38.8	38.4	39.0	36.7	34.8	38.1

(a) Construct a histogram for these data by hand. Describe the distribution.

(b) Use your calculator to construct a histogram. Do you get the same graph? If not, how can you make your calculator match the histogram you constructed in part (a)?

Section 1.3: Describing Quantitative Data with Numbers

Before You Read: Section Summary

In this section, you will learn how to use numerical summaries to describe the center and variability of a distribution of quantitative data. You will also learn how to identify outliers in a distribution. Being able to accurately describe a distribution and calculate and interpret numerical summaries forms the foundation for our statistical study. By the end of this section, you will want to make sure you are comfortable selecting appropriate numerical summaries, calculating them (by hand or by using technology), and interpreting them for a set of quantitative data.

Learning Targets:

____ I can calculate and interpret measures of center (mean and median) for a distribution of quantitative data

____ I can calculate and interpret measures of variability (range, *IQR*, and standard deviation) for a distribution of quantitative data

____ I can explain how outliers and skewness affect measures of center and variability

____ I can identify outliers using the 1.5 x *IQR* Rule

____ I can make and interpret boxplots of quantitative data

____ I can use boxplots and numerical summaries to compare distributions of quantitative data

While You Read: Key Vocabulary and Concepts

mean:
median:
range:
standard Deviation:
quartiles:
interquartile range (*IQR*):
five-number summary:
boxplot:

After You Read: Check for Understanding

Concept 1: Exploratory Data Analysis

Statistical tools and ideas can help you examine data in order to describe their main features. This examination is called an ***exploratory data analysis (EDA)***. To organize our exploration, we want to:

- **Examine each variable by itself**...then move on to study relationships among the variables.
- **Always always always always always *plot your data*....**always.
- **Begin with a graph or graphs**...construct and interpret an appropriate graph of the data.
- **Add numeric summaries**...for quantitative data, calculate and interpret appropriate measures of center and measures of spread.
- **Don't forget your SOCV!** (**S**hape, **O**utliers, **C**enter, **V**ariability)...for quantitative data, note these important features!

Concept 2: Measures of Center

The mean and median measure the center of a set of data in different ways. While the mean, or average, is the most common measure of center, it is not always the most appropriate. Extreme values can "pull" the mean towards them. The median, or middle value, is resistant to extreme values and is sometimes a better measure of center. In symmetric distributions, the mean and median will be approximately equal, while in a skewed distribution the mean tends to be closer to the "tail" than the median. Always consider the shape of the distribution when deciding which measure to use to describe your data!

Check for Understanding: ____ *I can calculate and interpret measures of center.*

Consider the following stemplot of the lengths of time (in seconds) it took students to complete a logic puzzle. Use it to answer the following questions. Note 2|2 = 22 seconds.

```
1 | 58
2 | 23
2 | 677778888999
3 | 2344
3 | 68
4 | 011223
4 | 6
5 | 00
```

1) Based only on the plot, how does the mean compare to the median? How do you know?

2) Calculate and interpret the mean. Show your work.

3) Calculate and interpret the median.

4) Which measure of center would be the more appropriate summary of the center of this distribution? Why?

Technology) Use your calculator to find the mean and median for this set of data and verify that you get the same results as (2) and (3).

Concept 3: Measures of Variability and Boxplots

Like measures of center, we have several different ways to measure variability for distributions of quantitative data. The easiest way to describe the variability of a distribution is to calculate the range (maximum – minimum). However, extreme values can cause this measure to be much greater than the variability of the majority of values. A measure of variability that is resistant to the effect of outliers is the interquartile range (*IQR*).

> To find the *IQR*, arrange the observations from smallest to largest and determine the median. Then find the median of the lower half of the data. This is the first quartile. The median of the upper half of the data is the third quartile. The distance between the first quartile and third quartile is the interquartile range).

Not only does the *IQR* provide a measure of variability, it also provides us with a way to identify outliers. According to the 1.5 x *IQR* rule, any value that falls more than 1.5 x *IQR* above the third quartile or below the first quartile is considered an outlier. The minimum, maximum, median, and quartiles make up the "five-number summary". This set of numbers describes the center and variability of a set of quantitative data and leads to a useful display – the boxplot (or box-n-whisker plot).

Another measure of variability that we will use to describe data is the standard deviation. The standard deviation measures roughly the average distance of the observations from their mean. The calculation can be quite time-consuming to do by hand, so we'll rely on technology to provide the standard deviation for us. However, be sure you understand how it is calculated!

When describing data numerically, always make sure to note a measure of center AND a measure of variability. If you choose the median as your measure of center, it is best to use the *IQR* to describe the variability. If you choose the mean to describe center, use the standard deviation to measure the variability.

Check for Understanding: ____ *I can calculate and interpret measures of variability.*

The length (in pages) of Mrs. Molesky's favorite books are noted below

242 346 314 330 340 322 284 342 368 170 344 318 318 374 332

(a) Use these data to construct a boxplot. Describe the center and variability using the five-number summary.

(b) Calculate and interpret the mean and standard deviation for these data.

Chapter Summary: Exploring Data

In this chapter, we learned that statistics is the art and science of data. When working with data, it is important to know whether the variables are categorical or quantitative, as this will determine the most appropriate display for the distribution. For categorical data, the display will help us describe the distribution. For quantitative data, the display will help us describe the shape of the distribution and suggest the most appropriate numeric measures of center and variability. Always begin with a graph of the distribution, then move to a numerical description. When exploring quantitative data, we want to be sure to interpret the shape, outliers, center, and variability. Look for an overall pattern to describe your data and note any striking departures from that pattern.

As you study, be sure to focus on *understanding*, not just mechanics. While it may be easy to "plug the data" into your calculator, simply making graphs and calculating values is not the point of statistics. Rather, focus on being able to explain HOW a graph or value is constructed and WHY you would choose a certain display or numeric summary. Get in this habit early...your calculator is a powerful tool, but it cannot replace your thinking and communication skills!

After You Read: What Have I Learned?
Complete the vocabulary puzzle, multiple-choice questions, and FRAPPY. Check your answers and performance on each of the learning targets. Be sure to get extra help on any targets that you identify as needing more work!

Learning Target	Got It!	Almost There	Needs Work
I can identify individuals and variables for a set of data			
I can classify variables as categorical or quantitative			
I can make and interpret bar graphs for categorical data			
I can identify what makes some graphs of categorical data misleading			
I can calculate marginal and joint relative frequencies from a two-way table			
I can calculate conditional relative frequencies from a two–way table			
I can use bar graphs to compare distributions of categorical data			
I can describe the association between two categorical variables			
I can make and interpret dotplots, stemplots, and histograms of quantitative data			
I can identify the shape of a distribution from a graph			
I can describe the overall pattern (shape, center, and variability) of a distribution and identify any major departures from the pattern (outliers)			
I can compare distributions of quantitative data using dotplots, stemplots, and histograms			
I can calculate and interpret measures of center (mean and median) for a distribution of quantitative data			
I can calculate and interpret measures of variability (range, *IQR*, and standard deviation) for a distribution of quantitative data			
I can explain how outliers and skewness affect measures of center and variability			
I can identify outliers using the 1.5 x *IQR* Rule			
I can make and interpret boxplots of quantitative data			
I can use boxplots and numerical summaries to compare distributions of quantitative data			

Chapter 1 Multiple Choice Practice

Directions. *Identify the choice that best completes the statement or answers the question. Check your answers and note your performance when you are finished.*

1. You measure the age (years), weight (pounds), and breed (beagle, golden retriever, pug, or terrier) of 200 dogs. How many variables did you measure?

(A) 1
(B) 2
(C) 3
(D) 200
(E) 203

2. You open a package of Lucky Charms cereal and count how many there are of each marshmallow shape. The distribution of the variable "marshmallow" is:

(A) The shape: star, heart, moon, clover, diamond, horseshoe, balloon.
(B) The total number of marshmallows in the package.
(C) Seven—the number of different shapes that are in the package.
(D) The seven different shapes and how many there are of each.
(E) Since "shape" is a categorical variable, it doesn't have a distribution.

3. A review of voter registration records in a small town yielded the following table of the number of males and females registered as Democrat, Republican, or some other affiliation.

	Male	Female
Democrat	300	600
Republican	500	300
Other	200	100

The proportion of males that are registered as Democrats is

(A) 300
(B) 30
(C) 0.33
(D) 0.30
(E) 0.15

4. For a physics course containing 10 students, the maximum point total for the quarter was 200. The point totals for the 10 students are given in the stemplot below.

```
11 | 6 8
12 | 1 4 8
13 | 3 7
14 | 2 6
15 |
16 |
17 | 9
```

Which of the following statements is NOT true?

(A) In a symmetric distribution, the mean and the median are equal.
(B) About fifty percent of the scores in a distribution are between the first and third quartiles.
(C) In a symmetric distribution, the median is halfway between the first and third quartiles.
(D) The median is always greater than the mean.
(E) The range is the difference between the largest and the smallest observation in the data set.

5. When drawing a histogram it is important to

(A) have a separate class interval for each observation to get the most informative plot.
(B) make sure the heights of the bars exceed the widths of the class intervals so that the bars are true rectangles.
(C) label the vertical axis so the reader can determine the counts or percent in each class interval.
(D) leave large gaps between bars. This allows room for comments.
(E) scale the vertical axis according to the variable whose distribution you are displaying.

6. A set of data has a mean that is much larger than the median. Which of the following statements is most consistent with this information?

(A) The distribution is symmetric.
(B) The distribution is skewed left.
(C) The distribution is skewed right.
(D) The distribution is bimodal.
(E) The data set probably has a few low outliers.

7. The following is a boxplot of the birth weights (in ounces) of a sample of 160 infants born in a local hospital.

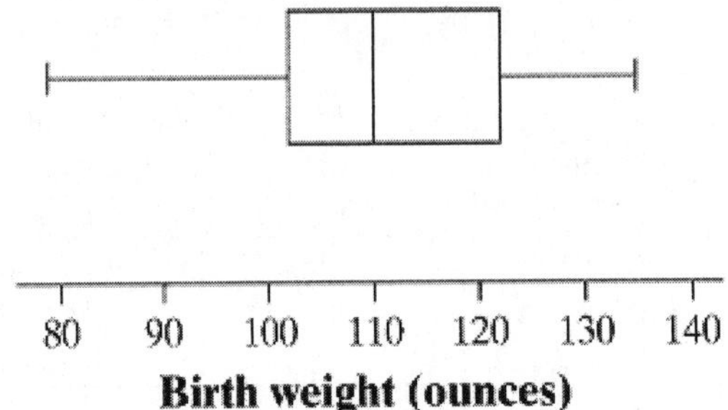

About 40 of the birth weights were below

(A) 92 ounces.
(B) 102 ounces.
(C) 112 ounces.
(D) 122 ounces.
(E) 132 ounces.

8. A sample of production records for an automobile manufacturer shows the following figures for production per shift:

705 700 690 705

The variance of the sample is approximately

(A) 8.66.
(B) 7.07.
(C) 75.00.
(D) 50.00.
(E) 20.00.

9. You catch 10 cockroaches in your bedroom and measure their lengths in centimeters. Which of these sets of numerical descriptions are *all* measured in centimeters?

(A) median length, variance of lengths, largest length
(B) median length, first and third quartiles of lengths
(C) mean length, standard deviation of lengths, median length
(D) mean length, median length, variance of lengths.
(E) both (B) and (C)

10. A policeman records the speeds of cars on a certain section of roadway with a radar gun. The histogram below shows the distribution of speeds for 251 cars.

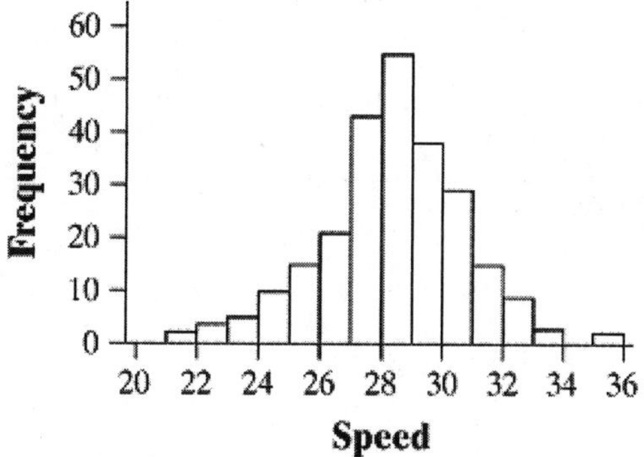

Which of the following measures of center and spread would be the best ones to use when summarizing these data?

(A) Mean and interquartile range.
(B) Mean and standard deviation.
(C) Median and range.
(D) Median and standard deviation.
(E) Median and interquartile range.

Check your answers below. If you got a question wrong, check to see if you made a simple mistake or if you need to study that concept more. After you check your work, identify the concepts you feel very confident about and note what you will do to learn the concepts in need of more study.

#	Answer	Concept	Right	Wrong	Simple Mistake?	Need to Study More
1	C	Variables				
2	D	Categorical variables				
3	D	Two-way table				
4	D	Distribution basics				
5	C	Constructing histograms				
6	C	Skewed distributions				
7	B	Interpreting boxplots				
8	D	Variance				
9	E	Summary statistics units				
10	B	Choosing statistics				

Chapter 1 Reflection

Summarize the "Big Ideas" in Chapter 1:

My strengths in this chapter:

Concepts I need to study more and what I will do to learn them:

FRAPPY! Free Response AP® Problem, Yay!

The following problem is modeled after actual Advanced Placement Statistics free response questions. Your task is to generate a complete, concise response in 15 minutes. After you generate your response, view two example solutions and determine whether you feel they are "complete", "substantial", "developing" or "minimal". If they are not "complete", what would you suggest to the student who wrote them to increase their score? Finally, you will be provided with a rubric. Score your response and note what, if anything, you would do differently to increase your own score.

SugarBitz candies are packaged in 15 oz. snack-size bags. The back-to-back plot below displays the weights (in ounces) of two samples of SugarBitz bags filled by different filling machines. The weights ranged from 14.1 oz. to 15.9 oz.

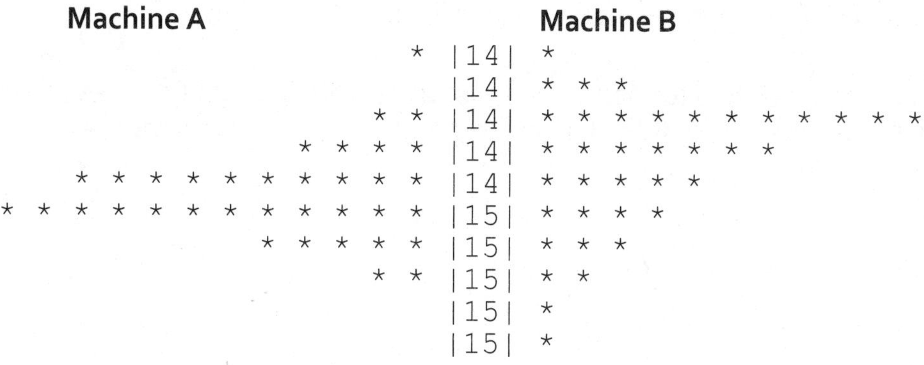

(a) Compare the distributions of weights of bags packaged by the two machines.

(b) The company wishes to be as consistent as possible when packing its snack bags. Which machine would be *least* likely to produce snack bags of SugarBitz that have a consistent weight? Explain.

(c) Suppose the company filled its bags using the machine you chose in part (b). Which measure of center, mean or median, would be closer to the advertised 15oz.? Explain why you chose this measure.

FRAPPY! Student Responses

Student Response 1:

a) Machine A has a slightly higher center than Machine B. Machine B has a much larger range. Machine A is approximately symmetric and Machine B is slightly skewed right. Neither machine has any extreme values.
b) Machine B would be least likely to produce bags containing 15 oz of SugarBitz because it has a much wider range than Machine A.
c) The company should report the mean weight of Machine B. Since the distribution is skewed to the right, the mean will be pulled higher towards the tail. Therefore, the mean will be higher than the median and will be closer to 15.

How would you score this response? Is it substantial? Complete? Developing? Minimal? Is there anything this student could do to earn a better score?

Student Response 2:

a) Machine A is normal and Machine B is skewed. Both have a single peak and wide ranges.
b) Machine B usually fills bags with about 14.6 oz of candy. Machine A usually fills bags with 15 oz of candy. Machine B is least likely to fill the bags with 15 oz. of candy.
c) The mean because it is about 15.

How would you score this response? Is it substantial? Complete? Developing? Minimal? Is there anything this student could do to earn a better score?

FRAPPY! Scoring Rubric

Use the following rubric to score your response. Each part receives a score of "Essentially Correct," "Partially Correct," or "Incorrect." When you have scored your response, reflect on your understanding of the concepts addressed in this problem. If necessary, note what you would do differently on future questions like this to increase your score.

Intent of the Question

The goals of this question are (1) to determine your ability to use graphical displays to compare and contrast two distributions and (2) to evaluate your ability to use statistical information to make a decision.

Solution

(a) Both distributions are single-peaked. However, Machine A's distribution is roughly symmetric while Machine B's is skewed to the right. The center of the weights for Machine A (median A = about 15) is slightly higher than that of Machine B (median B = about 14.5). There is more variability in the weights produced by Machine B. Machine A has one low value (14.1) that does not fall with the majority of weights. However, it does not appear to be extreme enough to be considered an outlier.

(b) Both machines produce bags of varying weight. However, Machine B has a higher variability as evidenced by a wider overall range. Machine B would be least likely to produce a consistent weight for the snack bags.

(c) The mean would be closer to the advertised 15 oz. weight. This is because in a skewed distribution, the mean is pulled away from the median in the direction of the tail. In Machine B's distribution, the peak is at about 14.5 oz so we would expect the mean to be higher and closer to 15 oz.

Scoring:

Parts (a), (b), and (c) are scored as essentially correct (E), partially correct (P), or incorrect (I).

Part (a) is essentially correct if you correctly identify similarities and differences in the shape, center, and spread for the two distributions.
Part (a) is partially correct if you correctly identify similarities and differences in two of the three characteristics for the two distributions.
Part (a) is incorrect if you only identify one similarity or difference of the three characteristics for the two distributions.

Part (b) is essentially correct if Machine B is chosen using rationale based on its measure of spread of the packaged weights.
Part (b) is partially correct if B is chosen, but the explanation does not refer to the variability in the weights.

Part (c) is incorrect if B is chosen and no explanation is provided OR if A is chosen.

Part (c) is essentially correct if the mean is chosen and the explanation addresses the fact that the mean will be greater than the median in a skewed right distribution.
Part (c) is partially correct if the mean is chosen, but the explanation is incomplete or incorrect.
Part (c) is incorrect if the mean is chosen, but no explanation is given OR if the median is chosen.
NOTE: If Machine A was chosen in part (b) and the solution to part (c) indicates either the mean or median would be appropriate due to the fact that they will be approximately equal in a symmetric, mound-shaped distribution, part (c) should be scored as essentially correct.

4 **Complete Response**
All three parts essentially correct

3 **Substantial Response**
Two parts essentially correct and one part partially correct

2 **Developing Response**
Two parts essentially correct and no parts partially correct
One part essentially correct and two parts partially correct
Three parts partially correct

1 **Minimal Response**
One part essentially correct and one part partially correct
One part essentially correct and no parts partially correct
No parts essentially correct and two parts partially correct

My Score:
What I did well:
What I could improve:
What I should remember if I see a problem like this on the AP Exam:

Chapter 1: Exploring Data

Across

3. The average distance of observations from their mean (two words)
5. The average squared distance of the observations from their mean
9. Displays the counts or percents of categories in a categorical variable through differing heights of bars
12. Tells you what values a variable takes and how often it takes these values
13. Displays a categorical variable using slices sized by the counts or percents for the categories
16. When specific values of one variable tend to occur in common with specific values of another
18. A measure of center, also called the average
19. A graphical display of quantitative data that involves splitting the individual values into two components
21. One of the simplest graphs to construct when dealing with a small set of quantitative data
22. Drawing conclusions beyond the data at hand
23. The shape of a distribution if one side of the graph is much longer than the other
24. What we call a measure that is relatively unaffected by extreme observations

Down

1. The objects described by a set of data
2. The midpoint of a distribution of quantitative data
4. A _____ distribution describes the distribution of values of a categorical variable among individuals who have a specific value of another variable.
6. A variable that places an individual into one of several groups or categories
7. A characteristic of an individual that can take different values for different individuals
8. When comparing two categorical variables, we can orgainze the data in a ___-___ _____.
9. A graphical display of the five-number summary
10. A graphical display of quantitative data that shows the frequency of values in intervals by using bars
11. A variable that takes numerical values for which it makes sense to find an average
14. The shape of a distribution whose right and left sides are approximate mirror images of each other
15. These values lie one-quarter, one-half, and three-quarters of the way up the list of quantitative data
17. A value that is at least 1.5 IQRs above the third quartile or below the first quartile
20. When exploring data, don't forget your ____

CHAPTER 2: MODELING DISTRIBUTIONS OF DATA

"It is not knowledge, but the act of learning, not possession but the act of getting there, which grants the greatest enjoyment." Carl Friedrich Gauss

Chapter Overview

In Chapter 1, we built a "toolbox" of graphical and numerical tools for describing distributions of data. We now have a clear strategy for exploring data. This chapter introduces you to a key concept in statistics – describing the location of an observation within a distribution. You will learn how to measure position by using percentiles as well as by using a standardized measure based on the mean and standard deviation. We'll discover that sometimes the overall pattern of a large number of observations is so regular it can be described by a smooth curve. Density curves will be introduced as a model to describe locations without having to rely on every one of the data values. One of the most common density curves, the Normal distribution, will be explored in Section 2.2. You will not only learn the properties of the Normal distributions, but also how to perform a number of calculations with them. The concepts introduced in this section will be revisited throughout the course, so be sure to master them in this chapter!

Sections in this Chapter

Plan Your Learning

Use the following *suggested* guide to help plan your reading and assignments in "The Practice of Statistics, 6th Edition." Note: your teacher may schedule a different pacing or assign different problems. Be sure to follow his or her instructions!

Read	**2.1: pp. 90-97**	**2.1: pp. 97-103**	**2.2: pp.109-119**
Do	1,3,7,9,11,13,15,19	21,25,29,31 MC 33-38	41,45,47,49,51

Read	**2.2: pp. 119-130**	**2.2: pp. 131-137**	**Chapter Summary**
Do	53,55,57,59,61,63	73,75,77,79,81 MC 85-90	Multiple-Choice FRAPPY!

Use this "Strive Guide" to help identify learning targets for each section, summarize key concepts, practice additional problems, and check your understanding as you read the text.

Section 2.1: Describing Location in a Distribution

Before You Read: Section Summary

In this section, you will learn how to describe the location of an individual value in a distribution. You have probably encountered measures of location before through the concept of percentiles. In this section, you will learn how to calculate and interpret percentiles as well as how to identify percentiles through a cumulative relative frequency graph. You will also learn a new way to describe location by using the mean and standard deviation that you studied in Chapter 1. Standardized scores, or *z*-scores, will be introduced as a way to describe location within a distribution and to compare observations from different distributions. Finally, you will be introduced to the concept of a density curve. This smooth curve models the overall shape of a distribution and can be used in place of the actual data to answer questions about the distribution. The concepts you learn in this section will very important throughout the course. Be sure to master the idea of a standardized score!

Learning Targets:

____ I can find and interpret the percentile of an individual value in a distribution.
____ I can estimate percentiles and individual values using a cumulative relative frequency graph.
____ I can find and interpret the standardized score (*z*-score) of an individual value in a distribution of data.
____ I can describe the effect of adding, subtracting, multiplying by, or dividing by a constant on the shape, center, and variability of a distribution of data.

While You Read: Key Vocabulary and Concepts

percentile:
cumulative relative frequency graph (ogive):
standardized score (*z*-score):
effect of adding (or subtracting) a constant:
effect of multiplying (or dividing) by a constant:

After You Read: Check for Understanding

Concept 1: Measuring Position - Percentiles

A common way to measure position within a distribution is to tell what percent of observations fall below the value in question. Percentiles are relatively easy to find – especially when our data is presented in order or in a dotplot or stemplot. To find a percentile, simply calculate the percent of values that fall below the value of interest. Note, some people calculate percentiles as the percent of values that are less than *or equal to* a given value. Either calculation is acceptable. In The Practice of Statistics, however, we'll stick with the first definition. Cumulative relative frequency graphs, or ogives, provide a graphical tool to find percentiles in a distribution.

Check for Understanding: ____ *I can find and interpret the percentile of an individual value in a distribution.* ____ *I can estimate percentiles and individual values using a cumulative relative frequency graph.*

The scores on Ms. Blockhus' chapter quiz are displayed in the stemplot and ogive below:

```
0|012
1|2248
2|11134599
3|00036
4|457
5|0
```

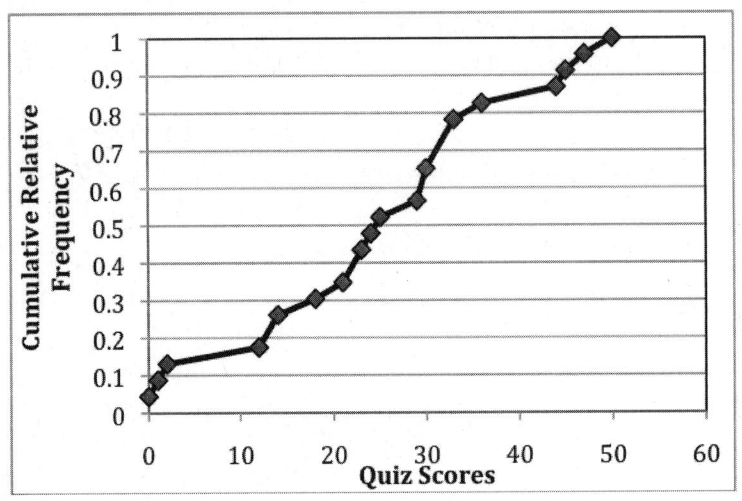

1) James earned a score of 33. Calculate and interpret his percentile.

2) Heather earned a score of 12. Calculate and interpret her percentile.

3) Using the cumulative relative frequency graph, at what percentile is a score of 38?

4) Using the cumulative relative frequency graph, estimate and interpret the quartiles of this distribution.

Concept 2: Measuring Position – z-Scores

When describing distributions, we learned that measures of center and spread are both very important characteristics. It follows, then, that measuring the position of an observation in a distribution should take into account both the center and the spread of the distribution. After all, saying a particular observation falls 5 points above the average doesn't mean much unless you know how varied the observations are. If the distribution has a small spread, an observation that is 5 points above the average might be an extreme value. However, if the distribution has a large spread, being 5 points above average might not be a big deal. We MUST consider center and spread when describing location. The standardized value, or *z*-score, of an observation does just that. The *z*-score simply tells us how many standard deviations above or below the mean a particular observation falls. This method of describing location not only allows us to describe individuals within a distribution, but also allows us to compare the positions of individuals in different distributions. We will be standardizing values a LOT in this course. Be sure to master the concept now!

Check for Understanding: ____*I can find and interpret the standardized score (z-score) of an individual value in a distribution of data.*

Use the scores from Ms. Blockhus' quiz.

0 | 0 1 2
1 | 2 2 4 8
2 | 1 1 3 4 5 9 9
3 | 0 0 0 3 6
4 | 4 5 7
5 | 0

1) Calculate the mean and standard deviation.

2) Paul earned a 45. Calculate and interpret his z-score.

3) Carl earned a 2. Calculate and interpret his z-score.

4) The scores on Ms. Blockhus' next quiz had a mean of 32 and a standard deviation of 6.5. Paul earned a 47 on this quiz. On which quiz did he perform better relative to the rest of the class? Explain.

Concept 3: Transforming Data

When we find *z*-scores, we are actually transforming our data to a standardized scale. That is, we subtract the mean and divide by the standard deviation, converting the observation from its original units to a standardized scale. Sometimes we transform data to switch between measurement units (inches to centimeters, Fahrenheit to Celsius, etc.). When we do this, it is important to know what happens to the center and spread of the transformed distribution. Adding (or subtracting) a constant to each of the observations in a set of data will have an effect on the center of the distribution, but not on the spread or the shape. Multiplying (or dividing) each of the observations by a constant will change the center and the spread of the distribution, but not its shape. Most important, while transforming data in these ways may change the center and spread of a distribution, the locations of individual observations remain unchanged! So, if you had a z-score of 1.5 on a quiz and your teacher decided to double everyone's score and give an additional 5 points, your z-score would STILL be 1.5!

Check for Understanding: ____ *I can describe the effect of adding, subtracting, multiplying by, or dividing by a constant on the shape, center, and variability of a distribution of data.*

Use the scores from Ms. Blockhus' quiz.

0 | 0 1 2
1 | 2 2 4 8
2 | 1 1 3 4 5 9 9
3 | 0 0 0 3 6
4 | 4 5 7
5 | 0

1) Describe the distribution of quiz scores.

2) Suppose Ms. Blockhus added 5 points to each student's score. How would the new distribution of scores compare to the distribution in question 1?

3) Suppose Ms. Blockhus decided to double the original scores instead of adding points. How would this new distribution of scores compare to the distribution in question 1?

Section 2.2: Density Curves and Normal Distributions

Before You Read: Section Summary

Sometimes the overall pattern of distribution is so regular that it can described by a smooth curve. Density curves are a handy tool for modeling distributions of data. This section is devoted entirely to the most common type of density curve – the Normal curve. In this section, you will learn the basic properties of the Normal distributions and how to use them to perform a variety of calculations. You will learn how to determine whether or not a distribution of data can be described as approximately Normal. For distributions that can be described by a Normal curve, you will learn how to determine the proportion of observations that fall into given intervals. You will also learn how to find the value corresponding to a specified percentile in a Normal distribution. Throughout this section, a variety of tools and methods will be presented to help you perform Normal calculations by hand and on your calculator. Remember, whenever you perform a calculation by hand or by calculator, it is important to interpret the results in the context of the situation! Make sure you are comfortable with not only performing the calculations in this section but also in describing what your results mean!

Learning Targets:

____ I can use a density curve to model distributions of quantitative data.

____ I can identify the relative locations of the mean and median of a distribution from a density curve.

____ I can use the 68-95-99.7 Rule to estimate (i) the proportion of values in a specified interval, or (ii) the value that corresponds to a given percentile in a Normal distribution.

____ I can find the proportion of values in a specified interval in a Normal distribution using Table A or technology.

____ I can find the value that corresponds to a given percentile in a Normal distribution using Table A or technology.

____ I can determine whether a distribution of data is approximately Normal from graphical and numerical evidence.

While You Read: Key Vocabulary and Concepts

density curve:
Normal distribution, Normal curve:
68-95-99.7 Rule:
standard Normal distribution:
Normal probability plot:

After You Read: Check for Understanding

Concept 1: Density Curves

You have learned that exploring quantitative data requires making a graph, describing the overall shape, and providing a numerical summary of the center and spread. When we have a large number of observations, we can describe the overall pattern using a smooth curve called a density curve. There are two key features of density curves you need to remember. First, since it is describing the distribution of values it is always on or above the horizontal axis. Second, since it is representing the distribution of all of the values, the area underneath it is exactly 1. Since the area under the entire curve is 1, the area under the curve and between any two values on the horizontal axis is the proportion of observations in the distribution that fall in that interval. One other key point is noted in this section. In a symmetric distribution, the mean and median will be the same. However, in a skewed distribution, the mean will be pulled away from the median in the direction of the tail. Make sure you note that fact, it will come in handy on the AP Exam!

Check for Understanding: ____ *I can use a density curve to model distributions of quantitative data.*

The scores on Ms. Chauvet's chapter test are displayed in the density curve below:

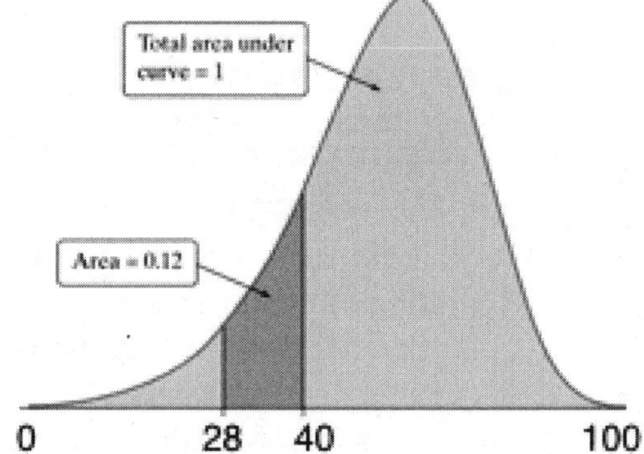

1) How do you know this is a legitimate density curve?

2) What percent of scores fell between 28 and 40?

3) How does the mean exam score compare to the median? Estimate where these values would fall on the density curve.

Concept 2: Normal Distributions

Normal distributions play an important role in statistics. We encounter them a LOT. That's because many distributions of real data and chance outcomes are symmetric, single-peaked, and bell-shaped and can be described by the family of Normal curves. This family of curves shares some important characteristics that help us perform calculations about observations in a distribution of data. First, in distributions that can be described as Normal, the mean is at the center of the Normal curve. Further, almost all of the observations in an approximately Normal distribution will fall within three standard deviations of the mean.

The 68-95-99.7 Rule allows us to be even more specific, noting that in Normal distributions, approximately 68% of the observations will fall within one standard deviation of the mean, approximately 95% of observations will fall within two standard deviations, and approximately 99.7% of observations will fall within three standard deviations. This fact allows us to perform calculations about the approximate proportion of observations in a distribution that fall into certain intervals without even knowing all of the individual observations!

Concept 3: Normal Distribution Calculations

As we explore situations in statistics, not all observations of interest will fall one, two, or three standard deviations from the mean. That is, we might be interested in knowing what proportion of observations are at least 1.72 standard deviations above the mean. In cases like this, the 68-95-99.7 rule can help us estimate the proportion, but we'll want to be more exact. Because all Normal distributions share the same properties, the standard Normal table and Normal calculations allow us to perform calculations for *any* observation in an approximately Normal distribution.

To perform a Normal calculation, start by expressing the problem in terms of a variable x. Sketch a picture of the distribution and shade the area of interest. Perform calculations by standardizing x and then using Table A or your calculator to find the required area under the Normal curve. Finally, be sure to write your conclusion in the context of the problem.

Check for Understanding: ____ *I can find the proportion of values in a specified interval in a Normal distribution using Table A or technology.*

Suppose the scores on all of the quizzes in Ms. Chauvet's class were Normally distributed with mean 83 and standard deviation 5.

1) Sketch the density curve that describes the distribution of scores on the quizzes.

2) What percent of scores are between 78 and 93? Use the 68-95-99.7 rule.

3) A score greater than 90 earns an "A." What percent of quizzes earn an "A"?

4) What percent of scores fall between 71 and 95?

5) What score would place a student in the 20th percentile in this class?

Concept 4: Assessing Normality

It is important to note that just because a distribution is symmetric, single-peaked, and bell-shaped, it is NOT necessarily Normal. Normal distributions are handy models for some distributions of data. However, before performing Normal calculations, you should be sure the distribution you are exploring really IS Normal! The Normal probability plot of a distribution of data is a helpful tool in assessing Normality of a distribution of data. If the points on a Normal probability plot fall in a relatively straight line, you have evidence that the data are approximately Normal and Normal calculations are justified. If the points taper off or display other systematic departures from a straight line, a Normal model probably isn't appropriate for the distribution. Make sure you are comfortable constructing Normal probability plots on your calculator and interpreting them in the context of the problem!

Check for Understanding: ____ *I can determine whether a distribution of data is approximately Normal from graphical and numerical evidence.*

Use these scores from Ms. Chauvet's quiz.

0 | 0 1 2
1 | 2 2 4 8
2 | 1 1 3 4 5 9 9
3 | 0 0 0 3 6
4 | 4 5 7
5 | 0

Use graphical and numerical methods to determine whether these data are approximately Normally distributed.

Chapter Summary: Modeling Distributions of Data

In this chapter, we expanded our toolbox for working with quantitative data. We learned how to describe the location of an individual within a distribution by determining its percentile or by calculating a standardized score (*z*-score) based on the mean and standard deviation of the distribution. We also learned that distributions with a clear overall pattern can be described using a density curve. A common density curve, the Normal curve, is a helpful model for describing many quantitative variables. Knowing how to justify that a distribution is approximately Normally distributed is an important skill. If you can show that a distribution is Normal, you can use the Normal distribution calculations to answer a number of questions about observations within the set of data. The 68-95-99.7 rule and the standard Normal table are both useful tools when performing calculations about observations in Normal distributions.

We have learned in these first two chapters that you should approach data analysis problems using four steps:

- Graph the data
- Look for an overall pattern and departures from this pattern
- Calculate and interpret numerical summaries
- If the data follow a regular overall pattern, describe the distribution with a smooth curve

The concepts introduced in this chapter will form the basis of much of our study of inference later on in the course. Standardizing data, justifying Normality, and performing Normal calculations are critical skills for statistical inference. Be sure to practice them as you will be using them a LOT!

After You Read: What Have I Learned?

Complete the vocabulary puzzle, multiple-choice questions, and FRAPPY. Check your answers and performance on each of the learning targets. Be sure to get extra help on any targets that you identify as needing more work!

Learning Target	Got It!	Almost There	Needs Work
I can find and interpret the percentile of an individual value in a distribution.			
I can estimate percentiles and individual values using a cumulative relative frequency graph.			
I can find and interpret the standardized score (*z*-score) of an individual value in a distribution of data.			
I can describe the effect of adding, subtracting, multiplying by, or dividing by a constant on the shape, center, and variability of a distribution of data.			
I can use a density curve to model distributions of quantitative data.			
I can use the 68-95-99.7 Rule.			
I can find the proportion of values in a specified interval in a Normal distribution.			
I can find the value that corresponds to a given percentile in a Normal distribution.			
I can determine whether a distribution of data is approximately Normal from graphical and numerical evidence.			

Chapter 2 Multiple Choice Practice

Directions. *Identify the choice that best completes the statement or answers the question. Check your answers and note your performance when you are finished.*

1. The 16th percentile of a Normally distributed variable has a value of 25 and the 97.5th percentile has a value of 40. Which of the following is the best estimate of the mean and standard deviation of the variable?

(A) Mean ≈ 32.5; Standard deviation ≈ 2.5
(B) Mean ≈ 32.5; Standard deviation ≈ 5
(C) Mean ≈ 32.5; Standard deviation ≈ 10
(D) Mean ≈ 30; Standard deviation ≈ 2.5
(E) Mean ≈ 30; Standard deviation ≈ 5

2. The proportion of observations from a standard Normal distribution that take values larger than 0.75 is about

(A) 0.2266
(B) 0.2500
(C) 0.7704
(D) 0.7764
(E) 0.8023

3. The density curve below takes the value 0.5 on the interval $0 < x < 2$ and takes the value 0 everywhere else. What percent of the observations lie between 0.4 and 1.08?

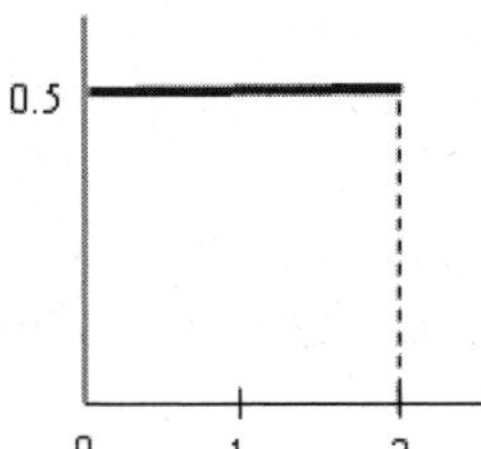

(A) 25%
(B) 34%
(C) 50%
(D) 68%
(E) 70%

4. The distribution of the heights of students in a large class is roughly Normal. The average height is 68 inches, and approximately 99.7% of the heights are between 62 and 74 inches. Thus, the standard deviation of the height distribution is approximately equal to

(A) 2
(B) 3
(C) 4
(D) 6
(E) 9

5. The mean age (at inauguration) of all U.S. Presidents is approximately Normally distributed with a mean of 54.6. Barack Obama was 47 when he was inaugurated, which is the 11th percentile of the distribution. George Washington was 57. What percentile was he in?

(A) 6.17
(B) 65.17
(C) 62.92
(D) 34.83
(E) 38.9

6. Which of the following statements are false?
   I. The standard Normal table can be used with z-scores from any distribution
   II. The mean is always equal to the median for any Normal distribution.
   III. Every symmetric, bell-shaped distribution is Normal
   IV. The area under a Normal curve is always 1, regardless of the mean and standard deviation.

(A) I and II
(B) I and III
(C) II and III
(D) III and IV
(E) None of the above gives the correct set of true statements.

7. High school textbooks don't last forever. The lifespan of all high school statistics textbooks is approximately Normally distributed with a mean of 9 years and a standard deviation of 2.5 years. What percentage of the books last more than 10 years?

(A) 11.5%
(B) 34.5%
(C) 65.5%
(D) 69%
(E) 84.5%

8. The distribution of the time it takes for different people to solve their Strive for a Five chapter crossword puzzle is strongly skewed to the right, with a mean of 10 minutes and a standard deviation of 2 minutes. The distribution of *z*-scores for those times is

(A) Normally distributed, with mean 10 and standard deviation 2.
(B) Skewed to the right, with mean 10 and standard deviation 2.
(C) Normally distributed, with mean 0 and standard deviation 1.
(D) Skewed to the right, with mean 0 and standard deviation 1.
(E) Skewed to the right, but the mean and standard deviation cannot be determined without more information.

9. The cumulative relative frequency graph below shows the distribution of lengths (in centimeters) of fingerlings at a fish hatchery. The third quartile for this distribution is approximately:

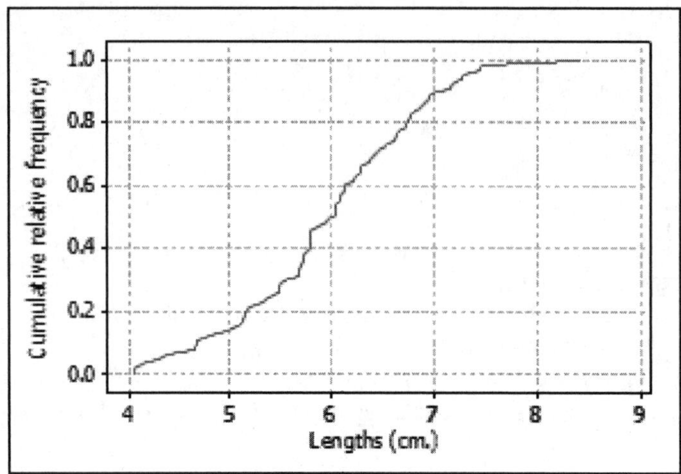

(A) 6.7 cm
(B) 7 cm
(C) 6 cm
(D) 5.5 cm
(E) 7.5 cm

10. The plot shown below is a Normal probability plot for a set of test scores. Which statement is true for these data?

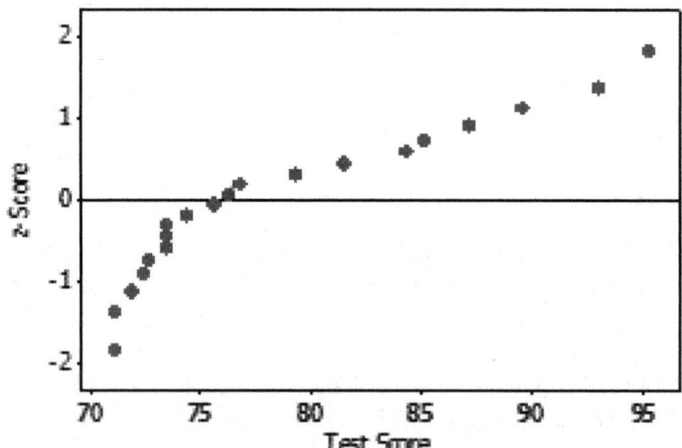

(A) The distribution is clearly Normal.
(B) The distribution is approximately Normal.
(C) The distribution appears to be skewed.
(D) The distribution appears to be uniform.
(E) There is insufficient information to determine the shape of the distribution.

Check your answers below. If you got a question wrong, check to see if you made a simple mistake or if you need to study that concept more. After you check your work, identify the concepts you feel very confident about and note what you will do to learn the concepts in need of more study.

#	Answer	Concept	Right	Wrong	Simple Mistake?	Need to Study More
1	E	68-95-99.7 Rule				
2	A	Standard Normal Table				
3	B	Area under a Density Curve				
4	A	68-95-99.7 Rule				
5	B	Standard Normal Calculations				
6	B	Properties of Normal Distributions				
7	B	Standard Normal Calculations				
8	D	Standardized Scores				
9	A	Cumulative Relative Frequency Graph				
10	C	Normal Probability Plots				

Chapter 2 Reflection

Summarize the "Big Ideas" in Chapter 2:

My strengths in this chapter:

Concepts I need to study more and what I will do to learn them:

FRAPPY! Free Response AP® Problem, Yay!

The following problem is modeled after actual Advanced Placement Statistics free response questions. Your task is to generate a complete, concise response in 15 minutes. After you generate your response, view two example solutions and determine whether you feel they are "complete", "substantial", "developing" or "minimal". If they are not "complete", what would you suggest to the student who wrote them to increase their score? Finally, you will be provided with a rubric. Score your response and note what, if anything, you would do differently to increase your own score.

Final exam grades are determined by the percent correct on the exam. A teacher's records indicate the performance on the exam is Normally distributed with mean 82 and standard deviation 5. The grades on her exam are assigned using the scale below.

Grade	Percent Correct
A	94 ≤ percent ≤ 100
B	85 ≤ percent < 94
C	76 ≤ percent < 85
D	65 ≤ percent < 76
F	0 ≤ percent < 65

(a) Use a sketch of a Normal distribution to illustrate the proportion of students who would earn a B. Calculate this proportion.

(b) Students who earn a B, C, or D are considered to "meet standards". Based on this grading scale, what percent of students will receive a score that places them in a category other than "meets standards"?

(c) What grade would the student who scored at the 25th percentile earn on this chapter? Justify your answer.

FRAPPY! Student Responses

Student Response 1:

a) $P(B) = 0.9918 - 0.7257 = 0.2661$. 26.61 % of students earn a B.
b) $P(\text{does not meet standards}) = P(F) = P(z<3.4) = 0.0003$
c) $z = 0.25$ so the score would be $82 + 0.25(5) = 83.25$

How would you score this response? Is it substantial? Complete? Developing? Minimal? Is there anything this student could do to earn a better score?

Student Response 2:

a) $P(B) = P(0.6 \leq z < 2.4) = 0.9793 - 0.5239 = 0.4554$
b) $P(A \text{ or } F) = P(z > 2.4) + P(z < -3.4) = 0.0085$
c) A *z*-score of -0.6745 corresponds to the 25th percentile. So, the score would be $82+(-0.6475)(5) = 78.63$.

How would you score this response? Is it substantial? Complete? Developing? Minimal? Is there anything this student could do to earn a better score?

FRAPPY! Scoring Rubric

Use the following rubric to score your response. Each part receives a score of "Essentially Correct," "Partially Correct," or "Incorrect." When you have scored your response, reflect on your understanding of the concepts addressed in this problem. If necessary, note what you would do differently on future questions like this to increase your score.

Intent of the Question

The goal of this question is to determine your ability to perform and interpret Normal calculations.

Solution

(a)

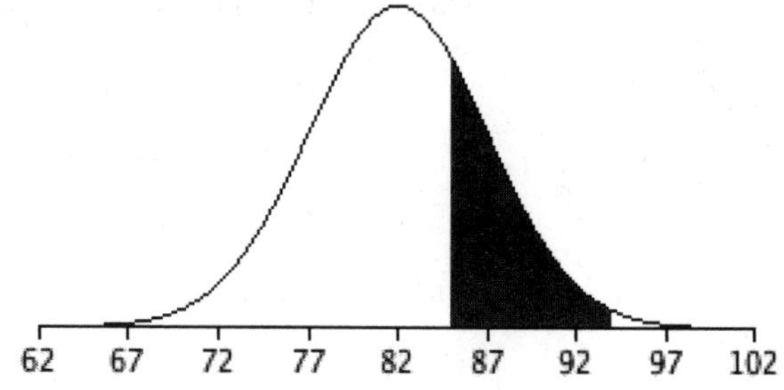

percent < 94)
$= P((85\text{-}82)/5 \le z < (94\text{-}82)/5)$
$= P(0.6 \le z < 2.4)$
$= 0.9918 - 0.7257$
$= 0.2661$

(b) $P(\text{A or F}) = P(\text{A}) + P(\text{F})$
$= P(z \ge (94\text{-}82)/5) + P(z < (65\text{-}82)/5)$
$= P(z \ge 2.4) + P(z < 3.4)$
$= 0.0082 + 0.0003$
$= 0.0085$

(c) A z-score of -0.6745 corresponds to the 25th percentile.
x = mean + z(std dev)
$x = 82 + (-0.6475)(5)$
$x = 78.63$

Scoring:

Parts (a), (b), and (c) are scored as essentially correct (E), partially correct (P), or incorrect (I).

Part (a) is essentially correct if (1) the appropriate probability is illustrated using a labeled Normal curve and (2) the proportion is correctly computed.
Part (a) is partially correct if only one of the above elements is correct

Part (b) is essentially correct if the response (1) recognizes the need to look at grades of A and F and (2) correctly computes the tail probabilities and adds them together.
Part (b) is partially correct if the response considers only an A or an F and calculates the corresponding tail area correctly OR recognizes the need to look at A an F but only calculates one of the tail areas correctly OR approximates the probabilities using the 68-95-99.7 rule OR computes the proportion that will "meet standards" OR states the correct answer without supporting work.

Part (c) is essentially correct if (a) the correct *z*-score is identified for the 25th percentile and (2) the correct corresponding score is calculated.
Part (c) is partially correct if only one of the above elements is correct.

NOTE: If the student makes an error in part (b) and correctly uses that probability in part (c) to compute a reasonable probability, part (c) is essentially correct.

4 **Complete Response**
All three parts essentially correct

3 **Substantial Response**
Two parts essentially correct and one part partially correct

2 **Developing Response**
Two parts essentially correct and no parts partially correct
One part essentially correct and two parts partially correct
Three parts partially correct

1 **Minimal Response**
One part essentially correct and one part partially correct
One part essentially correct and no parts partially correct
No parts essentially correct and two parts partially correct

My Score:
What I did well:
What I could improve:
What I should remember if I see a problem like this on the AP Exam:

Chapter 2: Modeling Distributions of Data

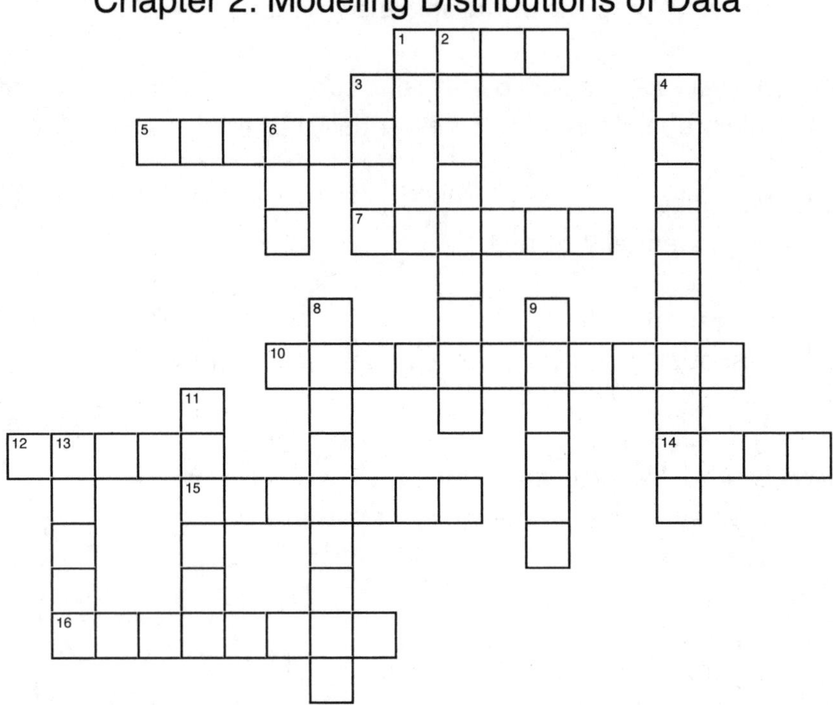

Across

1. The balance point of a density curve, if it were made of solid material
5. The standardized value of an observation
7. These common density curves are symmetric and bell-shaped
10. A Normal ______ plot provides a good assessment of whether a data set is approximately Normally distributed
12. Another name for a cumulative relative frequency graph
14. The standard Normal table tells us the area under the standard Normal curve to the ___ of z
15. A ___ curve is a smooth curve that can be used to model a distribution
16. This Normal distribution has mean 0 and standard deviation 1

Down

2. The ____ rule is also known as the 68-95-99.7 rule for Normal distributions
3. To standardize a value, subtract the ___ and divide by the standard deviation
4. The value with p percent of the observations less than it
6. The area under any density curve is always equal to
8. We ___ data when we change each value by adding a constant and/or multiplying by a constant.
9. If a Normal probability plot shows a _____ pattern, the data are approximately Normal
11. The point that divides the area under a density curve in half
13. This mathematician first applied Normal curves to data to errors made by astronomers and surveyors

CHAPTER 3: DESCRIBING RELATIONSHIPS

"You can only predict things after they've happened." Eugene Ionesco

Chapter Overview

Our statistics toolbox now contains a variety of ways to explore a single quantitative variable. Further, we have learned ways to explore one or more categorical variables. Often in our studies, though, we will need to explore and describe the relationship between two quantitative variables. In this chapter, we will learn how to analyze patterns in "bivariate" relationships by plotting them and calculating summary statistics about them. Further, we will learn how to describe them using mathematical models that can be used to make predictions based on the relationship between the variables. Investigating the relationship between two variables is a key component of statistical study and is the final skill necessary for our data exploration toolbox. Be sure to master the concepts and methods in this chapter!

Sections in this Chapter
Section 3.1: Scatterplots and Correlation
Section 3.2: Least-Squares Regression

Plan Your Learning

Use the following *suggested* guide to help plan your reading and assignments in "The Practice of Statistics, 6th Edition." Note: your teacher may schedule a different pacing or assign different problems. Be sure to follow his or her instructions!

Read	**Introduction** **3.1: pp.152-159**	**3.1: pp. 160-170**	**3.2: pp.176-182**	**3.2: pp. 183-187**
Do	1,3,5,9,11	13,15,17,19,23 MC 29-34	37,39,41,43,45	47,49,51,53

Read	**3.2: pp.188-193**	**3.2: pp. 194-201**	**Chapter Summary**
Do	55,57,59,67	63,65 MC 71-78	Multiple Choice FRAPPY!

Section 3.1: Scatterplots and Correlation

Before You Read: Section Summary

Many statistical studies examine more than one variable. So far, we have learned methods to graph and describe relationships between categorical variables. In this chapter, we'll learn that the approach to data analysis that we learned for a single quantitative variable can also be applied to explore the relationship between two quantitative variables. That is, we'll learn how to plot our data and add numerical summaries. We'll then learn how to describe the overall patterns and departures from patterns that we see. Finally, we'll learn how to create a mathematical model to describe the overall pattern. This section will focus primarily on displaying the relationship between two quantitative variables and describing its form, direction, and strength. Like the previous chapters, you will find that technology can be used to do most of the difficult calculations. However, be sure you understand *how* the calculator is determining its results and *what* those results mean!

Learning Targets:

____ I can distinguish between explanatory and response variables for quantitative data.

____ I can make a scatterplot to display the relationship between two quantitative variables.

____ I can describe the direction, form, and strength of a relationship displayed in a scatterplot and identify unusual features.

____ I can interpret the correlation.

____ I can understand the basic properties of correlation, including how the correlation is influenced by outliers.

____ I can distinguish correlation from causation.

While You Read: Key Vocabulary and Concepts

response variable:
explanatory variable:
scatterplot:
direction, form, strength, unusual features:
positive association, negative association, no association:
correlation *r*:

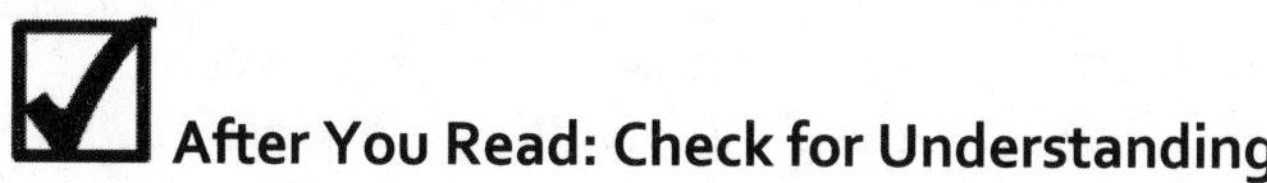

After You Read: Check for Understanding

Concept 1: Explanatory and Response Variables

The purpose of many studies involving two quantitative variables is to develop a model so that we can use one variable to make a prediction for the other. Because of that, it is important to clearly identify which variable in a situation is *explanatory* and which is the *response*. The explanatory variable is the one we think explains the relationship or "predicts" changes in the response variable. It is important to know the difference as identifying the explanatory and response variable will determine how we display the data and how we calculate a summary model. Note: you may have learned about *independent* and *dependent* variables in an earlier math or science class. Those are just different names for explanatory and response variables. We'll avoid using independent and dependent here, though, because those terms have a different meaning later in the course.

Check for Understanding: ____ *I can distinguish between explanatory and response variables for quantitative data.*

Identify the explanatory and response variable in the following situations:

How does stress affect your test performance? In a recent study, researchers studied students' test anxiety and subsequent performance on a standardized test.

Is brain size related to memory? A 1995 study measured the volume of each subject's hippocampus and then administered a short verbal retention assessment.

Concept 2: Scatterplots

As we learned in Chapters 1 and 2, plotting data should always be our first step in a data exploration. The most useful graph for exploring bivariate relationships is the scatterplot. Making a scatterplot is pretty easy. 1) Determine which variable goes on which axis. (Hint: eXplanatory goes on the *x*-axis!) 2) Label and scale the axes. 3) Plot individual data values. Once you have the scatterplot constructed, take some time to describe what you see. What is the overall form of the relationship? Is it linear? Nonlinear? What direction does the relationship take? Is it positive or negative? How strong is the relationship? Do the points follow the pattern closely, or are they widely scattered? Finally, are there any outliers?

Check for Understanding: ____ *I can make a scatterplot to display the relationship between two quantitative variables.* ____ *I can describe the direction, form, and strength of a relationship displayed in a scatterplot and identify unusual features.*

Use the following data to construct a scatterplot and describe the form, direction, and strength of the relationship between anxiety and exam performance. Note: Higher anxiety scores indicate higher levels of test anxiety.

Anxiety	23	14	14	0	7	20	20	15	21	4
Exam Score	43	59	48	77	50	52	46	60	51	70

Scatterplot:

Concept 3: Measuring Linear Relationships - Correlation

Scatterplots are great tools for displaying the direction, form, and strength of the relationship between two quantitative variables. Often, we will want to know whether or not the relationship is linear and, if so, how strong the linear relationship is. However, our eyes aren't the most accurate judges of the strength of linear relationships. The correlation r provides a numerical summary of the strength of the linear relationship that can be easily interpreted. Some key points to remember about r: It is always a number between -1 and 1. Perfect linear relationships are defined when $r = 1$ or $r = -1$. Positive relationships have a positive correlation and vice versa. Finally, the closer $|r|$ is to 1, the stronger the linear relationship between the quantitative variables. Note, however, just because the linear relationship is strong, it is possible that curvature still exists! Scatterplots that display little to no pattern will have a correlation close to 0.

Check for Understanding: ____ *I can interpret the correlation.*

Use the data from the anxiety vs. exam score example to calculate and interpret the correlation coefficient. What does this value tell you about the relationship?

Section 3.2: Least-Squares Regression

Before You Read: Section Summary

In the last section, we learned that we can display the relationship between two quantitative variables using a scatterplot. Further, we can use a scatterplot to describe the direction, form, and strength of the relationship. The correlation coefficient *r* allows us to describe the situation by telling us how strong the *linear* relationship between the variables is. In this section, we'll learn how to summarize the overall pattern of a linear relationship by finding the equation of the least squares regression line. This line can be used to not only model the linear relationship, but also to make predictions based on the overall pattern. Like correlation, our calculator will do most of the work for us. Your job is to be able to interpret and apply the results!

Learning Targets:

____ I can make predictions using regression lines, keeping in mind the dangers of extrapolation.
____ I can calculate and interpret a residual.
____ I can interpret the slope and *y* intercept of a regression line.
____ I can determine the equation of a least-squares regression line using technology or computer output.
____ I can construct and interpret residual plots to assess whether a regression model is appropriate.
____ I can interpret the standard deviation of the residuals and r^2 and use these values to assess how well a least-squares regression line models the relationship between two variables.
____ I can describe how the least-squares regression line, standard deviation of the residuals, and r^2 are influenced by outliers.
____ I can find the slope and *y* intercepts of the least-squares regression line from the means and standard deviations of *x* and *y* and their correlation.

While You Read: Key Vocabulary and Concepts

regression line:
extrapolation:
slope:
y-intercept:
least-squares regression line:
residual:

residual plot:
standard deviation of the residuals, *s:*
coefficient of determination r^2:
outliers and influential points:
association vs. causation:

After You Read: Check for Understanding

Concept 1: Least-Squares Regression Line

The main concept in this section is that of the least-squares regression line. When a scatterplot suggests a linear relationship between quantitative explanatory and response variables, we can summarize the pattern by "fitting" a line to the points. This line can then be used to predict values of the response variable for given values of the explanatory variable. Be careful, though. We should use caution not to make predictions too far outside of our observed *x*-values. Extrapolation can be dangerous, as we don't know whether or not the pattern continues outside our observations!

The equation of the least-squares regression line can be calculated by hand, if you know the mean and standard deviation of the variables and the correlation *r*. However, you might want to rely on technology to provide the equation for you. Focus your energy on interpreting the slope and *y*-intercept in context and on using the model to make predictions. The y-intercept is often meaningless in the context of our situations. It tells us what response value we'd predict to see if our explanatory value was zero. The slope is the key value of interest in describing the relationship between two quantitative variables. It tells us how much of an increase (or decrease) we expect to see, on average, in our predicted y-values for each one-unit increase in our x-values. Get familiar with that concept, as we will see it again in future chapters!

By the end of this section, you should be able to construct and interpret a least-squares regression model, justify its use, and use it to make predictions. Be sure to focus on interpreting the different components of the model!

Concept 2: Regression Lines-Prediction and Extrapolation

A regression line is a line the models the data. That is, it summarizes the overall pattern and provides an equation that represents the relationship between our explanatory and response variable. This equation can be used to predict the response for a given value of the explanatory variable. Use caution not to extrapolate when making predictions, though, as we do not know if the relationship between the variables extends far beyond the observed values of *x*!

Concept 3: Least-Squares Regression Line

Chances are any scatterplot you construct or encounter will not display a perfectly straight line. In most cases, the observed points will be, well, scattered. Since most of our observed relationships are not perfectly linear, predictions of *y* made from our regression line will often be different than observed *y* values, resulting in a prediction error. That is, there will be some amount of vertical distance between the regression line and the observed value. This vertical difference (actual *y* – predicted *y*) is called a residual. The regression line that "best fits" our observed data is the one that minimizes the squared residuals. This "line of best fit" that minimizes that prediction error is called the least-squares regression line.

Familiarize yourself with the formulas that can be used to determine the slope and intercept of the least-squares regression line. We will rely on technology to generate this equation, but you should recognize that we can construct the equation by hand given the mean and standard deviation of *x* and *y* as well as the correlation *r* between them.

Once you have the equation of the least-squares regression line, you should be able to interpret it and use it. The most important feature to note when interpreting is the slope. You should be able to explain what the slope means in the context of the variables you are analyzing. That is, the slope represents the expected change in the predicted *y* value for each one-unit increase of the *x* value. Be sure to get familiar with this interpretation as you may be asked to provide it on the AP Exam!

Check for Understanding: ____ *I can make predictions using regression lines, keeping in mind the dangers of extrapolation.* ____ *I can calculate and interpret a residual.* ____ *I can interpret the slope and y intercept of a regression line.* ____ *I can determine the equation of a least-squares regression line using technology or computer output.*

Using the following data, determine the least-squares regression line to predict exam scores from anxiety scores. Note: Higher anxiety scores indicate higher levels of test anxiety.

Anxiety	23	14	14	0	7	20	20	15	21	4
Exam Score	43	59	48	77	50	52	46	60	51	70

a) What is the equation of the least-squares regression line?

b) Interpret the slope of the least-squares regression line in the context of the situation.

c) What exam score can we predict for an anxiety score of 15?

d) What is the residual for an anxiety score of 15?

e) Would you use your least-squares regression line to predict an exam score for a person who had an anxiety score of 35? Why or why not?

Concept 4: Assessing How Well the Least-Squares Regression Line Fits the Data

In Section 3.1, we learned that our eyes aren't always the best judge of linear relationships. While correlation r gives us a better understanding of the strength of the linear relationship, we still need to assess how well the least-squares regression line fits the observed data. If it fits well, it may be a useful prediction tool. If it doesn't fit well, we may want to search for a model that fits it better.

One way to assess how well the least-squares regression line fits our data is to make a residual plot. Plotting the residuals gives us more information about the relationship between quantitative variables and helps us assess how well a linear model fits the data. If the residual plot displays a pattern, a better (perhaps nonlinear) model might exist!

We can also assess the fit of the least-squares regression line by interpreting the coefficient of determination r^2. r^2 is a measure of how well the regression model explains the response. Specifically, it is interpreted as the fraction of variation in the values of y that is explained by the least-squares regression line of y on x. For example, if $r^2 = 0.82$, we can say that 82% of the variation in y is due to the linear relationship between y and x. 18% is due to factors other than x.

Check for Understanding: ____ *I can construct and interpret residual plots to assess whether a regression model is appropriate.* ____ *I can interpret the standard deviation of the residuals and r^2 and use these values to assess how well a least-squares regression line models the relationship between two variables.*

Consider the equation of the least squares regression line of exam score on anxiety.

1) Construct and interpret the residual plot for the least-squares regression line.

2) What is the value of r? What is the value of r^2? Interpret each of these in the context of the problem.

Concept 5: Interpreting Computer Regression Output

As noted already, we will often rely on technology to generate the equation of the least-squares regression line. You are probably familiar with using your calculator to produce the equation. Make sure you can also interpret computer output to identify the slope and intercept of the regression line as well as other important values such as correlation and the coefficient of determination. There is a strong possibility you will need to read computer output on the AP Exam!

Check for Understanding: ____ *I can determine the equation of a least-squares regression line using technology or computer output.*

A study was performed to determine the effect of temperature on a pond's algae level. Temperature was measured in degrees F, and algae level was measured in parts per million. Consider the computer output below.

```
Predictor    Coef        Stdev     t-ratio   p
Constant     42.8477     5.750     77.40     0.000
Temp         0.47620     0.5911    13.70     0.000

s = 0.4224          R-sq= 91.7%          R-sq(adj)=91.2%
```

1) Write the equation of the least squares regression line. Identify any variables used.

2) Interpret the slope of the least-squares regression line.

3) Identify and interpret the correlation coefficient.

4) Identify and interpret the standard deviation of the residuals.

Chapter Summary: Modeling Distributions of Data

In this chapter, we expanded our toolbox for working with quantitative data. We learned how to analyze and describe the relationship between two quantitative variables. Using scatterplots, we can display the relationship and describe the direction, strength, and form of the overall pattern. Correlation provides a numerical summary of the strength of the linear relationship between the variables and the equation of the least-squares regression line provides a model that can be used to make predictions. Residual plots, the standard deviation of the residuals, and the coefficient of determination help us assess the fit of the least-squares regression line and may suggest whether or not a linear model is appropriate. Finally, we learned that outliers and influential points can affect our interpretations and regression results. Just like we did with a single quantitative variable, we should be able to identify departures from the overall pattern and explain their influence on our analysis.

Perhaps the most important note for this chapter, though, is that while we now have some tools to help us describe the relationship between two quantitative variables, association does not always imply causation!

After You Read: What Have I Learned?
Complete the vocabulary puzzle, multiple-choice questions, and FRAPPY. Check your answers and performance on each of the learning targets. Be sure to get extra help on any targets that you identify as needing more work!

Learning Target	Got It!	Almost There	Needs Work
I can distinguish between explanatory and response variables for quantitative data.			
I can make a scatterplot to display the relationship between two quantitative variables.			
I can describe the direction, form, and strength of a relationship displayed in a scatterplot and identify unusual features.			
I can interpret the correlation.			
I can understand the basic properties of correlation, including how the correlation is influenced by outliers.			
I can distinguish correlation from causation.			
I can make predictions using regression lines, keeping in mind the dangers of extrapolation.			
I can calculate and interpret a residual.			
I can interpret the slope and *y* intercept of a regression line.			
I can determine the equation of a least-squares regression line using technology or computer output.			
I can construct and interpret residual plots to assess whether a regression model is appropriate.			
I can interpret the standard deviation of the residuals and r^2 and use these values to assess how well a least-squares regression line models the relationship between two variables.			
I can describe how the least-squares regression line, standard deviation of the residuals, and r^2 are influenced by outliers.			
I can find the slope and *y* intercepts of the least-squares regression line from the means and standard deviations of *x* and *y* and their correlation.			

Chapter 3 Multiple Choice Practice

Directions. *Identify the choice that best completes the statement or answers the question. Check your answers and note your performance when you are finished.*

1. A study is conducted to determine if one can predict the academic performance of a first year college student based on their high school grade point average. The explanatory variable in this study is

(A) academic performance of the first year student.
(B) grade point average.
(C) the experimenter.
(D) number of credits the student is taking.
(E) the college.

2. If two variables are positively associated, then

(A) larger values of one variable are associated with larger values of the other.
(B) larger values of one variable are associated with smaller values of the other.
(C) smaller values of one variable are associated with larger values of the other.
(D) smaller values of one variable are associated with both larger /smaller values of the other.
(E) there is no pattern in the relationship between the two variables.

3. The correlation coefficient measures

(A) whether there is a relationship between two variables.
(B) the strength of the relationship between two quantitative variables.
(C) whether or not a scatterplot shows an interesting pattern.
(D) whether a cause and effect relation exists between two variables.
(E) the strength of the linear relationship between two quantitative variables.

4. Consider the following scatterplot, which describes the relationship between stopping distance (in feet) and air temperature (in degrees Centigrade) for a certain 2,000-pound car travelling 40 mph.

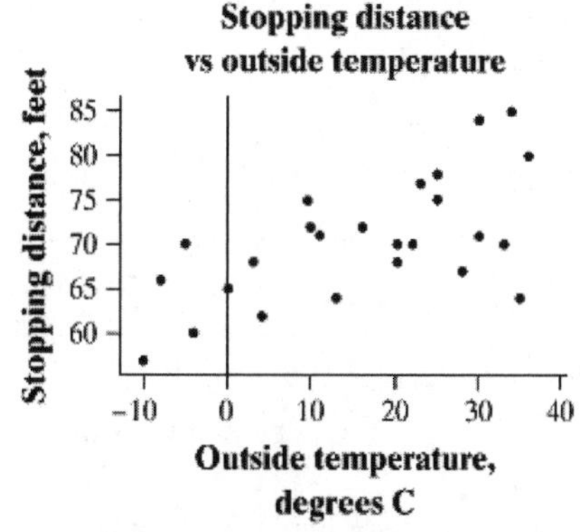

Do these data provide strong evidence that warmer temperatures actually *cause* a greater stopping distance?

(A) Yes. The strong straight-line association in the plot shows that temperature has a strong effect on stopping distance.
(B) No. $r \neq +1$
(C) No. We can't be sure the temperature is responsible for the difference in stopping distances.
(D) No. The plot shows that differences among stopping distances are not large enough to be important.
(E) No. The plot shows that stopping distances go down as temperature increases

5. If stopping distance was expressed in yards instead of feet, how would the correlation *r* between temperatures and stopping distance change?

(A) *r* would be divided by 12.
(B) *r* would be divided by 3.
(C) *r* would not change.
(D) *r* would be multiplied by 3.
(E) *r* would be multiplied by 12.

6. If another data point were added with an air temperature of 0° C and a stopping distance of 80 feet, the correlation would

(A) decrease, since this new point is an outlier that does not follow the pattern in the data.
(B) increase, since this new point is an outlier that does not follow the pattern in the data.
(C) stay nearly the same, since correlation is resistant to outliers.
(D) increase, since there would be more data points.
(E) Whether this data point causes an increase or decrease cannot be determined without recalculating the correlation.

7. Which of the following is true of the correlation *r*?

(A) It is a resistant measure of association.
(B) $-1 \leq r \leq 1$.
(C) If *r* is the correlation between *X* and *Y*, then -*r* is the correlation between *Y* and *X*.
(D) Whenever all the data lie on a perfectly straight line, the correlation *r* will always be equal to +1.0.
(E) All of the above.

Consider the following scatterplot of amounts of CO (carbon monoxide) and NOX (nitrogen oxide) in grams per mile driven in the exhausts of cars. The least-squares regression line has been drawn in the plot.

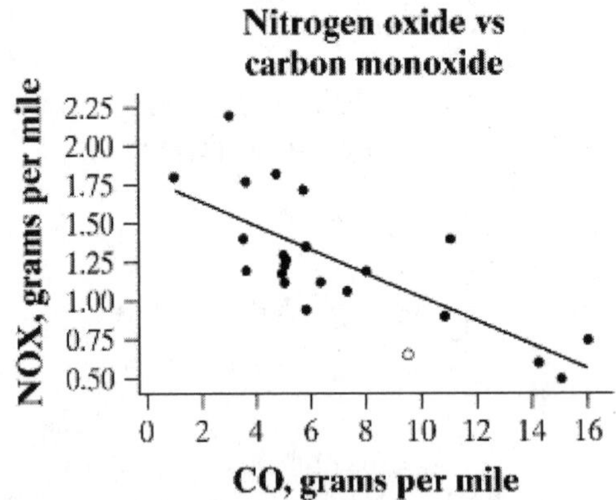

8. Based on the scatterplot, the least-squares line would predict that a car that emits 2 grams of CO per mile driven would emit approximately how many grams of NOX per mile driven?

(A) 4.0
(B) 1.25
(C) 2.0
(D) 1.7
(E) 0.7

9. In the scatterplot, the point indicated by the open circle

(A) has a negative value for the residual.
(B) has a positive value for the residual.
(C) has a zero value for the residual.
(D) has a zero value for the correlation.
(E) is an outlier.

10. Which of the following is correct?

(A) The correlation *r* is the slope of the least-squares regression line.
(B) The square of the correlation is the slope of the least-squares regression line.
(C) The square of the correlation is the proportion of the data lying on the least-squares regression line.
(D) The coefficient of determination is the fraction of variability in *y* that can be explained by least-squares regression of *y* on *x*.
(E) The sum of the squared residuals from the least-squares line is 0.

11. Which of the following statements concerning residuals from a LSRL is true?

(A) The sum of the residuals is always 0.
(B) A plot of the residuals is useful for assessing the fit of the least-squares regression line.
(C) The value of a residual is the observed value of the response minus the value of the response that one would predict from the least-squares regression line.
(D) An influential point on a scatterplot is not necessarily the point with the largest residual.
(E) All of the above.

A fisheries biologist studying whitefish in a Canadian lake collected data on the length (in centimeters) and egg production for 25 female fish. A scatter plot of her results and computer regression analysis of egg production versus fish length are given below. *Note that Number of eggs is given in thousands (i.e., "40" means 40,000 eggs).*

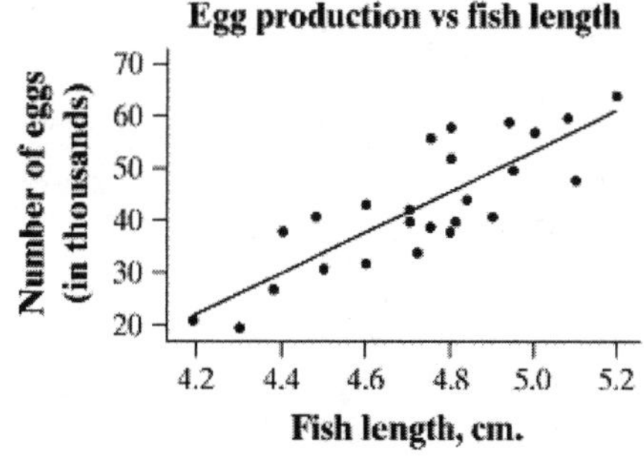

Predictor	Coef	SE Coef	T	P
Constant	-142.74	25.55	-5.59	0.000
Fish length	39.250	5.392	7.28	0.000

S = 6.75133 R-Sq = 69.7% R-Sq(adj) = 68.4%

12. Which of the following statements is a correct interpretation of the slope of the regression line?

(A) For each 1-cm increase in the fish length, the predicted number of eggs increases by 39.25.
(B) For each 1-cm increase in the fish length, the predicted number of eggs decreases by 142.74.
(C) For each 1-unit increase in the number of eggs, the predicted fish length increases by 39.25 cm.
(D) For each 1-unit increase in the number of eggs, the predicted fish length decreases by 142.74cm.
(E) For each 1-cm increase in the fish length, the predicted number of eggs increases by 39,250.

13. What percent of variability in the number of eggs is explained by the least-squares regression of *number of eggs* on *fish length*?

    (A) 25.55
    (B) 5.392
    (C) 6.75133
    (D) 69.7
    (E) Cannot be determined without the original data.

14. A study of the effects of television measured how many hours of television each of 125 grade school children watched per week during a school year and their reading scores. The study found that children who watch more television tend to have lower reading scores than children who watch fewer hours of television. The study report says that, "Hours of television watched explained 25% of the observed variation in the reading scores of the 125 subjects." The correlation between hours of TV and reading score must be

    (A) $r = 0.25$.
    (B) $r = -0.25$.
    (C) $r = -0.5$.
    (D) $r = 0.5$.
    (E) Can't tell from the information given.

15. A study gathers data on the outside temperature during the winter in degrees Fahrenheit and the amount of natural gas a household consumes in cubic feet per day. Call the temperature x and gas consumption y. The house is heated with gas, so x helps explain y. The least-squares regression line for predicting y from x is: $\hat{y} = 1344 - 19x$. When the temperature goes up 1 degree, what happens to the gas usage predicted by the regression line?

    (A) It goes up 19 cubic feet.
    (B) It goes down 19 cubic feet.
    (C) It goes up 1344 cubic feet.
    (D) It goes down 1344 cubic feet.
    (E) Can't tell without seeing the data.

Check your answers below. If you got a question wrong, check to see if you made a simple mistake or if you need to study that concept more. After you check your work, identify the concepts you feel very confident about and note what you will do to learn the concepts in need of more study.

#	Answer	Concept	Right	Wrong	Simple Mistake?	Need to Study More
1	B	Explanatory vs. Response				
2	A	Definition of Association				
3	E	Definition of Correlation				
4	C	Association vs. Causation				
5	C	Correlation				
6	A	Correlation				
7	B	Correlation				
8	D	Predicting with the LSRL				
9	A	Residuals				
10	D	Coefficient of Determination				
11	E	Residuals				
12	E	Slope of the LSRL				
13	D	Coefficient of Determination				
14	C	Coefficient of Determination				
15	B	Slope of the LSRL				

Chapter 3 Reflection

Summarize the "Big Ideas" in Chapter 3:

My strengths in this chapter:

Concepts I need to study more and what I will do to learn them:

FRAPPY! Free Response AP® Problem, Yay!

The following problem is modeled after actual Advanced Placement Statistics free response questions. Your task is to generate a complete, concise response in 15 minutes. After you generate your response, view two example solutions and determine whether or not you feel they are "complete," "substantial," "developing" or "minimal". If they are not "complete," what would you suggest to the student who wrote them to increase their score? Finally, you will be provided with a rubric. Score your response and note what, if anything, you would do differently to increase your own score.

A recent study was interested in determining the optimal location for fire stations in a suburban city. Ideally, fire stations should be placed so the distance between the station and residences is minimized. One component of the study examined the relationship between the amount of fire damage *y* (in thousands of dollars) and the distance between the fire station and the residence *x* (in miles). The results of the regression analysis are below.

```
Predictor     Coef     SE Coef         T          P
Constant      10.28    1.42            7.237      0.000
X             4.92     0.39            12.525     0.000

s = 2.232     R-Sq = 0.9235            R-Sq(adj) = 0.9176
```

(a) Write the equation of the least squares regression line. Define any variables used. Interpret the slope of the equation in context.

(b) A home located 3 miles from the fire station received $22,300 in damage. Use your equation in part (a) to calculate and interpret the residual for this observation.

(c) Identify and interpret the correlation coefficient.

FRAPPY! Student Responses

Student Response 1:

a) $\hat{y}$ = 10.28 + 4.92*x*
For each additional mile between the fire station and residence, we predict about 4,920 additional dollars in damages.

b) $\hat{y}$ = 10.28 + 4.92(3) = 25.04. Residual = 25.04 – 22.3 = 2.74. Our model overpredicted the amount of damage for this observation by $2,740.

c) r^2 = 0.9235. There is a strong, positive linear relationship between the distance between a fire station and residence and the resulting damage in a fire.

How would you score this response? Is it substantial? Complete? Developing? Minimal? Is there anything this student could do to earn a better score?

Student Response 2:

a) $\widehat{firedamage}$ = 4.92*distance* + 10.28
We predict about $4,920 additional damage for each increase of one mile between the fire station and residence that is on fire.

b) $\widehat{damage}$ = 2.92(3) + 10.28 = 25.04
residual = 22.3 – 25.04 = - 2.74. Our model overpredicts the damage amount by $2,740.

c) r = 0.96. There is a very strong, positive, linear relationship between a residence's damage from a fire and its distance from a fire station.

How would you score this response? Is it substantial? Complete? Developing? Minimal? Is there anything this student could do to earn a better score?

FRAPPY! Scoring Rubric

Use the following rubric to score your response. Each part receives a score of "Essentially Correct," "Partially Correct," or "Incorrect." When you have scored your response, reflect on your understanding of the concepts addressed in this problem. If necessary, note what you would do differently on future questions like this to increase your score.

Intent of the Question

The goal of this question is to determine your ability to interpret computer regression output and explain key concepts of linear regression.

Solution

(a) $\widehat{firedamage} = 10.28 + 4.92 distance$ OR $\hat{y} = 10.28 + 4.92x$ with x and y defined as distance and damage.

For each additional mile between the fire station and residence, we predict about $4920 additional dollars in damages.

(b) $\widehat{damage} = 4.92(3) + 10.28 = 25.04$

residual = 22.3 – 25.04 = - 2.74.

The model overpredicts the damage amount by $2740.

(c) Since $r^2 = 0.9325$, $r = 0.96$. There is a very strong, positive, linear relationship between a residence's damage from a fire and its distance from a fire station.

Scoring

Parts (a), (b), and (c) are scored as essentially correct (E), partially correct (P), or incorrect (I).

Part (a) is essentially correct if the response (1) correctly identifies the least-squares regression equation in context or with variables defined and (2) correctly interprets the slope

Part (a) is partially correct if the response fails to define the variables in context or reverses the coefficients OR if the slope is not correctly defined in context (eg, predicts 4.92 dollars instead of $4920).

Part (b) is essentially correct if (1) the correct residual is calculated and (2) the interpretation is correct.

Part (b) is partially correct if only one of the above elements is correct.

Part (c) is essentially correct if the correlation coefficient is correctly identified and interpreted correctly with all three elements (strong, positive, linear).

Part (c) is partially correct if one of the elements (strong, positive, linear) is missing OR if r^2 is used instead of r.

FRAPPY! Reflection

4 Complete Response
All three parts essentially correct

3 Substantial Response
Two parts essentially correct and one part partially correct

2 Developing Response
Two parts essentially correct and no parts partially correct
One part essentially correct and two parts partially correct
Three parts partially correct

1 Minimal Response
One part essentially correct and one part partially correct
One part essentially correct and no parts partially correct
No parts essentially correct and two parts partially correct

My Score:
What I did well:
What I could improve:
What I should remember if I see a problem like this on the AP Exam:

Chapter 3: Describing Relationships

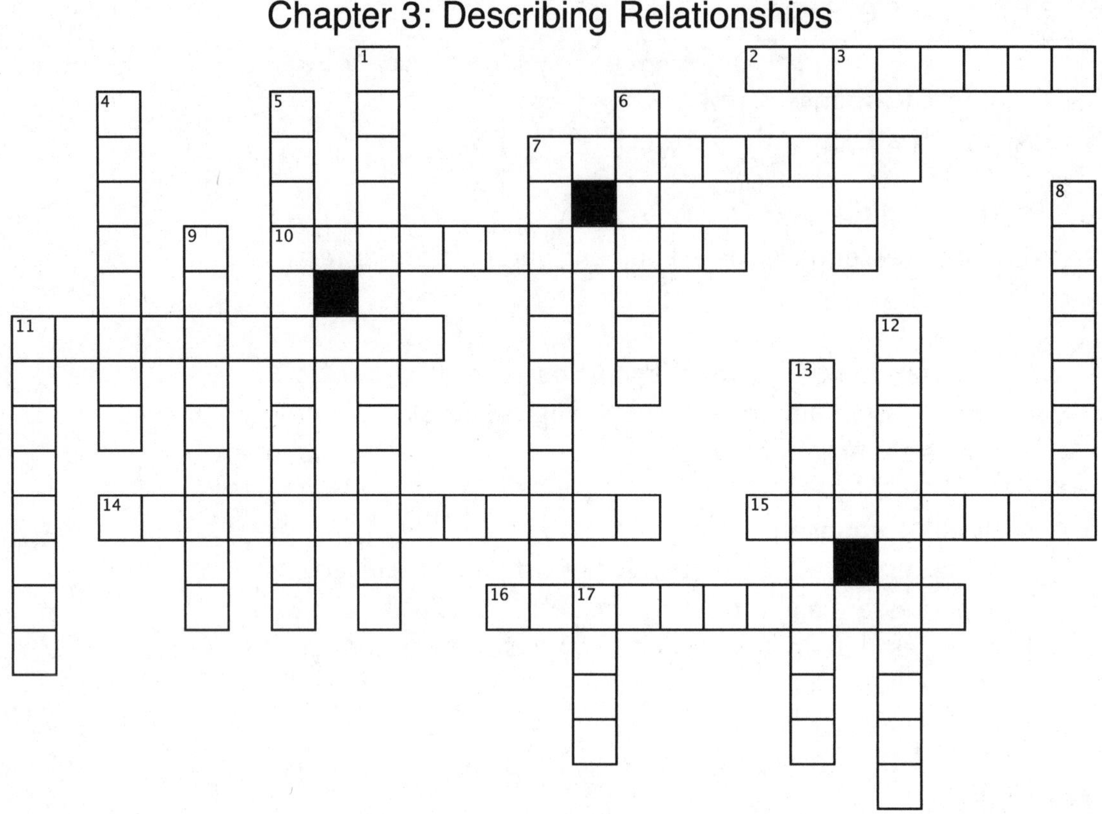

Across

2. the difference between an observed value of the response and the value predicted by a regression line
7. Important note: Association does not imply ______.
10. graphical display of the relationship between two quantitative variables
11. line that describes the relationship between two quantitative variables
14. the coefficient of ______ describes the fraction of variability in y values that is explained by least squares regression on x.
15. A _____ association is defined when above average values of one variable are accompanied by below average values of the other.
16. individual points that substantially change the correlation or slope of the regression line

Down

1. the use of a regression line to make a prediction far outside the observed x values
3. the amount by which y is predicted to change when x increases by one unit
4. The ______ of a relationship in a scatterplot is determined by how closely the point follow a clear form.
5. the ____-_____ regression line is also known as the line of best fit (2 words)
6. an individual value that falls outside the overall pattern of the relationship
7. value that measures the strength of the linear relationship between two quantitative variables
8. A ______ association is defined when above average values of the explanatory are accompanied by above average values of the response
9. y-hat is the _______ value of the y-variable for a given x
11. variable that measures the outcome of a study
12. variable that may help explain or influence changes in another variable
13. The _____ of a scatterplot indicates a positive or negative association between the variables.
17. The ____ of a scatterplot is usually linear or nonlinear.

CHAPTER 4: DESIGNING STUDIES

"Not everything that can be counted counts; and not everything that counts can be counted" ~George Gallup

Chapter Overview

The first three chapters introduced us to some of the basics of exploratory data analysis. In this chapter, we'll learn about the second major topic in statistics –planning and conducting a study. Since it is difficult to perform a descriptive statistical analysis without data, we need to learn appropriate ways to produce data. We will start by learning the difference between a population and a sample. Then, we will study sampling techniques and learn how to identify potential sources of bias. Our goal is to collect data that is representative of the population we wish to study. Therefore, it is important that our data collection techniques do not systematically over- or under-represent any segment of the population. In the second section, we will learn the difference between observational studies and experiments. We will learn a number of different ways to design experiments so we can establish relationships between variables. Finally, we will conclude this chapter by reviewing some cautions about using studies wisely. This chapter contains a lot of vocabulary and concepts. Be sure to study them, as proper application of terms is important for strong statistical communication!

Sections in this Chapter

Section 4.1: Sampling and Surveys
Section 4.2: Experiments
Section 4.3: Using Studies Wisely

Plan Your Learning

Use the following *suggested* guide to help plan your reading and assignments in "The Practice of Statistics, 6th Edition." Note: your teacher may schedule a different pacing or assign different problems. Be sure to follow his or her instructions!

Read	**4.1: pp. 220-228**	**4.1: pp. 229-232**	**4.1: pp. 232-236**	**4.2: pp. 241-247**
Do	1,3,5,7,11,13,15	17,19,21,22,23	25,27,29,31,33 MC 35-40	43,45,47,49,51,53

Read	**4.2: pp. 247-252**	**4.2: pp. 252-256**	**4.2: pp. 256-262**	**4.3: pp. 269-275**
Do	57, 59, 61, 63	55,65,67,69	71,75,77,79 MC 83-90	93,95,97,99

Read	**4.3: pp. 275-280**	**Chapter Summary**
Do	103,105,107 MC 117-118	Multiple-Choice FRAPPY!

Section 4.1: Sampling and Surveys

Before You Read: Section Summary

Often in statistics, our goal is to draw a conclusion about a population based on information gathered from a sample. In order for us to make a valid inference about the population, we must feel confident that the sample obtained will be representative of the group as a whole. There are a number of different ways to select samples from a population—some better than others. In this section, you will explore ways in which you can sample badly, how to sample well, and cautions to consider when sampling. You will be introduced to a number of different sampling methods. Be sure to familiarize yourself with how to select samples using each of the methods and how to explain potential advantages and disadvantages of each.

Learning Targets:

____ I can identify the population and sample in a statistical study.
____ I can identify voluntary response sampling and convenience sampling and explain how these sampling methods can lead to bias.
____ I can describe how to select a simple random sample with technology or a table of random digits.
____ I can describe how to select a sample using stratified random sampling and cluster sampling, distinguish stratified random sampling from cluster sampling, and give an advantage of each method.
____ I can explain how undercoverage, nonresponse, question wording, and other aspects of a sample survey can lead to bias.

While You Read: Key Vocabulary and Concepts

population:
sample:
census:
sample survey:
convenience sampling:
bias:
voluntary response sampling:
random sampling:
simple random sample (SRS):

strata:
stratified random sampling:
cluster:
cluster sampling:
undercoverage:
nonresponse:
response bias:

 **After You Read: Check for Understanding**

Concept 1: Population, Samples, and Inference

The purpose of sampling is to provide us with information about a population without actually gathering the information from every single element of the whole group. A sample is a part of the population from which we collect information. This information is then used to infer something about the population. We can have more faith in our inference if we are confident that the sampling method used is likely to produce a sample that is representative of the population of interest. Some sampling methods may introduce a level of bias into the situation that could cause us to make an incorrect inference about the population. It is important that you can clearly identify the population of interest and describe whether or not the sampling method is unbiased.

Check for Understanding: ____ *I can identify the population and sample in statistical study.*

A publisher is interested in determining the reading difficulty of mathematics textbooks. Reading difficulty is determined by the length of sentences and the length of words used in the text. Researchers randomly select 10 paragraphs out of the most popular Algebra 1 textbooks and calculate the average sentence length and average word length for that type of mathematic textbook. Identify the population and sample.

Concept 2: How to Sample Well

The key to sampling is to use a method that helps ensures that the sample is as representative of the population as possible. Because sampling methods like voluntary response and convenience samples can systematically favor certain outcomes in a population, we say they are biased. An unbiased sampling method is one that does not favor any element of the population. We rely on the use of chance to select unbiased samples. The easiest way to do this is to select a simple random sample (SRS) from the population of interest. One way we could select an SRS would be to write the name of each individual in the population on a piece of paper, put the pieces in a hat, mix them well, and draw out the necessary number of slips for our sample. We could also select an SRS by labeling each individual in the population with a number of the same length (ie. one-, two-, three- digit number depending on the size of the population) and then generate random numbers from technology or a random number table until you reach the desired sample size. The labeled individuals that match the generated numbers, ignoring repeats, make up the sample.

Check for Understanding: ____ *I can describe how to select a simple random sample with technology or a table of random digits.*

An alphabetized list of student names is found below. Use the random digit table provided to choose an SRS of 4 individuals. Clearly indicate how you are using the table to select the SRS.

Abney
Andreasen
Bearden
Bready
Buckley
Brock
Bush
Chauvet
Costello
Derksen
Greenberg
Knott
Lacey
Martin
McDonald
Osters
Preble
Ripp
Rohnkol
Sterken
Tyson
Wilcock
Yankay

19223 95034 05756 28713 96409 12531 42544 82853

73676 47150 99400 11327 27754 40548 82421 76290

While Simple Random Samples give each group of *n* individuals in the population an equal chance to be selected, they are not always the easiest to obtain. Several other sampling methods are available that can be used instead of an SRS. If the population consists of several groups of individuals that are likely to produce similar responses within groups, but systematically different responses between groups, consider taking a stratified random sample. To do this, select an SRS from each stratum (group) to obtain the sample. Another method, cluster sampling, divides the population into smaller groups (clusters) that mirror the overall population and selects an SRS of those clusters. You should be able to describe how to select a sample using each of these methods as you may be asked to do so on the AP exam!

Concept 3: What Can Go Wrong?

When designing a sample survey, random sampling helps avoid bias. But there are other kinds of mistakes in the sampling process that can lead to inaccurate information about the population. For instance, if the sampling method is designed in a way that leaves out certain segments of the population, the sample survey suffers from undercoverage. When selected individuals cannot be contacted or refuse to participate, the survey suffers from nonresponse. If people give incorrect or misleading responses, a survey suffers from response bias. Finally, if a question is worded in a way that favors certain responses, the survey suffers from question wording bias. Be sure to design sample surveys to avoid these issues and make sure you can identify them in existing studies.

Check for Understanding: ____ *I can explain how undercoverage, nonresponse, question wording, and other aspects of a sample survey can lead to bias.*

An SRS of 2,400 recent high school graduates was asked, "Did you ever cheat on an exam during your senior year?" Of the 2,400 respondents, 64 percent admitted to cheating on at least one test.

What is one practical problem with this survey?

What is the likely direction of the bias?

Section 4.2: Experiments

Before You Read: Section Summary

Sample surveys allow us to gather information about the population without actually doing anything to that population. Such observational studies provide a snapshot of the population, but cannot be used to establish any sort of cause-effect relationship. Observational studies only allow us to describe the population, compare groups, or examine basic relationships between variables. In this section, we will move beyond sampling to study the elements of experimental design. Experiments allow us to produce data in a way that can lead to conclusions about causation. There are a lot of vocabulary terms in this section. It is easy to confuse sampling terms and experimental design terms. Make sure you understand each vocabulary term or concept!

Learning Targets:

____ I can explain the concept of confounding and how it limits the ability to make cause-and effect conclusions.

____ I can distinguish between an observational study and an experiment, and identify the explanatory and response variables in each type of study.

____ I can identify experimental units and treatments in an experiment.

____ I can describe the placebo effect and the purpose of blinding in an experiment.

____ I can describe how to randomly assign treatments in an experiment using slips of paper, technology, or a table of random digits.

____ I can explain the purpose of comparison, random assignment, control, and replication in an experiment.

____ I can describe a completely randomized design for an experiment.

____ I can describe a randomized block design and a matched pairs design for an experiment and explain the purpose of blocking in an experiment.

While You Read: Key Vocabulary and Concepts

observational study:
response variable:
explanatory variable:
confounding:
experiment:
placebo:
treatment:

experimental unit:
subjects:
factors:
levels:
control group:
placebo effect:
random assignment:
double-blind/single-blind:
random assignment:
control:
replication:
completely randomized design:
block:
randomized block design:
matched pairs design:

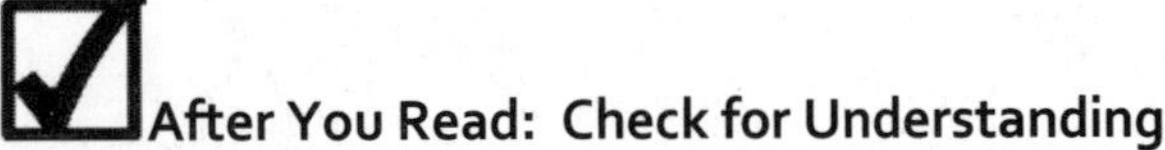

After You Read: Check for Understanding

Concept 1: Observational Studies vs. Experiments

Sample surveys are examples of observational studies. Their goal is to describe the population and examine relationships between variables. Often, however, we wish to determine whether or not a cause-effect relationship exists between an explanatory and a response variable. Observational studies cannot be used to establish this relationship because of confounding between the explanatory variable and one or more other variables.

Check for Understanding: ____ *I can explain the concept of confounding and how it limits the ability to make cause-and effect conclusions.*

Does shoe size affect spelling ability? A recent study was conducted in a suburban school district to answer this question. 30 students from grades 1 through 8 were randomly selected. Each student was administered a spelling test and had his or her feet measured. Test scores were plotted against shoe size and a strong, positive relationship was observed.

1. Was this an observational study or an experiment?

2. What are the explanatory and response variables?

3. Suggest a possible confounding variable in this setting. Explain carefully how it may confound the results.

Concept 2: How to Experiment Well

If we wish to establish cause and effect, we must conduct an experiment in which treatments are imposed on experimental units and other potential influences on the response variable are controlled as much as possible. The basic idea behind an experiment is that we obtain experimental units, apply treatments, and measure the results. When conducted properly, experiments can provide good evidence of causation. Proper experimental design incorporates three principles. First, we must control for the influences of variables that might affect the response. We want to ensure that any changes are due to the treatment alone. Second, we must randomly assign experimental units to treatments. This helps ensure equivalent groups of units. Random assignment helps ensure that the effects of variables will be felt equally by all groups in the experiment, so that the only systematic difference between groups are the treatments themselves. Finally, replication is necessary to ensure enough experimental units are in each group to convince us the differences in the effects of the treatments can be separated from chance differences between the groups. If a difference between groups is observed that is too large to have occurred by chance alone, we say it is statistically significant. We will learn how to establish statistical significance in a later chapter. For now, focus on how to design a quality experiment!

Check for Understanding: ____ *I can identify experimental units and treatments in an experiment.* ____ *I can describe a completely randomized design for an experiment.*

Mr. Tyson teaches statistics to 150 students. He is interested in knowing whether or not listening to classical music while studying results in higher test scores than listening to no music. He wishes to design an experiment to answer this question.

1. What are the experimental units, explanatory variable, treatments and response variable?

2. What other potential variables could confound the results in this situation? How could they affect the results? How could we avoid their effects?

3. Describe a completely randomized design for Mr. Tyson's experiment.

Concept 3: Other Types of Experimental Design

A completely randomized experiment is the simplest design that can give good evidence of cause-effect relationships. In some cases, however, we can add elements to this basic design to control for the effects of other variables. For example, placebos (fake treatments) can be given to control for the placebo effect—the phenomenon that occurs when subjects respond to getting any treatment, whether it is real or not. We can use blinding when there is a concern that knowing who receives what treatment might affect the results. Double-blind experiments ensure that neither the subjects nor the people who interact with the subjects and measure their responses know who receives which treatment.

When groups of subjects share a common characteristic that might systematically affect their responses to the treatments, we can use blocking to control for the effects of this variable. For example, suppose researchers are conducting an experiment to compare the effectiveness of a new medicine for treating high blood pressure with the most commonly prescribed drug. If they believe that older and younger people may respond differently to such medications, the researchers can separate the subjects into blocks of older and younger people and randomly assign treatments within each block. This randomized block design helps isolate the variation in responses due to age, which makes it easier for researchers to find evidence of a treatment effect. If we are only comparing two treatments, we can sometimes conduct a matched pairs design by creating blocks of two similar individuals and then randomly assigning each subject to a treatment. Another type of matched pairs design involves assigning two treatments to each subject in a random order.

In each type of experimental design, our goal is to ensure control, random assignment, and replication. If these principles are addressed in our design, we can establish good evidence of causal relationships. Make sure you can describe how you would conduct an experiment using each of these designs as you may be asked to do so on the AP exam!

Check for Understanding: ____ *I can describe a randomized block design and a matched pairs design for an experiment and explain the purpose of blocking in an experiment.*

Refer to the previous Check for Understanding. Describe an experimental design involving blocking that will help answer Mr. Tyson's question. Explain why this design is preferable to a completely randomized design.

Section 4.3: Using Studies Wisely

Before You Read: Section Summary

A common mistake on the AP exam is to confuse inference about the population and inference about cause and effect. In this section, we learn about the difference between the two as well as some of the challenges of establishing causation. While it is important to know how to design good surveys and experiments, it is just as important to understand how to properly use their results!

Learning Targets:

- ____ I can explain the concept of sampling variability when making an inference about a population and how sample size affects sampling variability.
- ____ I can explain the meaning of statistically significant in the context of an experiment and use simulation to determine if the results of an experiment are statistically significant.
- ____ I can identify when it is appropriate to make an inference about a population and when it is appropriate to make an inference about cause and effect.
- ____ I can evaluate if a statistical study has been carried out in an ethical manner.

While You Read: Key Vocabulary and Concepts

sampling variability:
sampling variability and sample size:
statistically significant:
scope of inference:
basic data ethics:

After You Read: Check for Understanding

Concept 1: Scope of Inference

When the members of a sample are selected at random from a population, we can use the sample results to infer things about the population. That is, well-designed samples allow us to make inferences about the population from which we sampled. Random sampling is what allows us to generalize our results with confidence. Larger random samples tend to provide more precise estimates of the true population value of interest than smaller random samples.

If our goal is to make an inference about cause and effect, we must use a randomized experiment. Unless the experimental units were randomly selected from a larger population of interest, we cannot extend our conclusions beyond individuals like those who took part in the experiment.

Check for Understanding: ____*I can explain the concept of sampling variability when making an inference about a population and how sample size affects sampling variability.*

In a random sample of 50 AP Statistics students, the average amount of change carried in their pockets is $0.42.

1. Do you think the true average amount of change carried by all AP Statistics students is $0.42?

2. Which would be more likely to give an estimate close to the true average amount of change carried by all AP Statistics students: a random sample of 50 students, or a random sample of 200 students?

Concept 2: The Challenges of Establishing Causation

Well-designed experiments can be used to establish causation. However, lack of realism in some experiments prevents us from seeing similar results outside the laboratory setting. In some cases, it is not practical, safe, or ethical to conduct an experiment. Even with strong evidence from observational studies, it is very difficult to establish a cause and effect conclusion.

Concept 3: Data Ethics (optional)

Because some sample surveys and experiments have potential to cause harm to the participants, it is important to consider data ethics when designing a study. Basic data ethics include having an institutional review board that monitors the well-being of the participants. Individuals must give their informed consent before data are collected and all individual results should be kept confidential.

Chapter Summary: Designing Studies

This chapter is an important one in your study of statistics. After all, we cannot describe or analyze data without collecting it first! Since one of the major goals of statistics is to make inferences that go beyond the data, it is critical that we produce data in a way that will allow for such inferences. Biased data production methods can lead to incorrect inferences. Random sampling allows us to make an inference about the population as a whole. Well-designed experiments in which we randomly assign treatments and control for other variables allow us to make inferences about cause and effect. We will learn how to perform these inferences in later chapters. Your goal in this chapter is to be able to describe good sampling and experimental design techniques and recognize when sampling or experimental design has been done poorly. There is almost always a question about sampling or experimental design on the free-response portion of the AP exam. Be sure to familiarize yourself with all of the vocabulary and concepts from this chapter so you can answer that question with confidence!

After You Read: What Have I Learned?

Complete the vocabulary puzzle, multiple-choice questions, and FRAPPY. Check your answers and performance on each of the learning targets. Be sure to get extra help on any targets that you identify as needing more work!

Learning Target	Got It!	Almost There	Needs Work
I can identify the population and sample in a statistical study.			
I can identify voluntary response sampling and convenience sampling and explain how these sampling methods can lead to bias.			
I can describe how to select a simple random sample with technology or a table of random digits.			
I can describe how to select a sample using stratified random sampling and cluster sampling, distinguish stratified random sampling from cluster sampling, and give an advantage of each method.			
I can explain how undercoverage, nonresponse, question wording, and other aspects of a sample survey can lead to bias.			
I can explain the concept of confounding and how it limits the ability to make cause-and-effect conclusions.			
I can distinguish between an observational study and an experiment, and identify the explanatory and response variables in each type of study.			
I can identify experimental units and treatments in an experiment.			
I can describe the placebo effect and the purpose of blinding in an experiment.			
I can describe how to randomly assign treatments in an experiment using slips of paper, technology, or a table of random digits.			
I can explain the purpose of comparison, random assignment, control, and replication in an experiment.			
I can describe a completely randomized design for an experiment.			
I can describe a randomized block design and a matched pairs design for an experiment and explain the purpose of blocking in an experiment.			
I can explain the concept of sampling variability when making an inference about a population and how sample size affects sampling variability.			
I can explain the meaning of statistically significant in the context of an experiment and use simulation to determine if the results of an experiment are statistically significant.			
I can identify when it is appropriate to make an inference about a population and when it is appropriate to make an inference about cause and effect.			
I can evaluate if a statistical study has been carried out in an ethical manner.			

Chapter 4 Multiple Choice Practice

Directions. *Identify the choice that best completes the statement or answers the question. Check your answers and note your performance when you are finished.*

1. A researcher is testing a company's new stain remover. He has contracted with 40 families who have agreed to test the product. He randomly assigns 20 families to the group that will use the new stain remover and 20 to the group that will use the company's current product. The most important reason for this random assignment is that

   (A) randomization makes the analysis easier since the data can be collected and entered into the computer in any order.
   (B) randomization eliminates the impact of any confounding variables.
   (C) randomization is a good way to create two groups of 20 families that are as similar as possible, except for the treatments they receive.
   (D) randomization ensures that the study is double-blind.
   (E) randomization reduces the impact of outliers.

2. A researcher observes that, on average, the number of traffic violations in cities with Major League Baseball teams is larger than in cities without Major League Baseball teams. The most plausible explanation for this observed association is that the

   (A) presence of a Major League Baseball team causes the number of traffic incidents to rise (perhaps due to the large number of people leaving the ballpark).
   (B) high number of traffic incidents is responsible for the presence of Major League Baseball teams (more traffic incidents means more people have cars, making it easier for them to get to the ballpark).
   (C) association is due to the presence of another variable (Major League teams tend to be in large cities with more people, hence a greater number of traffic incidents).
   (D) association makes no sense, since many people take public transit or walk to baseball games.
   (E) observed association is purely coincidental. It is implausible to believe the observed association could be anything other than accidental.

3. A researcher is testing the effect of a new fertilizer on crop growth. He marks 30 plots in a field, splits the plots in half, and randomly assigns the new fertilizer to one half of the plot and the old fertilizer to the other half. After 4 weeks, he measures the crop yield and compares the effects of the two fertilizers. This design is an example of

   (A) matched pairs experiment
   (B) completely randomized comparative experiment
   (C) cluster experiment
   (D) double-blind experiment
   (E) this is not an experiment

4. A large suburban school wants to assess student attitudes towards their mathematics textbook. The administration randomly selects 15 mathematics classes and gives the survey to every student in the class. This is an example of a

   (A) multistage sample
   (B) stratified sample
   (C) cluster sample
   (D) simple random sample
   (E) convenience sample

5. Eighty volunteers who currently use a certain brand of medication to reduce blood pressure are recruited to try a new medication. The volunteers are randomly assigned to one of two groups. One group continues to take their current medication, the other group switches to the new experimental medication. Blood pressure is measured before, during, and after the study. Which of the following best describes a conclusion that can be drawn from this study?

   (A) We can determine whether the new drug reduces blood pressure more than the old drug for anyone who suffers from high blood pressure.
   (B) We can determine whether the new drug reduces blood pressure more than the old drug for individuals like the subjects in the study.
   (C) We can determine whether the blood pressure improved more with the new drug than with the old drug, but we can't establish cause and effect.
   (D) We cannot draw any conclusions, since the all the volunteers were already taking the old drug when the experiment started.
   (E) We cannot draw any conclusions, because there was no control group.

6. To determine employee satisfaction at a large company, the management selects an SRS of 200 workers from the marketing department and a separate SRS of 50 workers from the sales department. This kind of sample is called a

   (A) simple random sample
   (B) simple random sample with blocking
   (C) multistage random sample
   (D) stratified random sample
   (E) random cluster sample

7. For a certain experiment you have 8 subjects, of which 4 are female and 4 are male. The name of the subjects are listed below:

   Males: Atwater, Bacon, Chu, Diaz. *Females: Johnson, King, Liu, Moore*

   There are two treatments, A and B. If a randomized block design is used with the subjects blocked by their gender, which of the following is *not* a possible group of subjects who receive treatment A?

   (A) Atwater, Chu, King, Liu
   (B) Bacon, Chu, Liu, Moore
   (C) Atwater, Diaz, Liu, King
   (D) Atwater, Bacon, Chu, Johnson
   (E) Atwater, Bacon, Johnson, King

8. An article in the student newspaper of a large university had the headline "A's swapped for evaluations?" Results showed that higher grades directly corresponded to a more positive evaluation. Which of the following would be a valid conclusion to draw from the study?

   (A) A teacher can improve his or her teaching evaluations by giving good grades.
   (B) A good teacher, as measured by teaching evaluations, helps students learn better, resulting in higher grades.
   (C) Teachers of courses in which the mean grade is higher apparently tend to have above-average teaching evaluations.
   (D) Teaching evaluations should be conducted before grades are awarded.
   (E) All of the above.

9. A new cough medicine was given to a group of 25 subjects who had a cough due to the common cold. 30 minutes after taking the new medicine, 20 of the subjects reported that their coughs had disappeared. From this information you conclude

(A) that the remedy is effective for the treatment of coughs.
(B) nothing, because the sample size is too small.
(C) nothing, because there is no control group for comparison.
(D) that the new treatment is better than the old medicine.
(E) that the remedy is not effective for the treatment of coughs.

10. 100 volunteers who suffer from anxiety take part in a study. 50 are selected at random and assigned to receive a new drug that is thought to be extremely effective in reducing anxiety. The other 50 are given an existing anti-anxiety drug. A doctor evaluates anxiety levels after two months of treatment to determine if there has been an larger reduction in the anxiety levels of those who take the new drug. This would be double blind if

(A) both drugs looked the same
(B) neither the subjects nor the doctor knew which treatment any subject had received
(C) the doctor couldn't see the subjects and the subjects couldn't see the doctor
(D) there was a third group that received a placebo
(E) all of the above

Check your answers below. If you got a question wrong, check to see if you made a simple mistake or if you need to study that concept more. After you check your work, identify the concepts you feel very confident about and note what you will do to learn the concepts in need of more study.

#	Answer	Concept	Right	Wrong	Simple Mistake?	Need to Study More
1	C	Why we randomize				
2	C	Confounding				
3	A	Matched Pairs				
4	C	Cluster Sampling				
5	B	Inference about the population				
6	D	Stratified random sampling				
7	D	Blocking				
8	C	Surveys vs Experiments				
9	C	Control				
10	B	Definition of Experiments				

Chapter 4 Reflection

Summarize the "Big Ideas" in Chapter 4:

My strengths in this chapter:

Concepts I need to study more and what I will do to learn them:

FRAPPY! Free Response AP® Problem, Yay!

The following problem is modeled after actual Advanced Placement Statistics free response questions. Your task is to generate a complete, concise response in 15 minutes. After you generate your response, view two example solutions and determine whether you feel they are "complete", "substantial", "developing" or "minimal". If they are not "complete", what would you suggest to the student who wrote them to increase their score? Finally, you will be provided with a rubric. Score your response and note what, if anything, you would do differently to increase your own score.

A large school district is interested in determining student attitudes about their co-curricular offerings such as athletics and fine arts. The district consists of students attending 4 elementary schools (2000 students total), 1 middle school (1000 students total), and 2 high schools (2000 students total).

The administration is considering two sampling plans. The first consists of taking a simple random sample of students in the district and surveying them. The second consists of taking a stratified random sample of students and surveying them.

(a) Describe how you would select a simple random sample of 200 students in the district.

(b) Describe how you would select a stratified random sample consisting of 200 students.

(c) Describe the statistical advantage of using a stratified random sample over the simple random sample in this study.

FRAPPY! Student Responses

Student Response 1:

a) Write the names of all 5,000 students on separate slips of paper. Place the slips into a large bin and mix them well. Draw slips of paper until you have 200.

b) Separate the students by level-elementary, middle, and high school. Label the students at each level and randomly select 66 elementary students, 66 middle school students, and 68 high school students.

c) By stratifying, we avoid surveying only elementary students or only high school students. This is important because student attitudes might be different at each level.

How would you score this response? Is it substantial? Complete? Developing? Minimal? Is there anything this student could do to earn a better score?

Student Response 2:

a) Label each student with a number from 0001 to 5000. Use your calculator to generate 200 random numbers. These numbers correspond to the individuals who will be surveyed.

b) Randomly select 1 elementary school, 1 middle school, and 1 high school. Randomly select 200 students from each school.

c) Stratifying is easier because we don't have to sample the entire population. It is less time consuming and gives better results.

How would you score this response? Is it substantial? Complete? Developing? Minimal? Is there anything this student could do to earn a better score?

FRAPPY! Scoring Rubric

Use the following rubric to score your response. Each part receives a score of "Essentially Correct," "Partially Correct," or "Incorrect." When you have scored your response, reflect on your understanding of the concepts addressed in this problem. If necessary, note what you would do differently on future questions like this to increase your score.

Intent of the Question

The goal of this question is to determine your ability to describe sampling methods and explain the advantages of stratifying over simple random sampling

Solution

(a) Write each student's name on a slip of paper. Place the slips of paper in a hat and mix well. Select 200 slips of paper and note the students in the sample. OR Label each student with a number from 0001 to 5000. Use a random number table or technology to produce random 4 digit numbers, ignoring repeats, until 200 are determined. These 200 numbers correspond to the individuals who will be surveyed.

(b) Because student attitudes may differ by level of school (elementary, middle, or high school), we should stratify by level. Label students at each level and randomly select 80 elementary students, 40 middle school students, and 80 high school students. This ensures each level is represented in the same proportion as the overall student enrollments

(c) Stratifying ensures no level is over or under represented in the sample. It is possible to select very few (or even no!) students from one level in a simple random sample. The opinions of students at one level may not reflect the opinions of all students in the district. Stratifying ensures each level is fairly represented.

Scoring:

Parts (a), (b), and (c) are scored as essentially correct (E), partially correct (P), or incorrect (I).

Part (a) is essentially correct if the response describes an appropriate method of selecting a simple random sample. This method should include labeling the individuals and employing a sufficient means of random selection that could be replicated by someone knowledgeable in statistics.
Part (a) is partially correct if random selection is used correctly, but the description does not provide sufficient detail for implementation.

Part (b) is essentially correct if the response describes selecting strata based on a reasonable variable (such as school level) and indicates randomly selecting individuals from each stratum to be a part of the survey. The method can result in an equal number of students from each level OR proportional representation based on the strata.
Part (b) is partially correct if a reasonable variable is identified, but the method is unclear or does not ensure proportional representation.

Part (c) is essentially correct if the response provides a reasonable statistical advantage of stratified random sampling based on the effects of an identified variable on the results in the context of the problem.
Part (c) is partially correct if the response provides a reasonable statistical advantage, but the communication is not clear or lacks context.

4 **Complete Response**
All three parts essentially correct

3 **Substantial Response**
Two parts essentially correct and one part partially correct

2 **Developing Response**
Two parts essentially correct and no parts partially correct
One part essentially correct and two parts partially correct
Three parts partially correct

1 **Minimal Response**
One part essentially correct and one part partially correct
One part essentially correct and no parts partially correct
No parts essentially correct and two parts partially correct

My Score:
What I did well:
What I could improve:
What I should remember if I see a problem like this on the AP Exam:

Chapter 4: Designing Studies

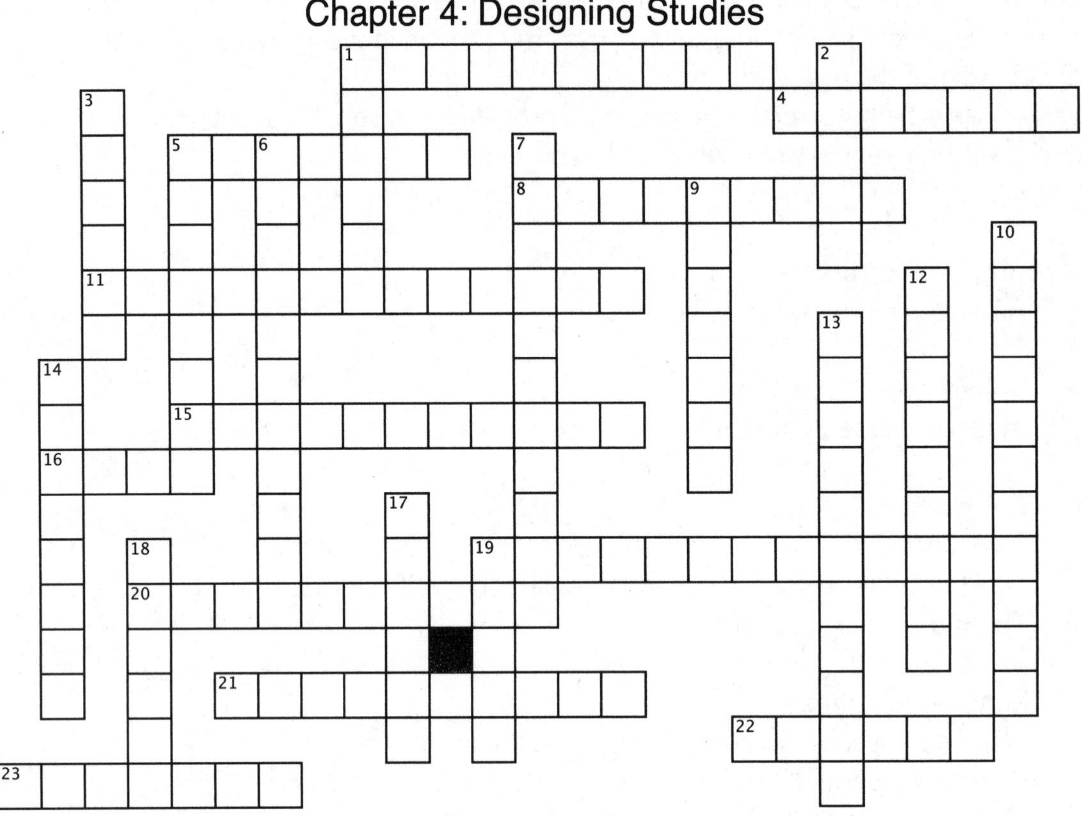

Across

1. a ______ random sample consists of separate simple random samples drawn from groups of similar individuals
4. a "fake" treatment that is sometimes used in experiments
5. the effort to minimize variability in the way experimental units are obtained and treated
8. the process of drawing a conclusion about the population based on a sample
11. this type of student can not be used to establish cause-effect relationships
15. the practice of using enough subjects in an experiment to reduce chance variation
16. a study that systematically favors certain outcomes shows this
19. this occurs when some groups in the population are left out of the process of choosing the sample
20. a study in which a treatment is imposed in order to observe a reasponse
21. the entire group of individuals about which we want information
22. a simple _____ sample consists of individuals from the population, each of which has an equally likely chance of being chosen
23. a _____ sample consists of a simple random sample of small groups from a population

Down

1. groups of similar individuals in a population
2. a group of experimental units that are similar in some way that may affect the response to the treatments
3. the rule used to assign experimental units to treatments is ____ assignment
5. smaller groups of individuals who mirror the population
6. this occurs when an individual chosen for the sample can't be contacted or refuses to participate
7. an observed effect that is too large to have occurred by chance alone
9. a lack of ____ in an experiment can prevent us from generalizing the results
10. a sample in which we choose individuals who are easiest to reach
12. a ____ response sample consists of people who choose themselves by responding to a general appeal
13. neither the subjects nor those measuring the response know which treatment a subject received (two words)
14. when units are humans, they are called
17. the part of the population from which we actually collect information
18. another name for treatments
19. the individuals on which an experiement is done are experimental ____

Chapter 5: Probability – What Are the Chances?

"The most important questions of life are, for the most part, really questions of probability." Pierre-Simon LaPlace

Chapter Overview

Now that we have learned how to collect data and how to analyze it graphically and numerically, we turn our study to probability, the mathematics of chance. Probability is the basis for the fourth and final theme in AP Statistics, inference. The next three chapters will provide you with the background in probability necessary to perform and understand the inferential methods we'll learn later in the course.

In this chapter, you will learn the definition of probability as a long-term relative frequency. You will study how to use simulation to answer probability questions as well as some basic rules to calculate probabilities of events. You will also learn two concepts that will reappear later in our studies: conditional probability and independence. Many of the ideas and methods you will learn in this chapter may be familiar to you. When it comes to statistics, your goal with probability is to be able to answer the question, "What would happen if we did this many times?" so you can make an informed statistical inference.

Sections in this Chapter

Section 5.1: Randomness, Probability, and Simulation
Section 5.2: Probability Rules
Section 5.3: Conditional Probability and Independence

Plan Your Learning

Use the following *suggested* guide to help plan your reading and assignments in "The Practice of Statistics, 6th Edition." Note: your teacher may schedule a different pacing or assign different problems. Be sure to follow his or her instructions!

Read	**5.1: pp. 298-304**	**5.1: pp. 304-308**	**5.2: pp. 313-318**	**5.2: pp. 318-325**
Do	1,3,5,7	9,11,15,21 MC 23-28	31,33,35,37,39	41,47,49,51,53 MC 55-58

Read	**5.3: pp. 330-338**	**5.3: pp. 338-347**	**Chapter Summary**
Do	61,63,65,67,69,71,77,79	81,83,87,89,91,93,99 MC 103-106	Multiple Choice FRAPPY!

Section 5.1: Randomness, Probability, and Simulation

Before You Read: Section Summary

This section introduces the basic definition of probability as a long-term relative frequency. That is, probability answers the question, "How often would we expect to see a particular outcome if we repeated a chance process many times?" The big idea that emerges when we study chance processes is that chance behavior is unpredictable in the short run but has a regular and predictable pattern in the long run. This idea seems pretty straightforward, yet there are a lot of common misconceptions about probability. Several are discussed in this section. Be sure to avoid falling for these common myths! The last topic in this section addresses the use of simulation to estimate probabilities. Simulation is a powerful tool for modeling chance behavior that can be used to illustrate many of the inference ideas you'll study later in the course.

Learning Targets:

____ I can interpret probability as a long-run relative frequency.

____ I can use simulation to model chance behavior.

While You Read: Key Vocabulary and Concepts

probability:
law of large numbers:
simulation:

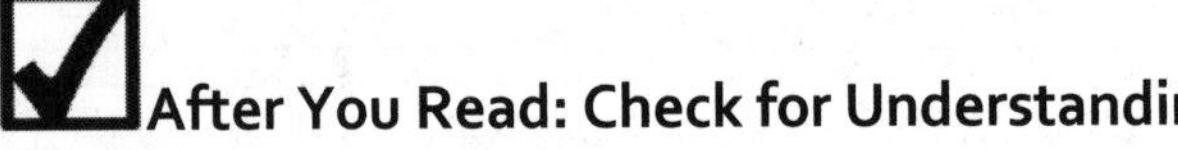

After You Read: Check for Understanding

Concept 1: The Idea of Probability

When we observe chance behavior over a long series of repetitions, a useful fact emerges. While chance behavior is unpredictable in the short term, a regular and predictable pattern becomes evident in the long run. The law of large numbers tells us that as we observe more and more repetitions of a chance behavior, the proportion of times a specific outcome occurs will "settle down" around a single value. This long-term proportion is the probability of the outcome occurring. The probability of an event is always described as a value between 0 and 1 -- 0 represents an event never occurs and 1 notes that it the event will occur on every repetition.

Check for Understanding: ____ *I can interpret probability as a long-run relative frequency.*

The probability of drawing a jack, queen, or king from a standard deck of playing cards is approximately 0.23.

a) Explain what this probability means in the context of drawing from a deck of cards.

b) Does this mean if we repeatedly draw a card, replace it, shuffle, and draw again 100 times that we will draw a jack, queen, or king 23 times? Why or why not?

Concept 2: Simulation

We first saw an example of simulation in Chapter 1. In this section, we learn how we can use simulation to estimate the probability of an event occurring. A three-step process can be used to perform a simulation of a question of interest by: 1) describing how to use a chance device to imitate a repetition of the chance behavior, 2) performing many repetitions of the simulation, and 3) using the results of the simulation to answer the original question. While simulations don't provide exact theoretical probabilities, the use of random numbers and other chance devices to imitate chance behavior can be a useful tool for estimating the likelihood of events.

Check for Understanding: ____ *I can use simulation to model chance behavior.*

A popular airline knows that, in general, 95% of individuals who purchase a ticket for a 10-seat commuter flight actually show up for the flight. In an effort to ensure a full flight, the airline sells 12 tickets for each flight. Design and carry out a simulation to estimate the probability that the flight will be overbooked, that is, more passengers show up than there are seats on the flight.

Section 5.2: Probability Rules

Before You Read: Section Summary

Now that you have the basic idea of probability down, you will learn how to describe probability models and use probability rules to calculate the likelihood of events. You will also learn how to organize information in two-way tables and Venn diagrams to help in determining probabilities. Understanding probability is important for understanding inference. Make sure you are comfortable with the definitions and rules in this section, as it will make your study of probability much easier!

Learning Targets:

____ I can give a probability model for a chance process with equally likely outcomes and use it to find the probability of an event.

____ I can use basic probability rules, including the complement rule and addition rule for mutually exclusive events.

____ I can use a two-way table or Venn diagram to model a chance process and calculate probabilities involving two events.

____ I can apply the general addition rule to calculate probabilities.

While You Read: Key Vocabulary and Concepts

sample space:
probability model:
event:
complement:
mutually exclusive (disjoint):
general addition rule:
intersection:
union:
Venn diagram:

After You Read: Check for Understanding

Concept 1: Probability Models and the Basic Rules of Probability

Chance behavior can be described using a probability model. This model provides two pieces of information: a list of possible outcomes (sample space) and the likelihood of each outcome. By describing chance behavior with a probability model, we can find the probability of an event—a particular outcome or collection of outcomes. Probability models must obey some basic rules of probability:

- For any event A, $0 \le P(A) \le 1$.
- If S is the sample space in a probability model, $P(S) = 1$.
- In the case of equally likely outcomes, $P(A)$ = (# outcomes in event A) / (# outcomes in S).
- Complement Rule: $P(A^C) = 1 - P(A)$.
- If A and B are mutually exclusive events, $P(A \text{ or } B) = P(A) + P(B)$.

After this section, you should be able to describe a probability model for chance behavior and apply the basic probability rules to answer questions about events.

Check for Understanding: ____*I can give a probability model for a chance process with equally likely outcomes and use it to find the probability of an event.* ____ *I can use basic probability rules, including the complement rule and addition rule for mutually exclusive events.*

Consider drawing a card from a shuffled fair deck of 52 playing cards.

1. How many possible outcomes are in the sample space for this chance process? What's the probability for each outcome?

Define the following events:
A: the card drawn is an Ace
B: the card drawn is a heart

2. Find P(A) and P(B).

3. What is $P(A^C)$?

4. Are events A and B mutually exclusive? Why or why not?

Concept 2: Two-Way Tables and Venn Diagrams

Often we'll need to find probabilities involving two events. In these cases, it may be helpful to organize and display the sample space using a two-way table or Venn diagram. This can be especially helpful when two events are not mutually exclusive. When dealing with two events A and B, it is important to be able to describe the union (or collection of all outcomes in A, B, or both) and the intersection (the collection of outcomes in both *A* and *B*). The general addition rule expands upon the basic rules presented in this section to help us find the probability of two events that are not mutually exclusive.

- If A and B are two events, $P(A \cup B) = P(A) + P(B) - P(A \cap B)$.

Check for Understanding: ____ *I can use a two-way table or Venn diagram to model a chance process and calculate probabilities involving two events.* ____ *I can apply the general addition rule to calculate probabilities.*

Consider drawing a card from a shuffled fair deck of 52 playing cards.
Define the following events:
A: the card drawn is an Ace
B: the card drawn is a heart

1. Use a two-way table to display the sample space.

2. Use a Venn diagram to display the sample space.

3. Find $P(A \cup B)$. Show your work.

Section 5.3: Conditional Probability and Independence

Before You Read: Section Summary

Two important concepts are introduced in this section: conditional probability and independence. These concepts will reappear throughout the remainder of your studies in statistics, so it is important that you understand what they mean! This section will also introduce you to several rules for calculating probabilities: the conditional probability formula, the general multiplication rule, and the multiplication rule for independent events. Not only do you want to know how to use these rules, but also when. As you perform probability calculations, make sure you can justify why you are using a particular rule!

Learning Targets:

____ I can calculate and interpret conditional probabilities.
____ I can determine if two events are independent.
____ I can use the general multiplication rule to calculate probabilities.
____ I can use a tree diagram to model a chance process involving a sequence of outcomes and to calculate probabilities.
____ I can use the multiplication rule for independent events to calculate probabilities, when appropriate.

While You Read: Key Vocabulary and Concepts

conditional probability:
Independent events:
general multiplication rule:
tree diagram:
multiplication rule for independent events:

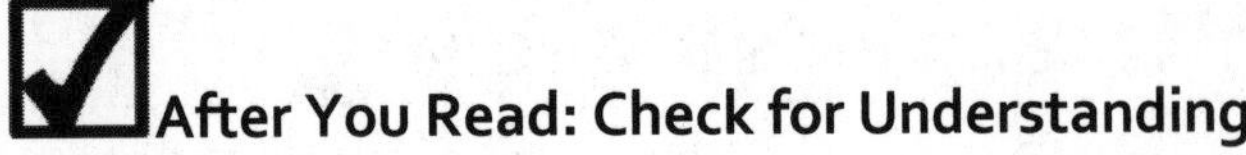

After You Read: Check for Understanding

Concept 1: Conditional Probability and Independence

A conditional probability describes the chance that an event will occur given that another event is already known to have happened. To note that we are dealing with a conditional probability, we use the symbol | to mean "given that." For example, suppose we draw one card from a shuffled deck of 52 playing cards. We could write "the probability that the card is an ace given that it is a red card as *P*(ace | red). Building on the concept of conditional probabilities, we say that when knowing that one event has occurred has no effect on the probability of another event occurring, the events are independent. That is, events A and B are independent if $P(A \mid B) = P(A)$ and $P(B \mid A) = P(B)$. Note that the events "get an ace" and "get a red card" described earlier are independent – if you know a randomly selected card is red, the probability it is an ace is 2/26. This is equal the same as the probability a randomly selected card (regardless of color) is an ace. That is, 2/26 = 4/52. Likewise, if you know a randomly selected card is an ace, the probability it is red is 1/2. This is the same as the probability a randomly selected card (regardless of value) is red. Therefore, the two events are independent.

Check for Understanding: ____ *I can calculate and interpret conditional probabilities.*
____ *I can determine if two events are independent.*

Is there a relationship between gender and candy preference? Suppose 200 high school students were asked to complete a survey about their favorite candies. The table below shows the gender of each student and their favorite candy.

	Male	**Female**	**Total**
Skittles	80	60	**140**
M & M's	40	20	**60**
Total	**120**	**80**	**200**

Define A to be the event that a randomly selected student is *male* and B to be the event that a randomly selected student likes *Skittles*. Are the events A and B independent? Justify your answer.

Concept 2: Tree Diagrams and Conditional Probability

When chance behavior involves a sequence of events, we can model it using a tree diagram. A tree diagram provides a branch for each outcome of an event along with the associated probabilities of those outcomes. Successive branches represent particular sequences of outcomes. To find the probability of an event, we multiply the probabilities on the branches that make up the event.

This leads us to the general multiplication rule: $P(A \cap B) = P(A) \bullet P(B \mid A)$.

If A and B are independent, the probability that both events occur is $P(A \cap B) = P(A) \bullet P(B)$.

Check for Understanding: ____ *I can use the general multiplication rule to calculate probabilities.* ____ *I can use a tree diagram to model a chance process involving a sequence of outcomes and to calculate probabilities.*

A study of high school juniors in three districts – Lakeville, Sheboygan, and Omaha – was conducted to determine enrollment trends in AP mathematics courses—Calculus or Statistics. 42% of students in the study came from Lakeville, 37% came from Sheboygan, and the rest came from Omaha. In Lakeville, 64% of juniors took Statistics and the rest took Calculus. 58% of juniors in Sheboygan and 49% of juniors in Omaha took Statistics while the rest took Calculus in each district. No juniors took both Statistics and Calculus. Describe this situation using a tree diagram and find the probability that a randomly selected student from in the study took Statistics.

Concept 3: Calculating Conditional Probabilities

By rearranging the terms in the general multiplication rule, we can determine a rule for conditional probabilities. That is, $P(B \mid A) = P(A \cap B) / P(A)$. Most conditional probabilities can be determined by using a two-way table, Venn diagram, or tree diagram. However, the formula can also be used if you know the appropriate probabilities in the situation.

Check for Understanding: ____ *I can use a tree diagram to model a chance process involving a sequence of outcomes and to calculate probabilities.*

Consider the situation from Concept 2.

Find *P*(student is from Lakeville | took Statistics).

Chapter Summary: Probability – What Are the Chances?

Probability describes the long-term behavior of chance processes. Since chance occurrences display patterns of regularity after many repetitions, we can use the rules of probability to determine the likelihood of observing particular results. At this point, you should be comfortable with the basic definition and rules of probability. In the next two chapters, you will study some further concepts in probability so we can build the foundation necessary for statistical inference.

Note that the AP exam may contain several questions about the probability of particular events. Make sure you understand how and when to apply each formula. More importantly, make sure you show your work when calculating probabilities so anyone reading your response understands exactly how you arrived at your answer!

After You Read: What Have I Learned?

Complete the vocabulary puzzle, multiple-choice questions, and FRAPPY. Check your answers and performance on each of the learning targets. Be sure to get extra help on any targets that you identify as needing more work!

Target	Got It!	Almost There	Needs Some Work
I can interpret probability as a long-run relative frequency.			
I can use simulation to model chance behavior.			
I can give a probability model for a chance process with equally likely outcomes and use it to find the probability of an event.			
I can use basic probability rules, including the complement rule and addition rule for mutually exclusive events.			
I can use a two-way table or Venn diagram to model a chance process and calculate probabilities involving two events.			
I can use the general addition rule to calculate probabilities.			
I can use a tree diagram to describe chance behavior.			
I can use the general multiplication rule to solve probability questions.			
I can calculate and interpret conditional probabilities.			
I can determine if two events are independent.			
I can use the general multiplication rule to calculate probabilities.			
I can use a tree diagram to model a chance process involving a sequence of outcomes and to calculate probabilities.			
I can use the multiplication rule for independent events to calculate probabilities, when appropriate.			

Chapter 5 Multiple Choice Practice

Directions. *Identify the choice that best completes the statement or answers the question. Check your answers and note your performance when you are finished.*

1. The probability that you will win a prize in a carnival game is about 1/7. During the last nine attempts, you have failed to win. You decide to give it one last shot. Assuming the outcomes are independent from game to game, the probability that you will win is:

(A) 1/7
(B) $(1/7) - (1/7)^9$
(C) $(1/7) + (1/7)^9$
(D) 1/10
(E) 7/10

2. A friend has placed a large number of plastic disks in a hat and invited you to select one at random. He informs you that half are red and half are blue. If you draw a disk, record the color, replace it, and repeat 100 times, which of the following is true?

(A) It is unlikely you will choose red more than 50 times.
(B) If you draw 10 blue disks in a row, it is more likely you will draw a red on the next try.
(C) The overall proportion of red disks drawn should be close to 0.50.
(D) The chance that the 100th draw will be red depends on the results of the first 99 draws.
(E) All of the above are true.

3. The two-way table below gives information on males and females at a high school and their preferred music format.

	CD	mp3	Vinyl	Totals
Males	146	106	48	**300**
Females	146	64	40	**250**
Totals	**292**	**170**	**88**	**550**

You select one student from this group at random. Which of the following statement is true about the events "prefers vinyl" and "Male"?

(A) The events are mutually exclusive and independent.
(B) The events are not mutually exclusive but they are independent.
(C) The events are mutually exclusive, but they are not independent.
(D) The events are not mutually exclusive, nor are they independent.
(E) The events are independent, but we do not have enough information to determine if they are mutually exclusive.

4. People with type O-negative blood are universal donors. That is, any patient can receive a transfusion of O-negative blood. Only 7.2% of the American population has O-negative blood. If 10 people appear at random to give blood, what is the probability that at least 1 of them is a universal donor?

(A) 0
(B) 0.280
(C) 0.526
(D) 0.720
(E) 1

5. A die is loaded so that the number 6 comes up three times as often as any other number. What is the probability of rolling a 4, 5, or 6?

(A) 2/3
(B) 1/2
(C) 5/8
(D) 1/3
(E) 1/4

6. You draw two candies at random from a bag that has 20 red, 10 green, 15 orange, and 5 blue candies without replacement. What is the probability that both candies are red?

(A) 0.1551
(B) 0.1600
(C) 0.2222
(D) 0.4444
(E) 0.8000

7. An event A will occur with probability 0.5. An event B will occur with probability 0.6. The probability that both A and B will occur is 0.1.

(A) Events A and B are independent.
(B) Events A and B are mutually exclusive.
(C) Either A or B always occurs.
(D) Events A and B are complementary.
(E) None of the above is correct.

8. Event A occurs with probability 0.8. The conditional probability that event B occurs, given that A occurs, is 0.5. The probability that both A and B occur is:

(A) 0.3
(B) 0.4
(C) 0.625
(D) 0.8
(E) 1.0

9. At Lakeville South High School, 60% of students have high-speed internet access, 30% have a mobile computing device, and 20% have both. The proportion of students that have neither high-speed internet access nor a mobile computing device is:

(A) 0%
(B) 10%
(C) 30%
(D) 80%
(E) 90%

10. Experience has shown that a certain lie detector will show a positive reading (indicates a lie) 10% of the time when a person is telling the truth and 95% of the time when a person is lying. Suppose that a random sample of 5 suspects is subjected to a lie detector test regarding a recent one-person crime. The probability of observing no positive readings if all suspects plead innocent and are telling the truth is:

(A) 0.409
(B) 0.735
(C) 0.00001
(D) 0.591
(E) 0.99999

Check your answers below. If you got a question wrong, check to see if you made a simple mistake or if you need to study that concept more. After you check your work, identify the concepts you feel very confident about and note what you will do to learn the concepts in need of more study.

#	Answer	Concept	Right	Wrong	Simple Mistake?	Need to Study More
1	A	Probability Basics				
2	C	Definition of Probability				
3	B	Mutually Exclusive/Independent				
4	C	Probability Calculations				
5	C	Probability Calculations				
6	A	Probability Calculations				
7	C	Probability Basics				
8	B	Conditional Probabilities				
9	C	General Addition Rule				
10	D	Conditional Probabilities				

Chapter 5 Reflection

Summarize the "Big Ideas" in Chapter 5:

My strengths in this chapter:

Concepts I need to study more and what I will do to learn them:

FRAPPY! Free Response AP® Problem, Yay!

The following problem is modeled after actual Advanced Placement Statistics free response questions. Your task is to generate a complete, concise response in 15 minutes. After you generate your response, view two example solutions and determine whether you feel they are "complete", "substantial", "developing" or "minimal". If they are not "complete", what would you suggest to the student who wrote them to increase their score? Finally, you will be provided with a rubric. Score your response and note what, if anything, you would do differently to increase your own score.

A simple random sample of adults in a metropolitan area was selected and a survey was administered to determine political views. The results are recorded below:

		Political Views		
Age	**Conservative**	**Moderate**	**Liberal**	**Total**
18-29	10	15	30	55
30-44	20	30	35	85
45-59	35	15	20	70
Over 60	20	15	10	45
Total	85	75	95	255

(a) What is the probability that a person chosen at random from this sample will have moderate political views?

(b) What is the probability that a person chosen at random from those in the sample who are between the ages of 30 and 44 will have moderate political views? Show your work.

(c) Based on your answers to (a) and (b), are political views and age independent for the population of adults in this metropolitan area? Why or why not?

FRAPPY! Student Responses

Student Response 1:

a) 75/255

b) P(moderate | 30-44) = 30/75 = 0.40

c) Yes, the two are independent because political views don't depend on age. There are moderates in every age category.

How would you score this response? Is it substantial? Complete? Developing? Minimal? Is there anything this student could do to earn a better score?

Student Response 2:

a) P(moderate) = 0.29

b) P(moderate | 30-44) = P(moderate and 30-44) / P(30-44) = 30/85 = 0.35

c) No, these are not independent because P(moderate) ≠ P(moderate | 30-44). In order to be independent, these probabilities should be the same and the condition of age should not affect the probability of political views.

How would you score this response? Is it substantial? Complete? Developing? Minimal? Is there anything this student could do to earn a better score?

FRAPPY! Scoring Rubric

Use the following rubric to score your response. Each part receives a score of "Essentially Correct," "Partially Correct," or "Incorrect." When you have scored your response, reflect on your understanding of the concepts addressed in this problem. If necessary, note what you would do differently on future questions like this to increase your score.

Intent of the Question

The goal of this question is to determine your ability to calculate probabilities and determine whether or not two events are independent.

Solution

(a) $P(\text{moderate}) = 75/255 = 0.2941$

(b) $P(\text{moderate} \mid \text{age } 30\text{-}44) = 30/85 = 0.3529$

(c) If moderate political views and age were independent, the probabilities in (a) and (b) would be the same. Since they are not equal, age and political views are not independent for the individuals in this sample.

Scoring:

Parts (a), (b), and (c) are scored as essentially correct (E), partially correct (P), or incorrect (I).

Part (a) is essentially correct if the probability is correct. Part (a) is partially correct if the correct formula is shown, but minor arithmetic errors are present. Otherwise it is incorrect.

Part (b) is essentially correct if the conditional probability is calculated correctly. Part (b) is partially correct if the conditioning is reversed and $P(\text{age } 30\text{-}44 \mid \text{moderate}) = 30/75 = 0.40$ is calculated.

Part (c) is essentially correct if the response indicates the two variables are not independent and justifies the conclusion based on an appropriate probability argument. Part (c) is partially correct if the response indicates the two variables are not independent, but the argument is weak or not based on an appropriate probability argument.

4 Complete Response
All three parts essentially correct

3 Substantial Response
Two parts essentially correct and one part partially correct

2 Developing Response
Two parts essentially correct and no parts partially correct
One part essentially correct and two parts partially correct
Three parts partially correct

1 Minimal Response
One part essentially correct and one part partially correct
One part essentially correct and no parts partially correct
No parts essentially correct and two parts partially correct

My Score:
What I did well:
What I could improve:
What I should remember if I see a problem like this on the AP Exam:

Chapter 5: Probability

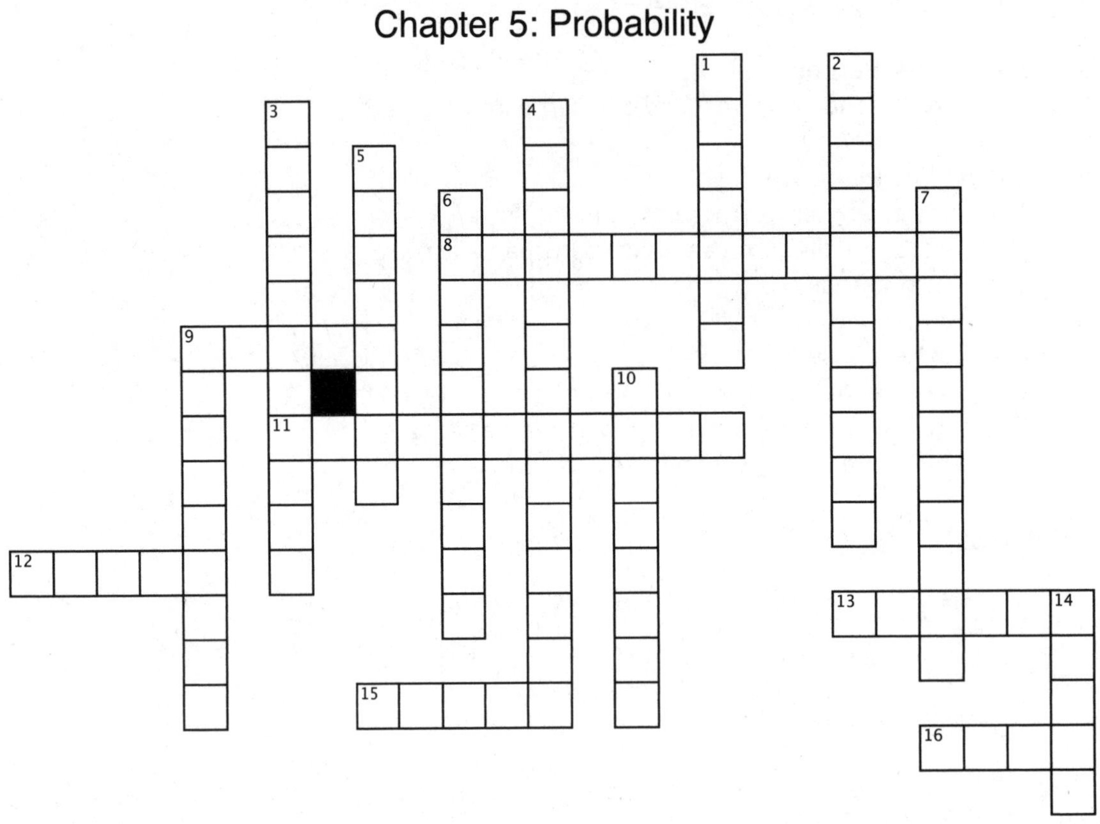

Across

8. The collection of outcomes that occur in both of two events.
9. A collection of outcomes from a chance process.
11. The proportion of times an outcome would occur in a very long series of repetitions.
12. _____ Theorem can be used to find probabilities that require going "backward" in a tree diagram.
13. In statistics, this doesn't mean "haphazard." It means "by chance."
15. The collection of outcomes that occur in either of two events.
16. A ____ diagram can help model chance behavior that involves a sequence of outcomes.

Down

1. The law of large _____ states that the proportion of times an outcome occurs in many repetitions will approach a single value.
2. The probability that one event happens given another event is known to have happened.
3. The set of all possible outcomes for a chance process (two words).
4. The probability that two events both occur can be found using the general _____ rule.
5. P(A or B) can be found using the general _____ rule.
6. The imitation of chance behavior, based on a model that reflects the situation.
7. The occurrence of one event has no effect on the chance that another event will happen.
9. Another term for disjoint: Mutually _____.
10. Two events that have no outcomes in common and can never occur together.
14. A probability ____ describes a chance process and consists of two parts.

Chapter 6: Random Variables

"We must become more comfortable with probability and uncertainty." Nate Silver

Chapter Overview

In the last chapter we learned the basic definition and rules of probability. We continue our study of probability by exploring situations that involve the assignment of a numerical value to each possible outcome of a chance process. The random variables that result form the foundation for inference procedures in later chapters. In this chapter, you will learn how to calculate probabilities of events involving random variables as well as how to describe their probability distributions. Specifically, you will learn formulas to determine the mean and standard deviation of individual random variables as well as the combination of several independent random variables. Finally, you'll explore two special random variables – binomial and geometric – and learn how to calculate probabilities of events in binomial and geometric settings. This chapter involves a lot of formulas, so you may want to familiarize yourself with the formula sheet provided on the AP exam. Like earlier chapters, you should focus less on memorizing formulas or calculator keystrokes and more on how to apply the formulas and interpret results.

Sections in this Chapter
Section 6.1: Discrete and Continuous Random Variables
Section 6.2: Transforming and Combining Random Variables
Section 6.3: Binomial and Geometric Random Variables

Plan Your Learning

Use the following *suggested* guide to help plan your reading and assignments in "The Practice of Statistics, 6th Edition." Note: your teacher may schedule a different pacing or assign different problems. Be sure to follow his or her instructions!

Read	**6.1: pp. 360-367**	**6.1: pp. 368-375**	**6.2: pp. 381-387**	**6.2: pp. 388-397**
Do	1, 3, 5, 7, 9, 11	13,19,21,23,27,29 MC 31-34	37,39,41,43,47	49,51,55,57,59,65,67 MC 73-74

Read	**6.3: pp. 402-412**	**6.3: pp. 412-422**	**6.3: pp. 422-427**	**Chapter Summary**
Do	77,79,81,83,85,89	91,93,95,99,101,105	107,109,111 MC 113-116	Multiple Choice FRAPPY!

Section 6.1: Discrete and Continuous Random Variables

Before You Read: Section Summary

A random variable takes numerical values that describe the outcomes of a chance process. In the last chapter, we learned that a probability model describes the possible outcomes for a chance process and the probability of each outcome. A random variable does the same thing, describing the possible values that the variable takes and the probability of each. Random variables fall into two categories: discrete and continuous. What differentiates the two is the set of values the random variable can take. If the set is limited to fixed values with gaps between, it is discrete. If the variable can take on any value in an interval, it is continuous. Regardless of the type, we are interested in describing the shape of the random variable's probability distribution, its center, and its variability. Knowing these characteristics will give us a sense of what to expect in repeated observations of the random variable as well as what can be considered likely and unlikely results. This idea forms the basis for inferential thinking in later chapters so you want to get used to thinking along those lines in this section!

Learning Targets:

- ____ I can use the probability distribution of a discrete random variable to calculate the probability of an event.
- ____ I can make a histogram to display the probability distribution of a discrete random variable and describe its shape.
- ____ I can calculate and interpret the mean (expected value) of a discrete random variable.
- ____ I can calculate and interpret the standard deviation of a discrete random variable.
- ____ I can use the probability distribution of a continuous random variable (uniform or Normal) to calculate the probability of an event.

While You Read: Key Vocabulary and Concepts

random variable:
probability distribution:
discrete random variable:
mean (expected value) of a discrete random variable:
standard deviation of a discrete random variable:
continuous random variable:

After You Read: Check for Understanding

Concept 1: Discrete and Continuous Random Variables

A random variable can be classified as either discrete or continuous depending on its possible values. If it takes a fixed (finite or infinite) set of possible values with gaps in between, then we call it a discrete random variable. If the random variable takes on *any* value in an interval of numbers, it is continuous. To describe a random variable, we follow the same process as describing a probability model. First, define the random variable, *X*, as a numerical outcome of a chance process. Next, indicate the possible values of the variable. Finally, give the probability that each value occurs using a table, formula, or graph.

Check for Understanding: ____ *I can use the probability distribution of a discrete random variable to calculate the probability of an event.* ____ *I can make a histogram to display the probability distribution of a discrete random variable and describe its shape.*

Consider two 4-sided dice, each having sides labeled 1, 2, 3, 4. Let *X* = the sum of the numbers that appear after a roll of the dice.

a) Is *X* a discrete or a continuous random variable? Sketch the probability distribution of *X* below. Describe what you see.

b) If somebody rolled the dice 10 times and got a sum less than 3 each time, would you be surprised? Why or why not?

Concept 2: Mean and Standard Deviation of Random Variables

In Chapter 1, we learned that when describing distributions of quantitative data, we should always note the shape, center, and variability. The same holds true for random variables. In order to make inferences, we need to know what is considered a "typical" value of the random variable being examined as well as how much variation around that value we can expect to see. As with distributions of quantitative data, the center of a random variable's probability distribution can be described by calculating the mean. However, in the case of random variables, the mean (or expected value) is a *weighted* average, taking into account the probability of each outcome occurring. Likewise, the standard deviation of a random variable takes into account the probability of each outcome occurring, giving more weight to those outcomes that are more likely. Make sure you get comfortable with the formulas for the mean and standard deviation of a discrete random variable so you can calculate and interpret the center and variability of the probability distribution.

Check for Understanding: ____ *I can calculate and interpret the mean (expected value) of a discrete random variable.* ____ *I can calculate and interpret the standard deviation of a discrete random variable.* ____ *I can use the probability distribution of a continuous random variable (uniform or Normal) to calculate the probability of an event.*

a) Suppose the random variable Y = *number of goals in a randomly selected high school hockey game* has the following probability distribution:

Goals:	0	1	2	3	4
Probability:	0.155	0.195	0.243	0.233	0.174

Sketch the probability distribution. Then calculate the mean and standard deviation of Y and interpret them in the context of the situation.

b) The weights of toddler boys follow an approximately Normal distribution with mean 34 pounds and standard deviation 3.5 pounds. Suppose you randomly choose one toddler boy and record his weight. What is the probability that the randomly selected boy weighs less than 31 pounds?

Section 6.2: Transforming and Combining Random Variables

Before You Read: Section Summary

This section introduces two distinct topics. First, you will explore transforming a single random variable. That is, you will learn how to describe the shape, center, and variability of the probability distribution of a random variable when a linear transformation (such as adding a constant to each value or multiplying each value by a constant) is applied. Second, you will learn how to combine two or more random variables. This topic is critical as many of the statistical inference problems we will explore involve observing the difference between two random variables. You will learn how to describe the mean and standard deviation of the sum and difference of independent random variables as well as how to calculate probabilities of observing particular outcomes in these situations. There are a lot of formulas to keep straight in this section, so you may wish to keep your AP formula sheet handy!

Learning Targets:

____ I can describe the effect of adding or subtracting a constant or multiplying or dividing by a constant on the probability distribution of a random variable.

____ I can calculate the mean and standard deviation of the sum or difference of random variables.

____ I can find probabilities involving the sum or difference of independent Normal random variables.

While You Read: Key Vocabulary and Concepts

effect on a probability distribution of adding/subtracting a constant:
effect on a probability distribution of multiplying/dividing by a constant:
mean (expected value) of a sum of random variables:
mean (expected value) of a difference of random variables:
independent random variables:
standard deviation of the sum of two independent random variables:
standard deviation of the difference of two independent random variables:

After You Read: Check for Understanding

Concept 1: Transforming a Random Variable

We learned how transformations affect the shape, center, and variability of distributions of quantitative data back in Chapter 2. Similar rules apply to random variables. That is, when we multiply (or divide) each value of a random variable by a constant b, the shape of the probability distribution does not change. However, measures of center are multiplied (divided) by b and measures of variability are multiplied (divided) by $|b|$. When we add (or subtract) a constant a to each value of a random variable, the shape and variability of the probability distribution do not change. However, the measures of center will increase (or decrease) by a.

Check for Understanding: ____ *I can describe the effect of adding or subtracting a constant or multiplying or dividing by a constant on the probability distribution of a random variable.*

A carnival game involves tossing a ball into numbered baskets with the goal of having your ball land in a high-numbered basket. The probability distribution of X = value of the basket on a randomly selected toss.

Value:	0	1	2	3
Probability:	0.3	0.4	0.2	0.1

The expected value of X is 1.1 and its standard deviation is 0.0943.

Suppose it costs \$2 to play and you earn \$1.50 for each point earned on your toss. That is, if you land in a basket labeled "2," you will earn \$3.00.

Define Y to be the amount of profit you make on a randomly selected toss. Describe the shape, center, and spread of the probability distribution of Y in the context of the situation.

Concept 2: Combining Random Variables

Many situations involve two or more random variables. Understanding how to describe the center and variability of the probability distribution for the sum or difference of two random variables is an important skill to have. When given two independent random variables, we can describe the mean and standard deviation of the sum (or difference) of the random variables using the formulas in this section. Basically, when adding or subtracting two or more random variables (whether they are independent or not), the mean of the sum or difference of those random variables will be the sum or difference of their means. However, to describe the spread of the sum or difference of *independent* random variables, we must perform two steps. First, we find the *variance* of the sum or difference of two or more independent random variables by *adding* their variances. Then, we take the square root of the variance to find the standard deviation. A common mistake is to simply add standard deviations. Remember to *always add* variances! Never subtract, and never combine standard deviations!

Check for Understanding: ____ *I can calculate the mean and standard deviation of the sum or difference of random variables.*

Students in Mr. Costello's class are expected to check their homework in groups of 4 at the beginning of class each day. Students must check it as quickly as possible, one at a time. The means and standard deviations of the time it takes to check homework for the 4 students in one group are noted below. Assume their times are independent.

	Mean	**Standard Deviation**
Alan	1.4 min	0.1 min
Barb	1.2 min	0.4 min
Corey	0.9 min	0.8 min
Doug	1.0 min	0.7 min

a) If each student checks one after the other, what are the mean and standard deviation of the total time necessary for these four students to check their homework on a randomly chosen day?

b) Suppose Alan and Doug like to race to see who can check their homework faster. What are the mean and standard deviation for the difference between their times (Doug – Alan)? Interpret these values in the context of the situation.

Concept 3: Combining Normal Random Variables

If our random variables of interest are Normally distributed, we can calculate the probability of observing particular outcomes using the skills we learned in Chapter 2. To do so, we rely on one important fact. When combining independent Normal random variables, the resulting distribution is also Normal! We can find the mean and standard deviation of the resulting distribution using the formulas we just learned. Then we can apply our knowledge from Chapter 2 to calculate and interpret probabilities about the situation.

Check for Understanding: ____ *I can find probabilities involving the sum or difference of independent Normal random variables.*

Mr. Molesky and Mr. Liberty are avid video game golfers. Both like to compare times to complete a particular course on their favorite game. Mr. Molesky's times are Normally distributed with a mean of 110 minutes and standard deviation of 10 minutes. Mr. Liberty's times are Normally distributed with mean 100 minutes and standard deviation 8 minutes.

a) Find the mean and standard deviation of the difference of their times (Molesky - Liberty). Assume their times are independent.

b) Find the probability that Mr. Molesky will finish his game before Mr. Liberty on any given day.

Section 6.3: Binomial and Geometric Random Variables

Before You Read: Section Summary

In the first two sections, you learned how to describe the probability distributions of discrete and continuous random variables as well as how to calculate probabilities for situations involving one or more random variables. In this section, you will focus on two special cases of discrete random variables: binomial and geometric. Binomial random variables count the number of successes that occur in a fixed number of independent trials of some chance process with a constant probability of success on each trial, while geometric random variables count the number of trials needed to get a success. Binomial random variables appear often on the AP exam, so you will want to pay particular attention to this topic. Again, try not to get bogged down in the formulas in this section. Familiarize yourself with the AP formula sheet and focus your efforts on being able to identify when to use binomial or geometric random variables and how to calculate and interpret probabilities involving them.

Learning Targets:

____ I can determine whether the conditions for a binomial setting have been met.
____ I can calculate and interpret probabilities involving binomial distributions.
____ I can calculate and interpret the mean and standard deviation of a binomial random variable.
____ I can use the Normal approximation to the binomial distribution to calculate probabilities, when appropriate.
____ I can find probabilities involving geometric random variables.

While You Read: Key Vocabulary and Concepts

binomial setting:
binomial random variable:
binomial coefficient:
binomial probability formula:
mean (expected value) of a binomial random variable:
standard deviation of a binomial random variable:
Normal approximation for binomial distributions:

10% condition:
large counts condition:
geometric setting:
geometric random variable:
geometric probability formula:
mean (expected value) of a geometric random variable:

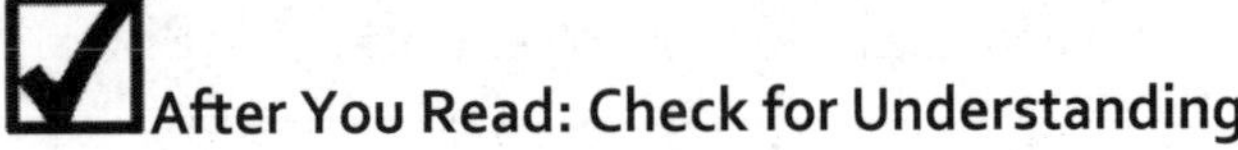

After You Read: Check for Understanding

Concept 1: Binomial Random Variables

When we observe a fixed number of repeated trials of the same chance process, we are often interested in how many times a particular outcome occurs. This is the basis for a binomial setting. That is, a binomial setting arises when we perform n independent trials of the same chance process and count the number of times a particular outcome occurs.

To be a binomial setting, four conditions must be met. First, we are interested in outcomes that can be classified in one of two ways – success or failure. The particular outcome of interest is considered a success, while anything else is considered a failure. Next, each observed trial of the chance process must be independent of the other trials. Third, the number of trials we observe must be fixed in advance. Last, the probability of success on each trial must be the same.

If these conditions are met, we can use the binomial probability formula to determine the likelihood of observing a certain number of successes in a fixed number of trials of the binomial random variable:

$$P(X=k)=\binom{n}{k}p^k(1-p)^{n-k}$$

Note that the formula uses the multiplication rule for independent events from Chapter 5 in multiplying the probabilities of successes and failures across the fixed number of trials. However, the formula also considers the number of ways in which we can arrange those successes across our trials.

Check for Understanding: ____ *I can determine whether the conditions for a binomial setting have been met.* ____ *I can calculate and interpret probabilities involving binomial distributions.*

Recall that there are 4 suits—spades, hearts, clubs, and diamonds—in a standard deck of playing cards. Suppose you play a game in which you draw a card, record the suit, replace it, shuffle, and repeat until you have observed 10 cards. Define X = number of hearts observed.

a) Show that X is a binomial random variable.

b) Find the probability of observing fewer than 4 hearts in this game.

Concept 2: Mean and Standard Deviation of a Binomial Distribution and the Normal Approximation

Like other discrete random variables, we can calculate the mean and standard deviation of binomial random variables. This will give us a better sense of what we'd expect to see in the long run as well as how much variability we can expect to observe in the observed number of successes. If a random variable X has a binomial probability distribution based on a chance process with n trials each having probability of success p, we can calculate the mean of X by multiplying np. We can find the standard deviation of X by taking the square root of the product $np(1 - p)$. Modified versions of these formulas are useful when trying to make inferences about the proportion of successes in a population. If we take an SRS of size n from a population (where n is less than 10% of the size of the population), then we can use a binomial distribution to model the number of successes in the sample.

Further, if n is so large that both np and $np(1 - p)$ are at least 10, we can use a Normal distribution to approximate binomial probabilities. As always, make sure you not only understand how to use the formulas, but also when to use them and how to interpret their results!

Check for Understanding: ____ *I can calculate and interpret the mean and standard deviation of a binomial random variable.* ____ *I can use the Normal approximation to the binomial distribution to calculate probabilities, when appropriate.*

Suppose 72% of students in the U.S. would give their teachers a positive rating if asked to score their effectiveness. A survey is conducted in which 500 students are randomly selected and asked to rate their teachers. Let X = the number of students in the sample who would give their teachers a positive rating.

a) Show that X can be approximated by a Normal distribution.

b) Use a Normal approximation to find the probability that 400 or more students would give their teacher a positive rating in this sample.

Concept 3: Geometric Random Variables

In a binomial setting, we are interested in knowing how many successes will occur in a fixed number of trials. Sometimes we are interested in knowing how long it will take until a success occurs. When we perform independent trials of a chance process with the same probability of success on each trial, and record how long it takes to get a success, we have a geometric setting. We can describe the number of trials it takes to get a success using a geometric random variable. As with other random variables, we can describe the probability distribution of a geometric random variable and calculate its mean and standard deviation. Using what we learned in Chapter 5, we can calculate the probability of observing the first success on the kth trial by multiplying the probabilities of $(k - 1)$ consecutive failures by the probability of a success:

$$P(Y = k) = (1 - p)^{k-1} p .$$

Check for Understanding: ____ *I can find probabilities involving geometric random variables.*

Suppose 20% of Super Crunch cereal boxes contain a secret decoder ring. Let X = the number of boxes of Super Crunch that must be opened until a ring is found.

a) Show that X is a geometric random variable.

b) Find the probability that you will have to open 7 boxes to find a ring.

c) Find the probability that it will take fewer than 4 boxes to find a ring.

d) How many boxes would you expect to have to open to find a ring?

Chapter Summary: Random Variables

In the last chapter we learned the basic definition and rules of probability. We continued our study of probability in this chapter by exploring situations that involve assigning a numerical value to each possible outcome of a chance process. The random variables that result form the foundation for inference procedures in later chapters. You learned how to calculate probabilities of events involving random variables as well as how to describe their probability distributions. You learned formulas to determine the mean and standard deviation of individual random variables as well as the combination of several independent random variables. Finally, you explored two special random variables – binomial and geometric – and learned how to calculate probabilities of events in binomial and geometric settings.

This chapter involved a lot of formulas, so you may want to familiarize yourself with the formula sheet provided on the AP exam. Like earlier chapters, you should focus less on memorizing formulas or calculator keystrokes and more on how to apply the formulas and interpret results. Make sure you understand how and when to apply each formula. More importantly, make sure you show your work when calculating probabilities so anyone reading your response understands exactly how you arrived at your answer!

After You Read: What Have I Learned?

Complete the vocabulary puzzle, multiple-choice questions, and FRAPPY. Check your answers and performance on each of the learning targets. Be sure to get extra help on any targets that you identify as needing more work!

Learning Target	Got It!	Almost There	Needs Some Work
I can use the probability distribution of a discrete random variable to calculate the probability of an event.			
I can make a histogram to display the probability distribution of a discrete random variable and describe its shape.			
I can calculate and interpret the mean (expected value) of a discrete random variable.			
I can calculate and interpret the standard deviation of a discrete random variable.			
I can use the probability distribution of a continuous random variable (uniform or Normal) to calculate the probability of an event.			
I can describe the effect of adding or subtracting a constant or multiplying or dividing by a constant on the probability distribution of a random variable.			
I can calculate the mean and standard deviation of the sum or difference of random variables.			
I can find probabilities involving the sum or difference of independent random variables.			
I can determine whether the conditions for a binomial setting are met.			
I can calculate and interpret probabilities involving binomial distributions.			
I can calculate and interpret the mean and standard deviation of a binomial random variable.			
I can find probabilities involving geometric random variables.			
I can use the Normal approximation to the binomial distribution to calculate probabilities, when appropriate.			

Chapter 6 Multiple Choice Practice

Directions. *Identify the choice that best completes the statement or answers the question. Check your answers and note your performance when you are finished.*

1. A marketing survey compiled data on the number of cars in households. If X = the number of cars in a randomly selected household, and we omit the rare cases of more than 5 cars, then X has the following probability distribution:

X	0	1	2	3	4	5
$P(X)$	0.24	0.37	0.20	0.11	0.05	0.03

What is the probability that a randomly chosen household has at least two cars?

(A) 0.19
(B) 0.20
(C) 0.29
(D) 0.39
(E) 0.61

2. What is the expected value of the number of cars in a randomly selected household?

(A) 2.5
(B) 0.1667
(C) 1.45
(D) 1
(E) Can not be determined

3. A dealer in Las Vegas selects 10 cards from a standard deck of 52 cards. Let Y be the number of diamonds in the 10 cards selected. Which of the following best describes this setting?

(A) Y has a binomial distribution with $n = 10$ observations and probability of success $p = 0.25$.
(B) Y has a binomial distribution with $n = 10$ observations and probability of success $p = 0.25$, provided the deck is shuffled well.
(C) Y has a binomial distribution with $n = 10$ observations and probability of success $p = 0.25$, provided that after selecting a card it is replaced in the deck and the deck is shuffled well before the next card is selected.
(D) Y has a geometric distribution with $n = 10$ observations and probability of success $p = 0.25$.
(E) Y has a geometric distribution with n = 52 observations and probability of success $p = 0.25$.

4. In the town of Lakeville, the number of cell phones in a household is a random variable W with the following probability distribution:

Value w_i	0	1	2	3	4	5
Probability p_i	0.1	0.1	0.25	0.3	0.2	0.05

The standard deviation of the number of cell phones in a randomly selected house is

(A) 1.32
(B) 1.7475
(C) 2.5
(D) 0.09
(E) 2.9575

5. A random variable Y has the following probability distribution:

Y	-1	0	1	2
$P(Y)$	$4C$	$2C$	0.07	0.03

The value of the constant C is:

(A) 0.10.
(B) 0.15.
(C) 0.20.
(D) 0.25.
(E) 0.75.

6. The variance of the sum of two random variables X and Y is

(A) $\sigma_X + \sigma_Y$.
(B) $(\sigma_X)^2 + (\sigma_Y)^2$.
(C) $\sigma_X + \sigma_Y$, but only if X and Y are independent.
(D) $(\sigma_X)^2 + (\sigma_Y)^2$, but only if X and Y are independent.
(E) None of these.

7. It is known that about 90% of the widgets made by Buckley Industries meet specifications. Every hour a sample of 18 widgets is selected at random for testing and the number of widgets that meet specifications is recorded. What is the approximate mean and standard deviation of the number of widgets meeting specifications?

(A) $\mu = 1.62$; $\sigma = 1.414$
(B) $\mu = 1.62$; $\sigma = 1.265$
(C) $\mu = 16.2$; $\sigma = 1.62$
(D) $\mu = 16.2$; $\sigma = 1.273$
(E) $\mu = 16.2$; $\sigma = 4.025$

8. A raffle sells tickets for $10 and offers a prize of $500, $1000, or $2000. Let *C* be a random variable that represents the prize in the raffle drawing. The probability distribution of *C* is given below.

Value c_i	$0	$500	$1000	$2000
Probability p_i	0.60	0.05	0.13	0.22

The expected profit when playing the raffle is

(A) $145.
(B) $585.
(C) $865.
(D) $635.
(E) $485.

9. Let the random variable *X* represent the amount of money Carl makes tutoring statistics students in the summer. Assume that *X* is Normal with mean $240 and standard deviation $60. The probability is approximately 0.6 that, in a randomly selected summer, Carl will make less than about

(A) $144
(B) $216
(C) $255
(D) $30
(E) $360

10. Which of the following random variables is geometric?

(A) The number of phone calls received in a one-hour period
(B) The number of times I have to roll a six-sided die to get two 5s.
(C) The number of digits I will read beginning at a randomly selected starting point in a table of random digits until I find a 7.
(D) The number of 7s in a row of 40 random digits.
(E) All four of the above are geometric random variables.

Check your answers below. If you got a question wrong, check to see if you made a simple mistake or if you need to study that concept more. After you check your work, identify the concepts you feel very confident about and note what you will do to learn the concepts in need of more study.

#	Answer	Concept	Right	Wrong	Simple Mistake?	Need to Study More
1	D	Discrete Random Variable				
2	C	Expected Value of Discrete Random Variables				
3	C	Binomial Settings				
4	A	Standard Deviation of Discrete Random Variables				
5	B	Probability Distribution				
6	D	Combining Random Variables				
7	D	Binomial Approximations				
8	B	Expected Value				
9	C	Normal Approximations				
10	C	Geometric Random Variables				

Chapter 6 Reflection

Summarize the "Big Ideas" in Chapter 6:

My strengths in this chapter:

Concepts I need to study more and what I will do to learn them:

FRAPPY! Free Response AP® Problem, Yay!

The following problem is modeled after actual Advanced Placement Statistics free response questions. Your task is to generate a complete, concise response in 15 minutes. After you generate your response, view two example solutions and determine whether you feel they are "complete", "substantial", "developing" or "minimal". If they are not "complete", what would you suggest to the student who wrote them to increase their score? Finally, you will be provided with a rubric. Score your response and note what, if anything, you would do differently to increase your own score.

A recent study revealed that a new brand of mp3 player may need to be repaired up to 3 times during its ownership. Let R represent the number of repairs necessary over the lifetime of a randomly selected mp3 player of this brand. The probability distribution of the number of repairs necessary is given below.

r_i	0	1	2	3
p_i	0.4	0.3	0.2	0.1

(a) Compute and interpret the mean and standard deviation of R.

(b) Suppose we also randomly select a phone that may require repairs over its lifetime. The mean and standard deviation of the number of repairs for this brand of phone are 2 and 1.2, respectively. Assuming that the phone and mp3 player break down independently of each other, compute and interpret the mean and standard deviation of the total number of repairs necessary for the two devices.

(c) Each mp3 repair costs \$15 and each phone repair costs \$25. Compute the mean and standard deviation of the total amount you can expect to pay in repairs over the life of the devices.

FRAPPY! Student Responses

Student Response 1:

a) mean = 0(0.4) + 1(0.3)+ 2(0.2) + 3(0.1) = 1
standard deviation = 1
We will have to repair our mp3 player exactly once or twice over its lifetime.

b) mean = 1 + 2 = 3
$1^2 + 1.2^2 = 2.44$ $\sqrt{(2.44)} = 1.562$ = standard deviation
We can expect to have to perform a total of about 3 repairs on the two devices. However, this can vary by up to 1.562 repairs or so over their lifetimes.

c) I would expect to pay \$15(1) + \$25(2) = \$65 in repairs over the lifetimes of the devices.

How would you score this response? Is it substantial? Complete? Developing? Minimal? Is there anything this student could do to earn a better score?

Student Response 2:

a) mean = 0(0.4) + 1(0.3)+ 2(0.2)+ 3(0.1) = 1
standard deviation = 1
This means we can expect to have to repair our mp3 player about once over its lifetime. But, we might have to repair it up to 2 times or maybe not at all. It is pretty unlikely we'd have to repair it three times.

b) mean = 1 + 2 = 3
standard deviation = 1 + 1.2 + 2.2
We can expect to have to perform a total of about 3 repairs on the two devices. However, this can vary by up to 2.2 repairs or so over their lifetimes.

c) The mean amount we can expect to pay in repairs is \$15(1) + \$25(2) = \$65. However, this amount will vary since we are not guaranteed to have to perform 3 repairs. The amount the cost can vary is $\sqrt{(15^2(1^2) + 25^2(1.2^2)} = \$33.54$.

How would you score this response? Is it substantial? Complete? Developing? Minimal? Is there anything this student could do to earn a better score?

FRAPPY! Scoring Rubric

Use the following rubric to score your response. Each part receives a score of "Essentially Correct," "Partially Correct," or "Incorrect." When you have scored your response, reflect on your understanding of the concepts addressed in this problem. If necessary, note what you would do differently on future questions like this to increase your score.

Intent of the Question

The goal of this question is to determine your ability to calculate and interpret the mean and standard deviation of a discrete random variable, combine random variables, and describe linear transformations of random variables.

Solution

(a) Mean: $\mu_R = 0(0.4)+1(0.3)+2(0.2)+3(0.1)=1$

Standard deviation: $\sigma_R = \sqrt{(0-1)^2(0.4)+(1-1)^2(0.3)+(2-1)^2(0.2)+(3-1)^2(0.1)}$ = 1

We can expect to have to repair our mp3 player once over its lifetime, but that can vary on average by about 1 repair.

(b) If T = the total number of repairs across the two devices, the mean of T will be μ_T = 1 + 2 = 3 and the standard deviation will be $\sigma_T = \sqrt{1^2+1.2^2}$ = 1.562.

We can expect to have to perform 3 repairs, total, on our devices over their lifetime, but that amount can vary on average by about 1.562 repairs.

(c) The total amount we can expect to pay in repairs will be \$15(1) + \$25(2) = \$65.

The standard deviation will be $\sqrt{15^2(1^2)+25^2(1.2)^2}$ = \$33.54.

We can expect to pay \$65 in repairs over the lifetime of the devices, but that can vary by an average of \$33.52.

Scoring:

Parts (a), (b), and (c) are scored as essentially correct (E), partially correct (P), or incorrect (I).

Part (a) is essentially correct if the mean and standard deviation are calculated correctly AND interpreted correctly.
Part (a) is partially correct if no interpretation or an incorrect interpretation is provided OR if only one calculation/interpretation is correct.

Part (b) is essentially correct if the mean and standard deviation are calculated correctly AND interpreted correctly.
Part (b) is partially correct if no interpretation or an incorrect interpretation is provided OR if only one calculation/interpretation is correct.

Part (c) is essentially correct if the mean and standard deviation are calculated correctly AND interpreted correctly.
Part (c) is partially correct if no interpretation or an incorrect interpretation is provided OR if only one calculation/interpretation is correct.

4 **Complete Response**
All three parts essentially correct

3 **Substantial Response**
Two parts essentially correct and one part partially correct

2 **Developing Response**
Two parts essentially correct and no parts partially correct
One part essentially correct and two parts partially correct
Three parts partially correct

1 **Minimal Response**
One part essentially correct and one part partially correct
One part essentially correct and no parts partially correct
No parts essentially correct and two parts partially correct

My Score:
What I did well:
What I could improve:
What I should remember if I see a problem like this on the AP Exam:

Chapter 6: Random Variables

1 2 3 4 5 6 7 8 9 10 11 12 13 14 15 16 17 18

Across

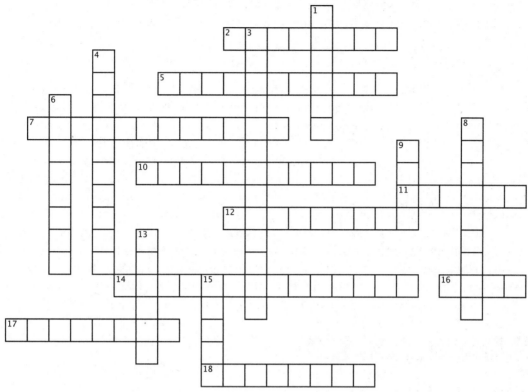

2. The average of the squared deviations of the values of a variable from its mean.
5. Random variables are _____ if knowing whether an event in X has occurred tells us nothing about the occurrence of an event involving Y.
7. The probability _____ of a random variable gives its possible values and their probabilities.
10. The number of ways of arranging k successes among n observations is the binomial _____.
11. The sum or difference of independent Normal random variables follows a _____ distribution.
12. When you combine independent random variable, you always add these.
14. A linear _____ occurs when we add/subtract and multiply/divide by a constant.
16. An easy way to remember the requirements for a geometric setting.
17. This setting arises when we perform several independent trials of a chance process and record the number of times an outcome occurs.
18. The mean of a discrete random variable is also called the _____ value.

Down

1. A _____ variable takes numerical values that describe the outcomes of some chance process.
3. When n is large, we can use a Normal _____ to determine probabilities for binomial settings.
4. A random variable that takes on all values in an interval of numbers.
6. A random variable that takes a fixed set of possible values with gaps between.
8. A _____ setting arises when we perform independed trials of the same chance process and record the number of trials until a particular outcome occurs.
9. An easy way to remember the requirements for a binmial setting.
13. Adding a constant to each value of a random variable has no effect on the shape or _____ of the distribution.
15. Multiplying each value of a random variable by a constant has no effect on the _____ of the distribution.

Chapter 7: Sampling Distributions

"Statistics may be defined as 'a body of methods for making wise decisions in the face of uncertainty.'" W.A. Wallis

Chapter Overview

In chapters 1-3, you learned how to explore a set of data. Chapter 4 introduced you to methods for producing data. Chapters 5 and 6 focused on the basics behind probability and random variables. In this chapter, you will study the final piece necessary to study statistical inference – sampling distributions. The foundation of statistical inference lies in the concept of the sampling distribution. In order to make a conclusion about a population based on information from a sample, you need to be able to answer the question, "What results would I expect to see if I sampled repeatedly from a population of interest?" Sampling distributions provide an answer to that question, allowing us to draw a conclusion about a population parameter based on an observed statistic from a sample. In this chapter, you will learn how to describe sampling distributions for sample proportions as well as sampling distributions for sample means. You will also learn an important theorem for sample means—the central limit theorem. The final chapters in the textbook will build upon your learning in this chapter to present formal methods for statistical inference. The better you understand sampling distributions, the easier your study of inference will be. Be sure to get a good grasp of these concepts before moving on!

Sections in this Chapter
Section 7.1: What Is a Sampling Distribution?
Section 7.2: Sample Proportions
Section 7.3: Sample Means

Plan Your Learning

Use the following *suggested* guide to help plan your reading and assignments in "The Practice of Statistics, 6th Edition." Note: your teacher may schedule a different pacing or assign different problems. Be sure to follow his or her instructions!

Read	**7.1: pp. 440 – 446**	**7.1: pp. 447 – 454**	**7.2: pp. 458 - 465**
Do	1,3,5,7,9	11,13,15,19,21,25 MC 26-30	35,37,41,43 MC 47-50

Read	**7.3: pp. 468 – 474**	**7.3: pp. 474 - 479**	**Chapter Summary**
Do	53,55,57,61	63,65,67,69,71 MC 73-76	Multiple Choice FRAPPY!

Section 7.1: What Is a Sampling Distribution?

Before You Read: Section Summary

This section will introduce you to the big ideas behind sampling distributions. First, you will learn how to distinguish between population parameters and statistics derived from samples. Next, you will explore the fact that statistics vary from sample to sample. This simple fact is the reason we study sampling distributions. By describing the shape, center, and variability of the sampling distribution of a statistic, we can determine the critical information necessary to perform statistical inference.

Learning Targets:

____ I can distinguish between a parameter and a statistic.
____ I can create a sampling distribution using all possible samples from a small population.
____ I can use the sampling distribution of a statistic to evaluate a claim about a parameter.
____ I can distinguish among the distribution of a population, the distribution of a sample and the sampling distribution of a statistic.
____ I can determine if a statistic is an unbiased estimator of a population parameter.
____ I can describe the relationship between sample size and the variability of a statistic.

While You Read: Key Vocabulary and Concepts

statistic:
parameter:
sampling variability:
sampling distribution:
population distribution:
distribution of sample data:
unbiased estimator:
variability of a statistic:

☑ After You Read: Check for Understanding

Concept 1: Parameters and Statistics

One of the most powerful skills we learn from statistics is the ability to answer a question about a population characteristic based on information gathered from a random sample. That is, we can use a statistic calculated from a sample to make a conclusion about a corresponding parameter in the population. However, we must note that the statistics we calculate from a sample may differ somewhat from the population characteristic we are trying to measure. Further, the statistic would likely differ from sample to sample. This sample-to-sample variability poses a problem when we try to generalize our findings to the population. However, based on what we learned in the last chapter, we can view a sample statistic as a random variable. That is, while we have no way of predicting exactly what statistic value we will get from a sample, we know how those values will behave in repeated random sampling. With the probability distribution of this random variable in mind, we can use a sample statistic to estimate the population parameter.

Check for Understanding: ____ *I can distinguish between a parameter and a statistic.*

For each of the following situations, identify the population of interest, the parameter, and the statistic.

a) A medical researcher is interested in exploring the effects of a new medicine on blood pressure. 500 males with high blood pressure are randomly selected and given the new drug. After two weeks, their blood pressure is measured and the average arterial pressure is calculated.

b) A study is conducted to determine whether or not the dangerous activity of texting while driving is a common practice. 1500 16- to 24-year-olds are randomly selected and asked whether or not they text while driving. Of the 1500 drivers, 12% indicate they text while driving.

Concept 2: Describing Sampling Distributions

To draw a conclusion about a population proportion *p*, we take a random sample and calculate the sample proportion $\hat{p}$. Likewise, to reach a conclusion about a population mean μ, we take a random sample and calculate the sample mean $\bar{x}$. Because of chance variation in random sampling, the values of our sample statistic will vary from sample to sample. The distribution of statistic values in all possible samples of the same size from a population is called the sampling distribution of the statistic. The sampling distribution describes the sampling variability and provides a foundation for performing inference. The variability of a sampling distribution is an important attribute as all inference calculations depend upon it! When trying to estimate a parameter, we want minimum sampling variability and no bias. Random sampling helps us avoid bias while larger samples help us minimize sampling variability.

Check for Understanding: ____ *I can distinguish among the distribution of a population, the distribution of a sample and the sampling distribution of a statistic.* ____ *I can create a sampling distribution using all possible samples from a small population.*

A breakfast cereal includes marshmallow shapes in the following distribution: 10% stars, 10% crescent moons, 20% rockets, 40% astronauts, 20% planets. We are interested in examining the proportion of rockets in a random sample of 2000 marshmallows from the cereal.

a) Sketch the population distribution of marshmallow shapes.

b) Suppose you were to collect a random sample of 2000 marshmallow shapes. Sketch the distribution of sample data you would expect to see. How many rockets would you expect to see in your sample?

c) Now, suppose you collected many samples of the same size. Sketch the sampling distribution of the proportion of rockets you think you would see in the samples.

Section 7.2: Sample Proportions

Before You Read: Section Summary

The objective of some statistical applications is to reach a conclusion about a population proportion p. For example, we may try to estimate an approval rating through a survey, or test a claim about the proportion of defective light bulbs in a shipment based on a random sample. Since p is unknown to us, we must base our conclusion on a sample proportion, $\hat{p}$. However, as we have noted, we know that the value of $\hat{p}$ will vary from sample to sample. The amount of variability will depend on the size of our sample. In this section, you will learn how to describe the shape, center, and spread of the sampling distribution of $\hat{p}$ in detail.

Learning Targets:

____ I can calculate the mean and standard deviation of the sampling distribution of a sample proportion $\hat{p}$ and interpret the standard deviation.
____ I can determine if the sampling distribution of $\hat{p}$ is approximately Normal.
____ I can use a Normal distribution to calculate probabilities involving $\hat{p}$, if appropriate.

While You Read: Key Vocabulary and Concepts

sampling distribution of $\hat{p}$:

mean of the sampling distribution of $\hat{p}$:

standard deviation of the sampling distribution of $\hat{p}$:

Normal approximation for $\hat{p}$:

After You Read: Check for Understanding

Concept 1: The Sampling Distribution of $\hat{p}$

If we take repeated samples of the same size n from a population with a proportion of interest p, the sampling distribution of $\hat{p}$ will have the following characteristics:

1) The shape of the sampling distribution of $\hat{p}$ will be approximately Normal as long as $np \geq 10$ and $n(1-p) \geq 10$.
2) The mean of the sampling distribution of $\hat{p}$ is $\mu_{\hat{p}} = p$.
3) The standard deviation of the sampling distribution of $\hat{p}$ is $\sigma_{\hat{p}} = \sqrt{\frac{p(1-p)}{n}}$

Note: The formula for the standard deviation is exactly correct only if we are sampling from an infinite population or *with replacement* from a finite population. When we are sampling without replacement from a finite population, the formula is approximately correct as long as the 10% condition is satisfied. That is, the sample size must be less than or equal to 10% of the population size.

Check for Understanding: ____*I can calculate the mean and standard deviation of the sampling distribution of a sample proportion* $\hat{p}$ *and interpret the standard deviation.*

____*I can determine if the sampling distribution of* $\hat{p}$ *is approximately Normal.*

Suppose your job at a potato chip factory is to check each shipment of potatoes for quality assurance. Further, suppose that a truckload of potatoes contains 95% that are acceptable for processing. If more than 10% are found to be unacceptable in a random sample, you must reject the shipment. To check, you randomly select and test 250 potatoes. Let $\hat{p}$ be the sample proportion of unacceptable potatoes.

a) What is the mean of the sampling distribution of $\hat{p}$?

b) Check the 10% condition and calculate the standard deviation of the sampling distribution of $\hat{p}$.

c) Check the Normal condition and sketch the sampling distribution of $\hat{p}$. Based on this sketch, do you think it would be likely to reject the truckload based on a random sample of 250 potatoes? Why or why not?

Concept 2: Using the Normal Approximation for $\hat{p}$

When the sample size *n* is large enough for *np* and *n*(1-*p*) to both be at least 10, the shape of the sampling distribution of $\hat{p}$ will be approximately Normal. In that case, we can use Normal calculations to determine the probability that an SRS will generate a value of $\hat{p}$ in a particular interval. This calculation is an important component of inference.

Check for Understanding: ____ *I can use a Normal distribution to calculate probabilities involving $\hat{p}$, if appropriate.*

A phone company is interested in exploring marketing possibilities for a new smartphone for teenagers. They ask an SRS of 1000 high school students whether they own a smartphone. Suppose 65% of all high school students own a smartphone. What is the probability that the random sample selected by the company will result in a $\hat{p}$-value within 3 percentage points of the true population proportion? Show all your work!

Section 7.3: Sample Means

Before You Read: Section Summary

When the goal of a statistical application is to reach a conclusion about a population mean μ we must rely on a sample mean $\bar{x}$. However, as we have noted, the value of $\bar{x}$ will vary from sample to sample. Like we observed with sample proportions, the amount of variability will depend on the size n of our sample. In this section, you will learn how to describe the shape, center, and variability of the sampling distribution of $\bar{x}$ in detail.

Learning Targets:

____ I can calculate the mean and standard deviation of the sampling distribution of a sample mean $\bar{x}$ and interpret the standard deviation.

____ I can explain how the shape of the sampling distribution of $\bar{x}$ is affected by the shape of the population distribution and the sample size.

____ I can use a Normal distribution to calculate probabilities involving $\bar{x}$, if appropriate.

While You Read: Key Vocabulary and Concepts

sampling distribution of $\bar{x}$:
mean of the sampling distribution of $\bar{x}$:
standard deviation of the sampling distribution of $\bar{x}$:
Central Limit Theorem:

After You Read: Check for Understanding

Concept 1: Sampling Distribution of $\bar{x}$

If we take repeated random samples of the same size n from a population with mean μ, the sampling distribution of $\bar{x}$ will have the following characteristics:

1) The shape of the sampling distribution depends upon the shape of the population distribution. If the population is Normally distributed, the sampling distribution of $\bar{x}$ will be Normally distributed. If the population distribution is non-Normal, the sampling distribution of $\bar{x}$ will become more and more Normal as n increases.
2) The mean of the sampling distribution is $\mu_{\bar{x}} = \mu$.
3) The standard deviation of the sampling distribution is $\sigma_{\bar{x}} = \dfrac{\sigma}{\sqrt{n}}$

4) Note: The formula for the standard deviation is exactly correct only if we are sampling from an infinite population or *with replacement* from a finite population. When we are sampling without replacement from a finite population, the formula is approximately correct as long as the 10% condition is satisfied. That is, the sample size must be less than or equal to 10% of the population size.

Check for Understanding: ____ *I can calculate the mean and standard deviation of the sampling distribution of a sample mean* $\bar{x}$ *and interpret the standard deviation.* ___ *I can use a Normal distribution to calculate probabilities involving* $\bar{x}$, *if appropriate.*

The times it takes 5th graders to complete a particular mathematics problem are Normally distributed with mean 2 minutes and standard deviation 0.8 minutes.

a) Find the probability that a randomly chosen 5th grader will take more than 2.5 minutes to complete the problem. Show your work.

b) Suppose you give the problem to an SRS of 20 students. Sketch the sampling distribution of $\bar{x}$. Then use this distribution to determine the probability that the mean time to complete the problem for the SRS of students is greater than 2.5 minutes. Show your work.

Concept 2: The Central Limit Theorem

When the population is Normally distributed, we know that the sampling distribution of $\bar{x}$ will be Normally distributed, so we can use Normal calculations. However, most population distributions are not Normally distributed. If our sampling distribution is skewed or non-Normal in some other way we cannot use Normal calculations to answer questions. Thankfully, a pretty remarkable fact about sample means helps us out: when the sample size *n* is large, the shape of the sampling distribution of $\bar{x}$ will be approximately Normal no matter what the shape of the population distribution may be! For our purposes, we'll define "large" to be any sample that is at least 30. So, if $n \geq 30$, we can be safe in assuming that the sampling distribution of $\bar{x}$ will be approximately Normal and we can proceed to perform Normal calculations. If $n < 30$, we must consider the shape of the population distribution.

Check for Understanding: ____ *I can use a Normal distribution to calculate probabilities involving* $\bar{x}$, *if appropriate.*

The blood cholesterol level of adult men has mean 188 mg/dl and standard deviation 41 mg/dl. A SRS of 250 men is selected and the mean blood cholesterol level in the sample is calculated.

Sketch the sampling distribution of $\bar{x}$ and calculate the probability that the sample mean will be greater than 193.

Chapter Summary: Sampling Distributions

This chapter introduced you to a key concept for inferential thinking – sampling distributions. Since we are interested in drawing conclusions about population proportions and means, it is important to know how statistics will behave in repeated random sampling. Being able to describe the sampling variability for sample statistics allows us to estimate and test claims about population parameters. This chapter provided us with some key facts about sample statistics that will help us as we begin our formal study of inference. First, statistics will vary from sample to sample. Second, if the sample size is large enough, we know that the distribution of sample statistic values will be approximately Normal. Third, the sampling distributions of $\hat{p}$ and $\bar{x}$ will be centered at p and μ, respectively. Finally, the variability of these sampling distributions can be computed (as long as the 10% condition is met). This variability will decrease as the sample size increases, so bigger random samples are more desirable.

One important fact about sample means was revealed in this chapter. When sampling from a Normal population, the sampling distribution of $\bar{x}$ will be Normal. However, as long as our sample size is at least 30, the shape of the sampling distribution of $\bar{x}$ will be approximately Normal--no matter what the population distribution looks like! The central limit theorem is a powerful fact that will be revisited in the coming chapters. Now that we have a grasp of the basic concept of sampling distributions, we are ready to begin the formal study of statistical inference!

After You Read: What Have I Learned?

Complete the vocabulary puzzle, multiple-choice questions, and FRAPPY. Check your answers and performance on each of the learning targets. Be sure to get extra help on any targets that you identify as needing more work!

Learning Target	**Got It!**	**Almost There**	**Needs Work**
I can distinguish between a parameter and a statistic.			
I can create a sampling distribution using all possible samples from a small population.			
I can use the sampling distribution of a statistic to evaluate a claim about a parameter.			
I can distinguish among the distribution of a population, the distribution of a sample and the sampling distribution of a statistic.			
I can determine if a statistic is an unbiased estimator of a population parameter.			
I can describe the relationship between sample size and the variability of a statistic.			
I can calculate the mean and standard deviation of the sampling distribution of a sample proportion $\hat{p}$ and interpret the standard deviation.			
I can determine if the sampling distribution of $\hat{p}$ is approximately Normal.			
I can use a Normal distribution to calculate probabilities involving $\hat{p}$, if appropriate.			
I can calculate the mean and standard deviation of the sampling distribution of a sample mean $\bar{x}$ and interpret the standard deviation.			
I can explain how the shape of the sampling distribution of $\bar{x}$ is affected by the shape of the population distribution and the sample size.			
I can use a Normal distribution to calculate probabilities involving $\bar{x}$, if appropriate.			

Chapter 7 Multiple Choice Practice

Directions. *Identify the choice that best completes the statement or answers the question. Check your answers and note your performance when you are finished.*

1. The variability of a statistic is described by
   (A) the spread of its sampling distribution.
   (B) the amount of bias present.
   (C) the vagueness in the wording of the question used to collect the sample data.
   (D) probability calculations.
   (E) the stability of the population it describes.

2. Below are dot plots of the values taken by three different statistics in 30 samples from the same population. The true value of the population parameter is marked with an arrow.

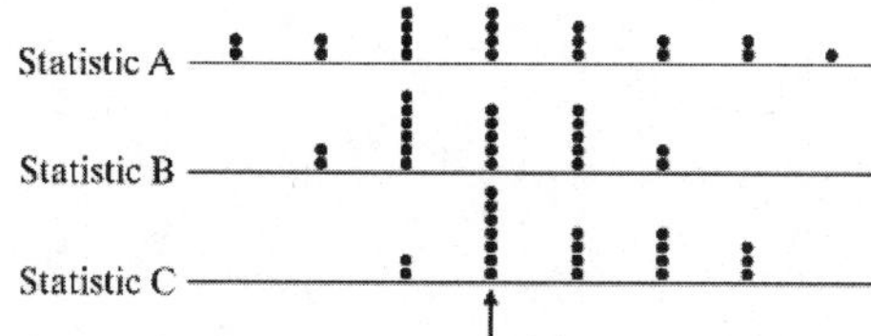

The statistic that has the largest *bias* among these three is
   (A) statistic A.
   (B) statistic B.
   (C) statistic C.
   (D) A and B have similar bias, and it is larger than the bias of C.
   (E) B and C have similar bias, and it is larger than the bias of A.

3. According to a recent poll, 27% of Americans prefer to read their news in a physical newspaper instead of online. Let's assume this is the parameter value for the population. If you take a simple random sample of 25 Americans and let $\hat{p}$ = the proportion in the sample who prefer a newspaper, is the shape of the sampling distribution of $\hat{p}$ approximately Normal?
   (A) No, because $p < 0.50$
   (B) No, because $np = 6.75$
   (C) Yes, because we can reasonably assume there are more than 250 individuals in the population.
   (D) Yes, because we took a simple random sample.
   (E) Yes, because $n(1-p) = 18.25$

4. The time it takes students to complete a statistics quiz has a mean of 20.5 minutes and a standard deviation of 15.4 minutes. What is the probability that a random sample of 40 students will have a mean completion time greater than 25 minutes?
   (A) 0.9678
   (B) 0.0322
   (C) 0.0344
   (D) 0.3851
   (E) 0.6149

5. A fair coin (one for which both the probability of heads and the probability of tails are 0.5) is tossed 60 times. The probability that more than 1/3 of the tosses are heads is closest to
   (A) 0.9951.
   (B) 0.33.
   (C) 0.109.
   (D) 0.09.
   (E) 0.0049.

6. The histogram below was obtained from data on 750 high school basketball games in a regional athletic conference. It represents the number of three-point baskets made in each game.

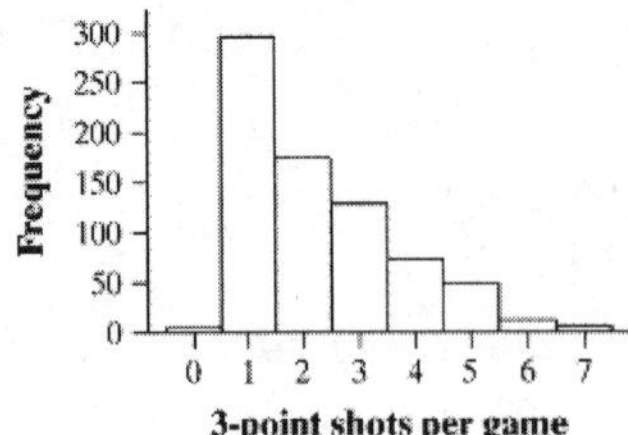

What is the range of sample sizes a researcher could take from this population without violating conditions required for performing Normal calculations with the sampling distribution of $\overline{x}$?

(A) $0 \leq n \leq 30$
(B) $30 \leq n \leq 50$
(C) $30 \leq n \leq 75$
(D) $30 \leq n \leq 750$
(E) $75 \leq n \leq 750$

7. The incomes in a certain large population of college teachers have a normal distribution with mean \$60,000 and standard deviation \$5000. Four teachers are selected at random from this population to serve on a salary review committee. What is the probability that their average salary exceeds \$65,000?

(A) 0.0228
(B) 0.1587
(C) 0.8413
(D) 0.9772
(E) essentially 0

8. A random sample of size 25 is to be taken from a population that is Normally distributed with mean 60 and standard deviation 10. The mean $\overline{x}$ of the observations in our sample is to be computed. The sampling distribution of $\overline{x}$

(A) is Normal with mean 60 and standard deviation 10.
(B) is Normal with mean 60 and standard deviation 2.
(C) is approximately Normal with mean 60 and standard deviation 2.
(D) has an unknown shape with mean 60 and standard deviation 10.
(E) has an unknown shape with mean 60 and standard deviation 2.

9. The scores of individual students on a college entrance examination have a left-skewed distribution with mean 18.6 and standard deviation 6.0. At Millard North High School, 36 seniors take the test. The sampling distribution of mean scores for random samples of 36 students is

(A) approximately Normal.
(B) symmetric and mound-shaped, but non-Normal.
(C) skewed right.
(D) neither Normal nor non-normal. It depends on the particular 36 students selected.
(E) exactly Normal.

10. The distribution of prices for home sales in Minnesota is skewed to the right with a mean of \$290,000 and a standard deviation of \$145,000. Suppose you take a simple random sample of 100 home sales from this (very large) population. What is the probability that the mean of the sample is above \$325,000?

(A) 0.0015
(B) 0.0027
(C) 0.0079
(D) 0.4046
(E) 0.4921

Check your answers below. If you got a question wrong, check to see if you made a simple mistake or if you need to study that concept more. After you check your work, identify the concepts you feel very confident about and note what you will do to learn the concepts in need of more study.

#	Answer	Concept	Right	Wrong	Simple Mistake?	Need to Study More
1	A	Sampling Variability				
2	C	Bias and Variability				
3	B	Normality Condition				
4	B	Normal Probability Calculation				
5	A	Normal Probability Calculation				
6	C	10% Condition and CLT				
7	A	Normal Probability Calculation				
8	B	Sampling Distribution for Means				
9	A	Sampling Distribution for Means				
10	C	Normal Probability Calculation				

Chapter 7 Reflection

Summarize the "Big Ideas" in Chapter 7:

My strengths in this chapter:

Concepts I need to study more and what I will do to learn them:

FRAPPY! Free Response AP® Problem, Yay!

The following problem is modeled after actual Advanced Placement Statistics free response questions. Your task is to generate a complete, concise response in 15 minutes. After you generate your response, view two example solutions and determine whether you feel they are "complete", "substantial", "developing" or "minimal". If they are not "complete", what would you suggest to the student who wrote them to increase their score? Finally, you will be provided with a rubric. Score your response and note what, if anything, you would do differently to increase your own score.

A television producer must schedule a selection of paid advertisements during each hour of programming. The lengths of the advertisements are Normally distributed with a mean of 28 seconds and standard deviation of 5 seconds. During each hour of programming, 45 minutes are devoted to the program and 15 minutes are set aside for advertisements. To fill in the 15 minutes, the producer randomly selects 30 advertisements.

a) Describe the sampling distribution of the sample mean length for random samples of 30 advertisements.

b) If 30 advertisements are randomly selected, what is the probability that the total time needed to air them will exceed the 15 minutes available? Show your work.

FRAPPY! Student Responses

Student Response 1:

(a) The sampling distribution will have a mean of 28 seconds and a standard deviation of 5/√(28) = 0.945 seconds.

(b) z=(30-28)/0.945 = 2.12. The probability the total time will exceed the 15 available minutes is 1 - 0.9826 = 0.0174.

How would you score this response? Is it substantial? Complete? Developing? Minimal? Is there anything this student could do to earn a better score?

Student Response 2:

(a) Since the sample size is only 28, we can't use the central limit theorem. However, because the advertisement times are Normally distributed, the sampling distribution of the average time will be Normal. The mean will be 28 seconds and the standard deviation is 5/√(30) = 0.913 seconds.

(b) Since 15 minutes = 900 seconds, the average time for the 30 ads can not exceed 900/30 = 30 seconds. Using the information from part (a), z = (30-28) / 0.913 = 2.19. Therefore, the probability the ads will exceed 900 seconds is 1 – 0.9857 = 0.0143. This is not very likely, so the producer should not be concerned about running over if he or she selects 30 ads randomly.

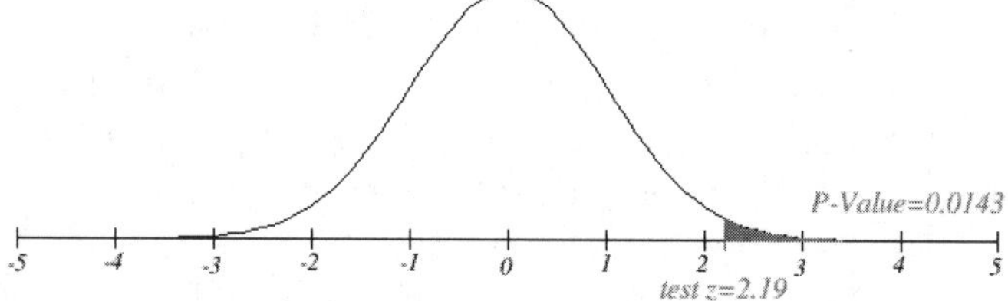

How would you score this response? Is it substantial? Complete? Developing? Minimal? Is there anything this student could do to earn a better score?

FRAPPY! Scoring Rubric

Use the following rubric to score your response. Each part receives a score of "Essentially Correct," "Partially Correct," or "Incorrect." When you have scored your response, reflect on your understanding of the concepts addressed in this problem. If necessary, note what you would do differently on future questions like this to increase your score.

Intent of the Question

The goal of this question is to determine your ability to describe the sampling distribution of a sample mean and use it to perform a probability calculation.

Solution

(a) The sampling distribution of the sample mean length for random samples of 30 advertisements has mean 28 seconds and standard deviation $\sigma_{\bar{x}} = \frac{5}{\sqrt{30}} = 0.913$ seconds. Because we are told that the population of advertisement lengths is Normally distributed, the shape of the sampling distribution will be Normal.

(b) The probability that a random sample of 30 advertisements will exceed the allotted time is equivalent to the probability that the sample mean length of the 30 advertisements is greater than 900/30 = 30 seconds. In part (a), we determined that the sampling distribution is Normal with mean = 28 and standard deviation = 0.913. Therefore,

$$P(\bar{x} > 30) = P\left(Z > \frac{30-28}{0.913}\right) = P(Z > 2.19) = 1 - 0.9857 = 0.0143$$

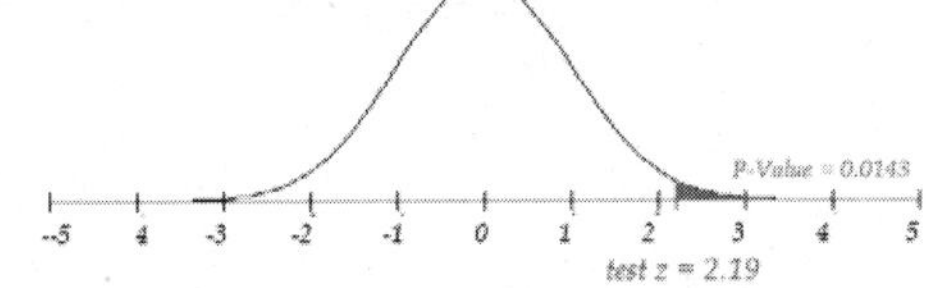

There is a 1.43% chance the randomly selected advertisements will exceed the allotted time.

Scoring:

Parts (a), (b), and (c) are scored as essentially correct (E), partially correct (P), or incorrect (I).

Part (a) is essentially correct if the response correctly identifies the shape (Normal), center (mean=28 seconds) and variability (standard deviation = 0.913 seconds) of the sampling distribution. The calculation of the standard deviation should be shown to earn an essentially correct score. Part (a) is partially correct if the solution only identifies 2 of the 3 components correctly or correctly identifies the standard deviation, but fails to show the calculation.

Part (b) is essentially correct if the response sets up and performs a correct probability calculation. Part (b) is partially correct if the response includes a correctly set up calculation, but fails to calculate the correct value OR if it sets up an incorrect, but plausible, calculation but carries it through correctly.

4 **Complete Response**
Both parts essentially correct

3 **Substantial Response**
One part essentially correct and one part partially correct

2 **Developing Response**
Both parts partially correct
One part essentially correct

1 **Minimal Response**
One part partially correct

My Score:
What I did well:
What I could improve:
What I should remember if I see a problem like this on the AP Exam:

Chapter 7: Sampling Distributions

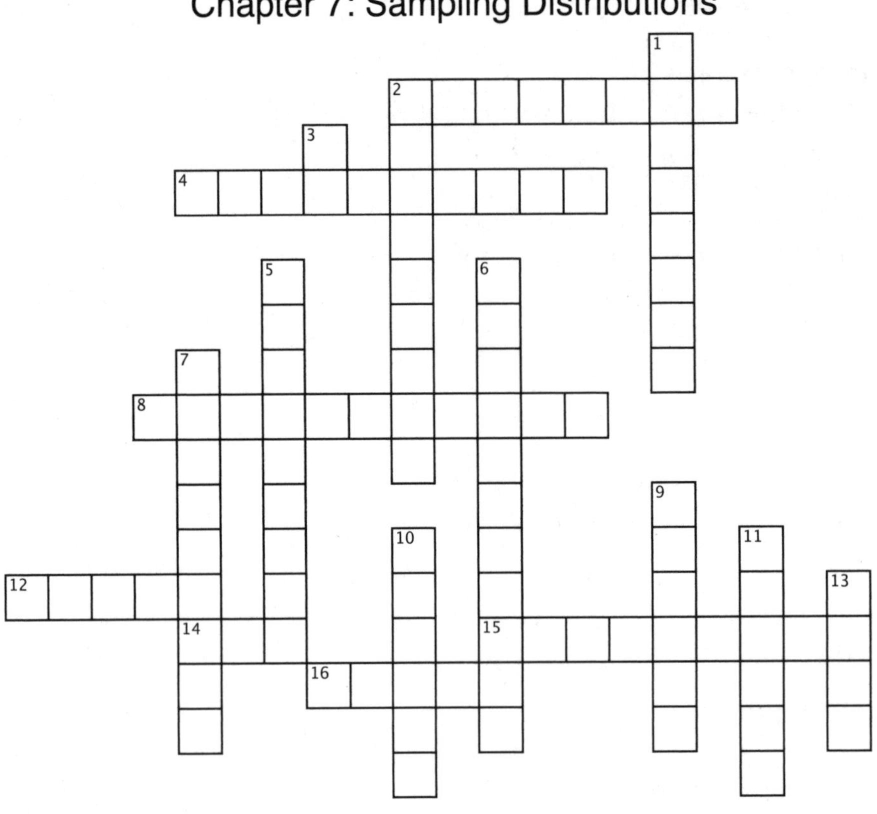

Across

2. _____ distribution: the distribution of values taken by the statistic in all possible samples of the same size from the population
4. ______ distribution: the distribution of all values of a variable in the population
8. _____ of a statistic is described by the spread of the sampling distribution
12. Greek letter used for the population standard deviation
14. the Normal approximation for the sampling distribution of a sample proportion can be used when both the number of successes and failures are greater than _____
15. sampling distributions and sampling variability provide the foundation for performing ______
16. central _____ theorem tells us if the sample size is large, the sampling distribution of the sample mean is approximately Normal, regardless of the shape of the population

Down

1. a statistic is an _____ estimator if the mean of the sampling distribution is equal to the true value of the parameter being estimated.
2. a number, computed from sample data, that estimates a parameter
3. Greek letter used for the population mean
5. standard ______ : measure of spread of a sampling distribution
6. sampling ______ notes the value of a statistic may be different from sample to sample
7. a number that describes a population
9. the rule of thumb for using the central limit theorem - the sample size should be greater than ______
10. when the sample size is large, the sampling distribution of a sample proportion is approximately ______
11. to draw a conclusion about a population parameter, we can look at information from a _____ sample
13. center of a sampling distribution

Chapter 8: Estimating with Confidence

***"Do not put your faith in what statistics say until you have carefully considered what they do not say."* William W. Watt**

Chapter Overview

Now that you have learned the basics of probability and sampling distributions, you are ready to begin your formal study of inference. In this chapter, you will use what you have learned about sampling distributions to construct confidence intervals for population proportions and population means. In later chapters, you'll learn how to determine confidence intervals for paired data, differences in means and proportions, and slopes of regression lines. Each of those procedures uses the same approach as the intervals you'll construct in this chapter, so you'll want to get a solid foundation in the basics here!

Sections in this Chapter
Section 8.1: Confidence Intervals: The Basics
Section 8.2: Estimating a Population Proportion
Section 8.3: Estimating a Population Mean

Plan Your Learning

Use the following *suggested* guide to help plan your reading and assignments in "The Practice of Statistics, 6th Edition." Note: your teacher may schedule a different pacing or assign different problems. Be sure to follow his or her instructions!

Read	**8.1: pp. 494 - 499**	**8.1: pp. 499 – 506**	**8.2: pp. 510 – 516**	**8.2: pp. 517 -521**
Do	1, 3, 5, 7, 9	11,15,17,19,21 MC 23-26	29,31,35,37,39	41,45,49 MC 55-58

Read	**8.3: pp. 525 – 534**	**8.3: pp. 534 - 541**	**Chapter Summary**
Do	61,63,65,67	69,73,77 MC 81-84	Multiple-Choice FRAPPY!

Section 8.1: Confidence Intervals: The Basics

Before You Read: Section Summary

In this section you will be introduced to the basic ideas behind constructing and interpreting a confidence interval for a population parameter. You will learn how we can take a point estimate for a population parameter and use what we know about sampling variability to construct an interval of plausible values for the parameter. You will focus on the big ideas in this section. Make sure you understand the different components of a confidence interval as well as the correct interpretation of both the interval and the confidence level. The next two sections will build upon the concepts presented here and focus on the details for estimating proportions and means.

Learning Targets:

____ I can identify an appropriate point estimator and calculate the value of a point estimate.
____ I can interpret a confidence interval in context.
____ I can determine the point estimate and margin of error from a confidence interval.
____ I can use a confidence interval to make a decision about the value of a parameter.
____ I can interpret a confidence level in context.
____ I can describe how the sample size and confidence level affect the margin of error.
____ I can explain how practical issues like nonresponse, undercoverage, and response bias can affect the interpretation of a confidence interval.

While You Read: Key Vocabulary and Concepts

point estimator:
point estimate:
confidence interval:
confidence level *C*:
margin of error:
critical value:

☑ After You Read: Check for Understanding

Concept 1: The Idea of a Confidence Interval

When our goal is to estimate a population parameter, we often must rely on a sample statistic to provide a "point estimate." However, as we learned in the last chapter, that estimate will vary from sample to sample. A confidence interval takes that variation in to account to provide an interval of plausible values, based on the statistic, for the true parameter. All confidence intervals have two main components: an interval based on the estimate that includes a margin of error and a confidence level *C* that reports the success rate of the method used to construct the interval in capturing the parameter in repeated constructions. For example, "C% confident" means C% of all samples of the same size from the population of interest would yield an interval that captures the true parameter. We can then interpret the interval itself to say "We are C% confident that the interval from a to b captures the true value of the population parameter."

Check for Understanding: ____ *I can interpret a confidence interval in context.* ____ *I can interpret a confidence level in context*

How much do the volumes of bottles of water vary? A random sample of 50 "20 oz." water bottles is collected and the contents are measured. A 90% confidence interval for the population mean μ is 19.10 to 20.74.

a) Interpret the confidence interval in context.

b) Interpret the confidence level in context.

c) Based on this interval, what can you say about the contents of the bottles in the sample? What can you say about the contents of bottles in the population?

Concept 2: Constructing a Confidence Interval

We will learn in the next section that to construct a confidence interval, we must work through several steps. First, you MUST check that the conditions for constructing the interval are met. That is, we must be assured that the data come from a random sample or randomized experiment. Next, the sampling distribution of the statistic must be approximately Normal. And, finally, the individual observations must be independent (which means checking the 10% condition if we're sampling without replacement from a finite population).

We construct the interval using the formula

statistic ± (critical value) ·(standard deviation of the statistic)

where the critical value is determined based on the confidence level *C* and the standard deviation is based on the sampling distribution of the statistic. We interpret the interval using the language we learned earlier in this section.

Our goal with confidence intervals is to provide as precise an estimate as possible. That is, we wish to construct a narrow interval that we are confident captures the parameter of interest. We can achieve this in two ways: by decreasing our confidence or by increasing our sample size.

Check for Understanding: ____ *I can describe how the sample size and confidence level affect the margin of error.* ____ *I can explain how practical issues like nonresponse, undercoverage, and response bias can affect the interpretation of a confidence interval.*

A large company is interested in developing a new bake ware product for consumers. In an effort to determine baking habits of adults, a researcher selects a random sample of 50 addresses in a large, midwestern, metropolitan area. She calls each selected home in the late morning to collect information on their baking habits. The proportion of adults who bake at least twice a week is calculated and a 95% confidence interval is constructed.

a) Explain what would happen to the length of the interval if the confidence level were decreased to 90%.

b) How would a 95% confidence interval based on a sample of size 100 compare to the original 95% confidence interval?

c) Describe one potential source of bias in this study that is not accounted for by the margin of error.

Section 8.2: Estimating a Population Proportion

Before You Read: Section Summary

In the last section, you learned the basic ideas behind confidence intervals. In the next two sections, you will learn how to construct and interpret confidence intervals for proportions and means. You will start by constructing them for proportions, focusing on the application of the four-step process to the procedure.

Learning Targets:

- ____ I can state and check the Random, 10%, and Large Counts conditions for constructing a confidence interval for a population proportion.
- ____ I can determine the critical value for calculating a *C*% confidence interval for a population proportion using a table or technology.
- ____ I can construct and interpret a confidence interval for a population proportion.
- ____ I can determine the sample size required to obtain a *C*% confidence interval for a population proportion with a specified margin of error.

While You Read: Key Vocabulary and Concepts

one-sample *z* interval for a population proportion:
standard error:
conditions for constructing a confidence interval about a proportion:
sample size for a desired margin of error when estimating *p*:

After You Read: Check for Understanding

Concept 1: Conditions for Estimating p

When constructing a confidence interval for *p*, it is critical that you begin by checking that the conditions are met. First, check to make sure that the sample was randomly selected or there was random assignment in an experiment. Because the construction of the interval is based on the sampling distribution of p-hat, next you must ensure that the condition for Normality is met. That is, check to see that $n\hat{p}$ and $n(1-\hat{p})$ are both at least 10. Finally, check for independence of measurements. If there is sampling without replacement, verify that the population of interest is at least 10 times as large as the sample. If all three of these conditions are met, you can safely proceed to construct and interpret a confidence interval for a population proportion *p*.

Concept 2: Constructing a Confidence Interval for p

To construct a confidence interval for a population proportion *p*, you should follow a four step process.

- **State** the parameter you want to estimate and at what confidence level.
- **Plan** which confidence interval you will construct and verify that the conditions have been met.
- **Do** the actual construction of the interval using

$$\hat{p} \pm z^* \sqrt{\frac{\hat{p}(1-\hat{p})}{n}}$$

where z^* is the critical value for the standard Normal curve with area C between $-z^*$ and z^*.

- **Conclude** by interpreting the interval in the context of the problem.

Check for Understanding: ____ *I can state and check the Random, 10%, and Large Counts conditions for constructing a confidence interval for a population proportion.* ____ *I can construct and interpret a confidence interval for a population proportion.*

According to a recent study, not everyone can roll their tongue. A researcher observed a random sample of 300 adults and found 68 who could roll their tongue. Use the four-step process to construct and interpret a 90% confidence interval for the true proportion of adults who can roll their tongue.

Concept 3: Choosing the Sample Size

As noted in section 8.1, our goal is to estimate the parameter as precisely as possible. We want high confidence and a low margin of error. To achieve that, we can determine how large a sample size is necessary before proceeding with the data collection. To calculate the sample size necessary to achieve a set margin of error at a confidence level C, we simply solve the following inequality for *n*:

$$z^* \sqrt{\frac{\hat{p}(1-\hat{p})}{n}} \le ME$$

where the sample proportion is estimated based on a previous study or set to 0.5 to maximize the possible margin of error.

Check for Understanding: ____ *I can determine the sample size required to obtain a C% confidence interval for a population proportion with a specified margin of error.*

A researcher would like to estimate the proportion of adults who can roll their tongues. However, unlike the previous example, she'd like the estimate to be within 2% at a 95% confidence level. How large a sample is needed?

Section 8.3: Estimating a Population Mean

Before You Read: Section Summary

In this section, you will continue your study of confidence intervals by learning how to construct and interpret a confidence interval for a mean. While the overall procedure is identical to that for a proportion, there is one major difference. When dealing with means and unknown population standard deviations, we must use a new distribution to determine critical values. You will be introduced to the *t*-distributions and learn how to use them to construct a confidence interval for a population mean.

Learning Targets:

____ I can determine the critical value for calculating a *C*% confidence interval for a population mean using a table or technology.

____ I can state and check the Random, 10%, and Normal/Large Sample conditions for constructing a confidence interval for a population mean.

____ I can construct and interpret a confidence interval for a population mean.

____ I can determine the sample size required to obtain a *C*% confidence interval for a population mean with a specified margin of error.

While You Read: Key Vocabulary and Concepts

t-distribution:
degrees of freedom:
standard error of the sample mean:
one-sample *t*-interval for a population mean:
conditions for constructing a confidence interval about a mean:
sample size for a desired margin of error when estimating μ:

After You Read: Check for Understanding

Concept 1: Conditions for Estimating μ

Like proportions, when constructing a confidence interval for μ, it is critical that you begin by checking that the conditions are met. First, check to make sure the sample was randomly selected or there was random assignment in an experiment. Because the construction of the interval is based on the sampling distribution of $\bar{x}$, next ensure that the condition for Normality is met. That is, check to see that the population distribution is Normal OR the sample size is at least 30. Finally, check for independence of measurements. If sampling without replacement was used, verify that the population of interest is at least 10 times as large as the sample. If all three of these conditions are met, you can safely proceed to construct and interpret a confidence interval about a population mean μ.

Concept 2: t-Distributions

When the population standard deviation is unknown, we can no longer model the test statistic with the Normal distribution. Therefore, we can't use critical z^* values to determine the margin of error in our confidence interval. Fortunately, it turns out that when the Normal condition is met, the test statistic calculated using the sample standard deviation s_x has a distribution similar in appearance to the Normal distribution, but with more area in the tails. That is, the statistic has the *t*-distribution with (n-1) degrees of freedom. As the sample size and, subsequently, the degrees of freedom increase, the t distribution approaches the standard Normal distribution. We calculate standardized t values the same way we calculate z values. However, we must refer to a t table and consider degrees of freedom when determining critical values. We will learn more about t distributions in Chapter 9. For now, we will focus on calculating the critical t^* for different sample sizes and confidence levels.

Check for Understanding:

____ *I can determine the critical value for calculating a C% confidence interval for a population mean using a table or technology.*

a) An 80% confidence interval from a sample with size n = 19

b) A 95% confidence interval from 248 degrees of freedom

c) A 99% confidence interval for a sample with size n = 30

Concept 3: Constructing a Confidence Interval for μ

To construct a confidence interval for a population mean μ when the population standard deviation is unknown, you should follow the four-step process.

- **State** the parameter you want to estimate and at what confidence level.
- **Plan** which confidence interval you will construct and verify that the conditions have been met.
- **Do** the actual construction of the interval.

$$\bar{x} \pm t^* \frac{s_x}{\sqrt{n}}$$

where t^* is the critical value for the t_{n-1} distribution with area C between $-t^*$ and t^*.

- **Conclude** by interpreting the interval in the context of the problem.

Check for Understanding: ____ *I can construct and interpret a confidence interval for a population mean.*

The amount of sugar in soft drinks is increasingly becoming a concern. To test sugar content, a researcher randomly sampled 8 soft drinks from a particular manufacturer and measured the sugar content in grams/serving. The following data were produced:

26 31 23 22 11 22 14 31

Use these data to construct and interpret a 95% confidence interval for the mean amount of sugar in this manufacturer's soft drinks.

Concept 4: Choosing the Sample Size

Similar to what we did with proportions, to calculate the sample size necessary to achieve a set margin of error for a population mean μ at a confidence level *C*, we simply solve the following inequality for *n*:

$$z^* \frac{\sigma}{\sqrt{n}} \leq ME$$

where σ is estimated based on a previous study.

Check for Understanding: ____ *I can determine the sample size required to obtain a C% confidence interval for a population mean with a specified margin of error.*

A researcher would like to estimate the mean amount of time it takes students to accomplish a particular task. A previous study indicates the time required varies in the population with a standard deviation of 4 seconds. He would like to estimate the true mean time within 0.5 seconds at 95% confidence. How large a sample is needed?

Chapter Summary: Estimating with Confidence

Statistical inference is the practice of drawing a conclusion about a population based on information gathered from a sample. This chapter introduced us to the practice of estimating a parameter based on a statistic. The underlying logic for confidence intervals is the same whether we are estimating a proportion or a mean. By using what we learned about sampling distributions, we can construct an interval around a point estimate that we are confident captures the parameter of interest. The confidence level itself tells what would happen if we used the construction method for the interval many times for samples of the same size. It is basically the capture rate for all of the constructed intervals. So, when we build a level C interval, we can interpret it by saying "We are *C*% confident the interval from a to b captures the true parameter of interest."

In the next chapter, we will learn how to test a claim about a parameter. Make sure you continue to practice confidence intervals, though, as they are just as important as the significance tests you are about to learn!

After You Read: What Have I Learned?

Complete the vocabulary puzzle, multiple-choice questions, and FRAPPY. Check your answers and performance on each of the learning targets. Be sure to get extra help on any targets that you identify as needing more work!

Learning Target	**Got It!**	**Almost There**	**Needs Some Work**
I can identify an appropriate point estimator and calculate the value of a point estimate.			
I can interpret a confidence interval in context.			
I can determine the point estimate and margin of error from a confidence interval.			
I can use a confidence interval to make a decision about the value of a parameter.			
I can interpret a confidence level in context.			
I can describe how the sample size and confidence level affect the margin of error.			
I can explain how practical issues like nonresponse, undercoverage, and response bias can affect the interpretation of a confidence interval.			
I can state and check the Random, 10%, and Large Counts conditions for constructing a confidence interval for a population proportion.			
I can determine the critical value for calculating a *C*% confidence interval for a population proportion using a table or technology.			
I can construct and interpret a confidence interval for a population proportion.			
I can determine the sample size required to obtain a *C*% confidence interval for a population proportion with a specified margin of error.			
I can determine the critical value for calculating a *C*% confidence interval for a population mean using a table or technology.			
I can state and check the Random, 10%, and Normal/Large Sample conditions for constructing a confidence interval for a population mean.			
I can construct and interpret a confidence interval for a population mean.			
I can determine the sample size required to obtain a *C*% confidence interval for a population mean with a specified margin of error.			

Chapter 8 Multiple Choice Practice

Directions. *Identify the choice that best completes the statement or answers the question. Check your answers and note your performance when you are finished.*

1. Gallup Poll interviews 1600 people. Of these, 18% say that they jog regularly. A news report adds: "The poll had a margin of error of plus or minus three percentage points." You can safely conclude
   (A) 95% of all Gallup Poll samples like this one give answers within ±3% of the true population value.
   (B) the percent of the population who jog is certain to be between 15% and 21%.
   (C) 95% of the population jog between 15% and 21% of the time.
   (D) we can be 3% confident that the sample result is true.
   (E) if Gallup took many samples, 95% of them would find that 18% of the people in the sample jog.

2. An agricultural researcher plants 25 plots with a new variety of corn. A 90% confidence interval for the average yield for these plots is found to be 162.72 ± 4.47 bushels per acre. Which of the following is the correct interpretation of the interval?
   (A) There is a 90% chance the interval from 158.28 to 167.19 captures the true average yield.
   (B) 90% of sample average yields will be between 158.28 and 167.19 bushels per acre.
   (C) We are 90% confident the interval from 158.28 to 167.19 captures the true average yield.
   (D) 90% of the time, the true average yield will fall between 158.28 and 167.19.
   (E) We are 90% confident the true average yield is 162.72.

3. I collect a random sample of size *n* from a population and from the data collected compute a 95% confidence interval for the mean of the population. Which of the following would produce a wider confidence interval, based on these same data?
   (A) Use a larger confidence level.
   (B) Use a smaller confidence level.
   (C) Use the same confidence level, but compute the interval *n* times. Approximately 5% of these intervals will be larger.
   (D) Increase the sample size.
   (E) Nothing can ensure that you will get a larger interval. One can only say the chance of obtaining a larger interval is 0.05.

4. A marketing company discovered the following problems with a recent poll:
I. Some people refused to answer questions
II. People without telephones could not be in the sample
III. Some people never answered the phone in several calls
Which of these sources is included in the ±2% margin of error announced for the poll?
   (A) Only source I.
   (B) Only source II.
   (C) Only source III.
   (D) All three sources of error.
   (E) None of these sources of error.

5. You are told that the proportion of those who answered "yes" to a poll about internet use is 0.70, and that the standard error is 0.0459. The sample size
   (A) is 50.
   (B) is 99.
   (C) is 100.
   (D) is 200.
   (E) cannot be determined from the information given.

6. The standardized test scores of 16 students have mean $\overline{x}$ = 200 and standard deviation s = 20. What is the standard error of $\overline{x}$?
   (A) 20
   (B) 10
   (C) 5
   (D) 1.25
   (E) 0.80

7. A newspaper conducted a statewide survey concerning the 2008 race for state senator. The newspaper took a random sample (assume it is an SRS) of 1200 registered voters and found that 620 would vote for the Republican candidate. Let *p* represent the proportion of registered voters in the state that would vote for the Republican candidate. A 90% confidence interval for *p* is
   (A) 0.517 ± 0.014.
   (B) 0.517 ± 0.022.
   (C) 0.517 ± 0.024.
   (D) 0.517 ± 0.028.
   (E) 0.517 ± 0.249.

8. After a college's football team once again lost a football game to the college's arch rival, the alumni association decided to conduct a survey to see if alumni were in favor of firing the coach. Let *p* represent the proportion of all living alumni who favor firing the coach. Which of the following is the smallest sample size needed to guarantee an estimate that's within 0.05 of *p* at a 95% confidence level?
   (A) 269
   (B) 385
   (C) 538
   (D) 768
   (E) 1436

9. An SRS of 100 postal employees found that the average time these employees had worked for the postal service was $\overline{x}$ = 7 years with standard deviation s_x = 2 years. Assume the distribution of the time the population of employees has worked for the postal service is approximately Normal. A 95% confidence interval for the mean time μ the population of postal service employees has spent with the postal service is
   (A) 7 ± 2.
   (B) 7 ± 1.984.
   (C) 7 ± 0.525.
   (D) 7 ± 0.4.
   (E) 7 ± 0.2.

10. Do students tend to improve their SAT Mathematics (SAT-M) score the second time they take the test? A random sample of four students who took the test twice earned the following scores.

Student	1	2	3	4
First Score	450	520	720	600
Second Score	440	600	720	630

Assume that the change in SAT-M score (second score – first score) for the population of all students taking the test twice is approximately Normally distributed with mean μ. A 90% confidence interval for μ is
   (A) 25.0 ± 118.03.
   (B) 25.0 ± 64.29.
   (C) 25.0 ± 47.56.
   (D) 25.0 ± 43.08.
   (E) 25.0 ± 33.24.

Check your answers below. If you got a question wrong, check to see if you made a simple mistake or if you need to study that concept more. After you check your work, identify the concepts you feel very confident about and note what you will do to learn the concepts in need of more study.

#	Answer	Concept	Right	Wrong	Simple Mistake?	Need to Study More
1	A	Interpreting Confidence				
2	C	Interpret a Confidence Interval				
3	A	Width of a Confidence Interval				
4	E	Biased Samples				
5	C	Standard Error of $\hat{p}$				
6	C	Standard Error of $\bar{x}$				
7	C	Confidence Interval for p				
8	B	Choosing Sample Size				
9	D	Confidence Interval for μ				
10	C	Confidence Interval for μ (paired)				

Chapter 8 Reflection

Summarize the "Big Ideas" in Chapter 8:

My strengths in this chapter:

Concepts I need to study more and what I will do to learn them:

FRAPPY! Free Response AP® Problem, Yay!

The following problem is modeled after actual Advanced Placement Statistics free response questions. Your task is to generate a complete, concise response in 15 minutes. After you generate your response, view two example solutions and determine whether you feel they are "complete", "substantial", "developing" or "minimal". If they are not "complete", what would you suggest to the student who wrote them to increase their score? Finally, you will be provided with a rubric. Score your response and note what, if anything, you would do differently to increase your own score.

A machine at a soft-drink bottling factory is calibrated to dispense 12 ounces of cola into cans. A simple random sample of 35 cans is pulled from the line after being filled and the contents are measured. The mean content of the 35 cans is 11.92 ounces with a standard deviation of 0.085 ounce.

a) Construct and interpret a 95% confidence interval to estimate the true mean contents of the cans being filled by this machine.

b) Based on your result from a), does the machine appear to be working properly? Justify your answer.

c) Interpret the confidence level of 95 percent in context.

FRAPPY! Student Responses

Student Response 1:

a) One samp-t-int = (11.89, 11.94)

b) There is a 95% chance the true mean of the amount the machine fills cans is captured in this interval.

c) If we took 100 samples, 95 of them would create an interval that captures the true mean.

How would you score this response? Is it substantial? Complete? Developing? Minimal? Is there anything this student could do to earn a better score?

Student Response 2:

a) Conditions: Random sample is given. The cans are independent of each other. Since 35 > 30, we can assume normality of the sampling distribution.

95% t-interval for the true mean contents:
$11.92 \pm 2.042(0.085/\sqrt{(35)}) = (11.89, 11.94)$

b) We are 95% confident the true mean contents of the cans filled by this machine falls between 11.89 and 11.94 oz. It appears the machine might be underfilling the cans since 12 oz is not in the interval.

c) If we were to take many samples of size 35 and construct intervals from their sample mean contents, 95% of the intervals would capture the true mean contents being dispensed by the filling machine.

How would you score this response? Is it substantial? Complete? Developing? Minimal? Is there anything this student could do to earn a better score?

FRAPPY! Scoring Rubric

Use the following rubric to score your response. Each part receives a score of "Essentially Correct," "Partially Correct," or "Incorrect." When you have scored your response, reflect on your understanding of the concepts addressed in this problem. If necessary, note what you would do differently on future questions like this to increase your score.

Intent of the Question

The goal of this question is to determine your ability to construct and interpret a confidence interval and correctly interpret the confidence level in the context of a problem.

Solution

(a) Conditions: Random – The cans were randomly selected.
Independent – There are more than 10(35) cans on the line.
Normal – $n = 35$ (greater than 30), so the sampling distribution of $\overline{x}$ will be approximately normal.

95% CI for μ: $11.92 \pm 2.042(0.085/\sqrt{(35)}) = (11.89, 11.94)$

(b) We are 95% confident that the interval from 11.89 ounces to 11.94 ounces captures the true mean contents of the cans filled by this machine. It appears the machine may be filling less than it is supposed to since 12 is not in the interval.

(c) 95% of intervals constructed from random samples of 35 cans from this machine will be successful in capturing the true mean contents.

Scoring

Parts (a), (b), and (c) are scored as essentially correct (E), partially correct (P), or incorrect (I).

Part (a) is essentially correct if the response correctly checks the conditions for a one-sample t confidence interval for a mean AND correctly calculates the interval. Part (a) is partially correct if the conditions are not properly checked but the interval is correct. Note: the construction of a z-interval receives a partial at most.

Part (b) is essentially correct if the response correctly interprets the confidence interval in context AND correctly notes the machine appears to be underfilling because 12 is not contained in the interval. Part (b) is partially correct if the interpretation lacks context OR fails to make a decision about the machine based on the interval.

Part (c) is essentially correct if the response correctly interprets the confidence level in context. Part (c) is partially correct if the interpretation lacks context.

4 Complete Response
All three parts essentially correct

3 Substantial Response
Two parts essentially correct and one part partially correct

2 Developing Response
Two parts essentially correct and no parts partially correct
One part essentially correct and two parts partially correct
Three parts partially correct

1 Minimal Response
One part essentially correct and one part partially correct
One part essentially correct and no parts partially correct
No parts essentially correct and two parts partially correct

My Score:
What I did well:
What I could improve:
What I should remember if I see a problem like this on the AP Exam:

Chapter 8: Estimating with Confidence

Across

4. _____ t procedures allow us to compare the responses to two treatments in a matched pairs design
8. a confidence interval consists of an estimate ± margin of _____
9. to find the standard error of the sample mean, divide the sample standard deviation by the _____ of the sample size (two-words)
10. to estimate with confidence, our estimate should be calculated from a ___ sample
11. methods for drawing conclusions about a population from sample data
12. a single value used to estimate a parameter is a _____ estimator
13. we can construct a narrow interval by _____ our confidence
14. as degrees of freedom increase, the t distribution approaches the _____ distribution
16. the spread of the t distributions is _____ than the spread of the standard Normal distribution

Down

1. inference procedures that remain fairly accurate even when a condition is violated
2. another condition for confidence intervals is that observations should be _____
3. particular t distributions are specified by degrees of _____
5. we can construct a narrow confidence interval by _____ our sample size
6. the margin of error consists of a _____ value and the standard error of the sampling distribution
7. a _____ interval provides an estimate for a population parameter
15. confidence _____: the success rate of the method in repeated sampling

Chapter 9: Testing a Claim

"A statistical analysis, properly conducted, is a delicate dissection of uncertainties, a surgery of suppositions." M.J. Moroney

Chapter Overview

In the last chapter, you learned how to estimate a parameter based on a sample statistic. In this chapter, you will learn how to use data from a random sample to test a claim about a population parameter. Significance tests are the second type of inference you will be introduced to in this course. By the end of this chapter, you should understand the logic behind a significance test, as well as how to use a four-step procedure to carry one out. Specific methods for testing a claim about a proportion and a mean will be explored. In the next few chapters, you'll learn how to test claims about the difference between proportions or means, as well as how to perform significance tests for slopes of regression lines and distributions of categorical variables. In each case, you will use the same logic and the same four-step procedure. Be sure to take the time in this chapter to get a solid understanding of both concepts!

Sections in this Chapter
Section 9.1: Significance Tests: The Basics **Section 9.2**: Tests About a Population Proportion **Section 9.3**: Tests About a Population Mean

Plan Your Learning

Use the following *suggested* guide to help plan your reading and assignments in "The Practice of Statistics, 6th Edition." Note: your teacher may schedule a different pacing or assign different problems. Be sure to follow his or her instructions!

Read	**9.1: pp. 552 - 560**	**9.1: pp. 560 - 563**	**9.2: pp. 568 – 572**	**9.2: pp. 572 - 581**
Do	1,3,5,7,9,13,14,15,19	21,23,25,27 MC 29-32	35,37,39,41	43,45,47,51,53,55 MC 59-62

Read	**9.3: pp. 585 – 594**	**9.3: pp. 595 - 605**	**Chapter Summary**
Do	65,67,69,73,77,79	81,85,87,93,95,97 MC 102-108	Multiple-Choice FRAPPY!

Section 9.1: Significance Tests: The Basics

Before You Read: Section Summary

In this section, you will learn the basic ideas and logic behind a significance test. You will be introduced to stating hypotheses, checking conditions, calculating a standardized test statistic, and drawing a conclusion. *P*-values will be introduced as a means of weighing the strength of evidence against a claim. Since we are basing our conclusion on a sample statistic that would vary from sample to sample, we must be aware of the fact that our conclusion could be wrong. This section ends with a discussion of the types of errors that could occur when performing a significance test and how to deal with them. Be sure to get a good understanding of the logic and format for a significance test. You will be using them throughout the rest of the course!

Learning Targets:

____ I can state appropriate hypotheses for a significance test about a population parameter.
____ I can interpret a *P*-value in context.
____ I can make an appropriate conclusion for a significance test.
____ I can interpret a Type I error and a Type II error in context, and give a consequences of each error in a given setting.

While You Read: Key Vocabulary and Concepts

null hypothesis:
alternative hypothesis:
one-sided and two-sided alternative hypotheses:
P-value:
"reject H_0" or "fail to reject H_0":
significance level:
Type I error:
Type II error:

After You Read: Check for Understanding

Concept 1: The Reasoning Behind Significance Tests

A significance test is a procedure that allows us to test a claim about a population parameter by studying a statistic from a random sample. If an observed statistic is "far" away from a hypothesized claim about a parameter, we have some evidence that the claim is wrong. Whether or not the statistic is "far enough" away depends on the sampling distribution of the statistics. That is, statistics will vary from sample to sample. What a significance test does is answers the question, "What are the chances we would observe a sample statistic at least this extreme, assuming the claim about the parameter was true?" If the chances are low, there is evidence that the claim may be wrong. If the chances are pretty good, then there is little evidence to suggest the claim is wrong. Of course, to even begin making that argument, you must start by setting up a null hypothesis (or claim about the parameter you are trying to find evidence against) and an alternative hypothesis (the claim about the parameter you are trying to find evidence for). Note: the null hypothesis will always be a statement of equality. That is, it says "Let's assume the parameter is equal to _____." The alternative hypothesis is set up to test whether or not the actual value of the parameter is greater than, less than, or simply not equal to the value in the null hypothesis.

Check for Understanding: ____ *I can state appropriate hypotheses for a significance test about a population parameter.*

Suppose you suspect a "chute" of playing cards is not fair. The chute supposedly contains 10 standard decks shuffled together. You are interested in knowing whether there are more hearts than usual. To test this, you deal 12 cards at random and calculate the proportion of hearts in your hand.

a) Describe the parameter of interest in this setting.

b) Write the appropriate null and alternative hypotheses for this situation.

c) Suppose your deal contains 7 hearts. What is the sample proportion of hearts? Is it possible to deal 7 hearts out of 12 cards if the chute really does contain standard decks? Is it likely? Why or why not?

Concept 2: Statistical Significance

To conclude that we have convincing evidence against a null hypothesis, our sample statistic must have a low likelihood of occurring under the assumption that the null hypothesis is true. To determine this, we must consider the sampling distribution of the statistic under the assumption that the null hypothesis is true and calculate the probability of observing a statistic at least as extreme as the one observed. If the conditions for inference are met, we can describe the sampling distribution and calculate a standardized test statistic. This test statistic can then be used to determine a *P*-value. If this *P*-value is low enough, we can say the data are statistically significant and we have evidence to reject the null hypothesis. The general rule of thumb is to use a significance level of 5%, although some situations may specify levels of 1% or even 10%. If the *P*-value is smaller than the chosen significance level, we have enough evidence to reject the null hypothesis and conclude that the alternative is true. The reasoning behind this is that if we observe an outcome that has an extremely low chance of occurring under a given assumption, then there is evidence that the assumption may be false. If we observe an outcome under that has a fairly likely chance of occurring under a given assumption, then we have little reason to doubt the assumption!

Check for Understanding: ____ *I can interpret a P-value in context.* ____ *I can make an appropriate conclusion for a significance test.*

Refer to the hypotheses you set in the previous check for understanding. Suppose you conduct a significance test for the proportion of hearts in the chute and find the *P*-value for obtaining 7 hearts in 12 randomly selected cards is 0.004.

a) Explain what it means for the null hypothesis to be true in this setting.

b) Interpret the *P*-value in context.

c) Do these data provide convincing evidence against the null hypothesis at a 5% significance level? Explain.

Concept 3: Type I and Type II Errors

Since our conclusion is based on what we would expect to see in a sampling distribution, there is a chance our sample statistic does not accurately reflect the true value of the parameter. That is, due to sampling variability, it is possible that the random sample that is collected will lead to an incorrect conclusion about the parameter. If we obtain a very small *P*-value and reject the null when, in fact, the null hypothesis is true, we commit a Type I error. The probability of committing a Type I error is equal to the significance level, α. If we obtain a relatively large *P*-value and fail to reject the null when, in fact, the null hypothesis is false, we commit a Type II error. You will not be expected to calculate the probability of Type II errors. However, you should be able to interpret both Type I and Type II errors in the context of the situation and explain the consequences of each.

Understanding: ____ *I can interpret a Type I error and a Type II error in context, and give a consequences of each error in a given setting.*

Refer to the previous checks for understanding.

a) Describe a Type I error in this setting.

b) Describe a Type II error in this setting.

Section 9.2: Tests About a Population Proportion

Before You Read: Section Summary

In this section, you will apply the logic of significance tests and the four-step process to test a claim about a population proportion. You will learn how to check the conditions for a significance test as well as how to calculate a standardized test statistic and *P*-value. Finally, you will learn how to write an appropriate conclusion in context. Your study will also include an introduction to two-sided tests to find evidence that a population proportion is different than a hypothesized value. You will then wrap up the section by studying the connection between confidence intervals and two-sided tests. Again, this section introduces you to concepts that will be used throughout the rest of the course. Make sure to get additional help if you struggle at all with the ideas behind significance testing!

Learning Targets:

____ I can state and check the Random, 10%, and Large Counts conditions for performing a significance test about a population proportion.

____ I can calculate the standardized test statistic and *P*-value for a test about a population proportion.

____ I can perform a significance test about a population proportion.

While You Read: Key Vocabulary and Concepts

conditions for performing a significance test about a proportion:
standardized test statistic:
significance tests – a four-step process:
one-sample *z* test for a proportion:

After You Read: Check for Understanding

Concept 1: One-Sample z Test for a Proportion

Significance tests and confidence intervals are both based on the same concept—sampling distributions. Therefore, when performing inference with either of these methods, it is important to follow a process that ensures the calculations are justified. To test a claim about a population parameter, the following four-step process should be used:

1) **STATE the parameter of interest and the hypotheses you would like to test.**
   When testing a claim about a population proportion, we start by defining hypotheses about the parameter. The null hypothesis assumes that the population proportion is equal to a particular value while the alternative hypothesis is that the population proportion is greater than, less than, or not equal to that value:

   $H_o: p = p_o$

   $H_a: p > p_o$ OR $p < p_o$ OR $p \neq p_o$

   State a significance level.

2) **PLAN: Choose the appropriate inference method and check the conditions.**
   We must check the Random, Normal, and Independent conditions before proceeding with inference. To ensure that the sampling distribution of $\hat{p}$ is approximately Normal, check to make sure np_o and $n(1 - p_o)$ are both at least 10. If sampling is done without replacement, verify that the population is at least 10 times as large as the sample.

3) **DO: If conditions are met, calculate a test statistic and *P*-value.**
   Compute the *z* test statistic

   $$z = \frac{\hat{p} - p_0}{\sqrt{\frac{p_0(1 - p_0)}{n}}}$$

   and find the *P*-value by calculating the probability of observing a *z* statistic at least this extreme in the direction of the alternative hypothesis.

4) **CONCLUDE by interpreting the results of your calculations in the context of the problem.**
   If the *P*-value is smaller than the stated significance level, you have significant evidence to reject the null hypothesis. If it is greater than or equal to the significance level, then you fail to reject the null hypothesis.

This four-step procedure can be used with *any* significance test! The only things that will change are the conditions that must be checked and the test statistic calculation. Be sure to get a good grasp of this process here as you will be using it a lot in the coming chapters!

Check for Understanding: ____ *I can state and check the Random, 10%, and Large Counts conditions for performing a significance test about a population proportion.* ____ *I can calculate the standardized test statistic and P-value for a test about a population proportion.*
____ *I can perform a significance test about a population proportion.*

A study of classic authors uncovered a distinguishable speech pattern that differed from author to author. Plato utilized this pattern in 21.4% of the passages in his works. The owner of a rare bookstore claims to have an original Plato work, but you suspect the speech pattern occurs too frequently to be an original Plato work. A random sample of passages from the work in question was taken and it was found that 136 of the 439 selected passages followed the speech pattern. Do these data provide convincing evidence that the work was not written by Plato? Conduct a one-sample *z* test using the four-step procedure.

Concept 2: Confidence Intervals Give More Information

Significance tests provide evidence that supports or rejects an assumption about a population parameter. If an observed sample statistic is far enough away from a parameter's assumed value, a significance test will suggest there is evidence to question that assumed value. However, the test does not say anything about what the actual parameter value may be. A confidence interval provides more information by suggesting a range of plausible values for the parameter. There is also a link between confidence intervals and *two-sided* tests. While not a perfect connection, a two-sided test at a significance level α and a $100(1 - \alpha)\%$ confidence interval give similar information about the population parameter.

Check for Understanding: ____ *I can use a confidence interval to make a conclusion for a two-sided test about a population parameter.*

A recent study suggested that 77% of teenagers have texted while driving. A random sample of 27 teenage drivers in Atlanta was taken and 15 admitted to texting while driving. Use a 99% confidence interval to determine whether there is convincing evidence that the population proportion of teens who text while driving is different than 77%.

Section 9.3: Tests About a Population Mean

Before You Read: Section Summary

In this section, you'll learn how to perform a significance test about a population mean μ. Just like you did for proportions, you'll learn how to set up hypotheses, check the appropriate conditions, calculate a standardized test statistic and *P*-value, and draw a conclusion in context. You will also learn how tests involving "paired data" can be performed using one-sample *t* procedures.

Learning Targets:

____ I can state and check the Random, 10%, and Normal/Large Sample conditions for performing a significance test about a population mean.

____ I can calculate the standardized test statistic and *P*-value for a test about a population mean.

____ I can perform a significance test about a population mean.

____ I can use a confidence interval to make a conclusion for a two-sided test about a population parameter.

____ I can interpret the power of a significance test and describe what factors affect the power of a test.

While You Read: Key Vocabulary and Concepts

conditions for performing a significance test about a mean:
t-distribution:
one-sample *t* test for a mean:
two-sided tests and confidence intervals:
power:
increasing the power of a significance test:

☑ After You Read: Check for Understanding

Concept 1: One Sample t Test for μ

The process for testing a claim about a population mean follows the same format as the process used for a population proportion. However, like confidence intervals for means, you usually need to base your calculations on the *t* distributions (unless σ is somehow known).

1) **STATE the parameter of interest and the hypotheses you would like to test.**
   The null hypothesis assumes that the population mean is equal to a particular value while the alternative hypothesis is that the population mean is greater than, less than, or not equal to that value:

   $$H_0: \mu = \mu_0$$
   $$H_a: \mu > \mu_0 \text{ OR } \mu < \mu_0 \text{ OR } \mu \neq \mu_0$$

   State a significance level.

2) **PLAN: Choose the appropriate inference method and check the conditions.**
   We must check the Random, Normal, and Independent conditions before proceeding with inference. To ensure that the sampling distribution of $\bar{x}$ is approximately Normal, check that the population distribution is Normal OR $n \geq 30$. If $n < 30$, it is sufficient if the sample data show no signs of strong skewness or outliers. If sampling is done without replacement, verify that the population is at least 10 times as large as the sample.

3) **DO: If conditions are met, calculate a standardized test statistic and *P*-value.**
   Compute the *t* test statistic

   $$t = \frac{\bar{x} - \mu_0}{\frac{s_x}{\sqrt{n}}}$$

   and find the *P*-value by calculating the probability of observing a *t* statistic at least this extreme in the direction of the alternative hypothesis in a *t* distribution with $n - 1$ degrees of freedom.

4) **CONCLUDE by interpreting the results of your calculations in the context of the problem.**
   If the *P*-value is smaller than the stated significance level, you have significant evidence to reject the null hypothesis. If it is larger than the significance level, then you fail to reject the null hypothesis.

As with proportions, a confidence interval for a population mean can be used to test a two-sided claim. Further, constructing a confidence interval gives a range of plausible values for μ, while a significance test only allows you to conclude that μ may be different from a particular value.

Check for Understanding: ____ *I can check conditions for carrying out a test about a population mean and ____ I can conduct a one-sample t test about a population mean μ.*

Humerus bones from the same species of animal tend to have approximately the same length-to-width ratios. When fossils of humerus bones are discovered, archeologists can often determine the species of animal by examining these ratios. It is known that the species Molekius Primatium exhibits a mean ratio of μ = 8.9. Suppose 41 fossils of humerus bones are unearthed at a site on Minnesota's Iron Range, where this species was known to have lived. Researchers are willing to view these as a random sample of all such humerus bones. The length-to-width ratios were calculated and are listed below. Test whether the population mean for the species that left these bones differs from 8.9 at α=.05.

9.73	10.89	9.07	9.2	9.33	9.98	9.84	9.59	9.48	8.71	9.57	9.29
9.94	8.07	8.37	6.85	8.52	8.87	6.23	9.41	6.66	9.35	8.86	9.93
8.91	11.77	10.48	10.39	9.39	9.17	9.89	8.17	8.39	8.8	10.02	8.38
11.67	8.3	9.17	12	9.38							

Concept 2: The Power of a Test

The power of a test is the probability it will wind convincing evidence for H_a when a specific alternative value of the parameter is true. You can think of this as the probability of avoiding a Type II error. Therefore, for a specific alternative value, Power = 1 – *P*(Type II error).

There are a few ways we can increase the power of a significance test. We can increase the sample size, increase the significance level, or increase the difference that is important to detect between the null and alternative parameter values. The design of our study can also increase power. Specifically, controlling for other variables and blocking in experiments or stratified random sampling can increase power.

Check for Understanding: ____ *I can interpret the power of a significance test and describe what factors affect the power of a test.*

A computer manufacturer requires each batch of processors it uses meets high quality standards. If the manufacturer finds convincing evidence that the processors use more than 70W of power in normal usage, the batch will be rejected. A quality control engineer inspects a random sample of processors from each shipment and performs a test of H_0: $\mu = 0.70$ versus H_a: $\mu > 0.70$, where μ is the true mean power usage (in watts) of the processors in a given batch.

a. The power of the test to detect that $\mu = 71$, based on a random sample of 30 processors and a significance level $\alpha = 0.05$, is 0.572. Interpret this value.

Determine if each of the following changes would increase or decrease the power of the test. Explain your answers.

b. Change the significance level to $\alpha = 0.01$.

c. Take a random sample of 50 processors, instead of 30.

d. The true mean is $\mu = 72$ instead of $\mu = 71$.

Chapter Summary: Testing a Claim

A significance test tells us whether or not a sample provides convincing evidence against a claim about a population parameter. The test answers the question, "How likely would it have been to observe this particular sample statistic (or one more extreme) if the claim about the parameter was true?" This probability, the *P*-value, gives us an idea just how surprising an observed statistic is under the assumption that the null hypothesis is true. If the *P*-value is small (less than 5% or some other chosen significance level), we have enough evidence to reject the null hypothesis, suggesting the actual parameter value may be greater than, less than, or different from the claim. If the *P*-value is greater than or equal to a specified significance level, we do not have convincing evidence against the claim about the parameter. Keep in mind, however, that since we are basing our decision on a likelihood calculated from a sampling distribution, it is possible that we may be wrong. Type I and Type II errors occur when we mistakenly reject a null hypothesis that is true or fail to reject a null hypothesis that is false, respectively. You should be able to define each of these errors in the context of the situation as well as explain the consequences of each.

This chapter not only introduced you to the logic behind significance tests, but also a four-step procedure for carrying them out. This procedure will be used throughout the rest of the course, so hopefully you are comfortable with it at this point! If not, be sure to practice a few more tests and focus on stating the parameter and hypotheses, checking the appropriate conditions, calculating a test statistic and *P*-value, and concluding in the context of the problem. If you follow those steps, you should have very few problems performing significance tests in the upcoming chapters!

After You Read: What Have I Learned?

Complete the vocabulary puzzle, multiple-choice questions, and FRAPPY. Check your answers and performance on each of the learning targets. Be sure to get extra help on any targets that you identify as needing more work!

Learning Target	**Got It!**	**Almost There**	**Needs Some Work**
I can state correct hypotheses for a significance test about a population parameter.			
I can interpret a *P*-value in context.			
I can make an appropriate conclusion for a significance test.			
I can interpret a Type I error and a Type II error in context, and give a consequence of each error in a given setting.			
I can state and check the Random, 10%, and Large Counts conditions for performing a significance test about a population proportion.			
I can calculate the standardized test statistic and *P*-value for a test about a population proportion.			
I can perform a significance test about a population proportion.			
I can state and check the Random, 10%, and Random/Large Sample conditions for performing a significance test about a population mean.			
I can calculate the standardized test statistic and *P*-value for a test about a population mean.			
I can perform a significance test about a population mean.			
I can use a confidence interval to make a conclusion for a two-sided test about a population parameter.			
I can interpret the power of a significance test and describe what factors affect the power of a test.			

Chapter 9 Multiple Choice Practice

Directions. *Identify the choice that best completes the statement or answers the question. Check your answers and note your performance when you are finished.*

1. The average yield of a certain crop is 10.1 bushels per plant. A biologist claims that a new fertilizer will result in a greater yield when applied to the crop. A random sample of 25 of plants given the fertilizer has an average yield of 10.8 bushels and a standard deviation of 2.1 bushels. The appropriate null and alternative hypotheses to test the biologist's claim are

   (A) H_0: $\mu = 10.8$ against H_a: $\mu > 10.8$.
   (B) H_0: $\mu = 10.8$ against H_a: $\mu \neq 10.8$.
   (C) H_0: $\mu = 10.1$ against H_a: $\mu > 10.1$.
   (D) H_0: $\mu = 10.1$ against H_a: $\mu < 10.1$.
   (E) H_0: $\mu = 10.1$ against H_a: $\mu \neq 10.1$.

2. An opinion poll asks a random sample of 200 adults how they feel about voting for an amendment in an upcoming election. In all, 150 say they are in favor of the amendment. Does the poll provide evidence that the proportion p of adults who are in favor of the amendment is greater than 60%? The null and alternative hypotheses are

   (A) H_0: $p = 0.6$ against H_a: $p > 0.6$.
   (B) H_0: $p = 0.6$ against H_a: $p \neq 0.6$.
   (C) H_0: $p = 0.6$ against H_a: $p < 0.6$.
   (D) H_0: $p = 0.6$ against H_a: $p = 0.75$.
   (E) H_0: $p = 0.75$ against H_a: $p < 0.6$.

3. A test of significance produces a P-value of 0.035. Which of the following conclusions is appropriate?

   (A) Accept H_a at the $\alpha = 0.05$ level
   (B) Reject H_a at the $\alpha = 0.01$ level
   (C) Fail to reject H_0 at the $\alpha = .05$ level
   (D) Reject H_0 at the $\alpha = 0.05$ level
   (E) Accept H_0 at the $\alpha = 0.01$ level

4. A Type II error is

   (A) rejecting the null hypothesis when it is true.
   (B) failing to reject the null hypothesis when it is false.
   (C) rejecting the null hypothesis when it is false.
   (D) failing to reject the null hypothesis when it is true.
   (E) more serious than a Type I error.

5. A researcher plans to conduct a significance test at the $\alpha = 0.05$ significance level. She designs her study to have a power of 0.85 for a particular alternative value of the parameter. The probability that the researcher will commit a Type II error for the particular alternative value of the parameter at which she computed the power is

   (A) 0.05.
   (B) 0.15.
   (C) 0.80.
   (D) 0.95.
   (E) equal to the 1 - (P-value) and cannot be determined until the data have been collected.

6. In hypothesis testing β is the probability of committing a Type II error in a test with significance level α. The probability of committing a Type I error is

(A) $1 - \beta$
(B) $1 - \alpha$
(C) $\beta - \alpha$
(D) α
(E) Can not be determined

7. A claimed psychic was presented with 200 cards face down and asked to determine if the card was one of five symbols: a star, cross, circle, square, or three wavy lines. The "psychic" was correct in 50 cases. To determine if he has ESP, we test the hypotheses H_0: $p = 0.20$, H_a: $p > 0.20$, where p represents the true proportion of cards for which the psychic would correctly identify the symbol in the long run. Assume the conditions for inference are met. The *P*-value of this test is

(A) between .10 and .05.
(B) between .05 and .025.
(C) between .025 and .01.
(D) between .01 and .001.
(E) below .001.

8. The most important condition for drawing sound conclusions from statistical inference is usually

(A) that the population standard deviation is known.
(B) that at least 30 people are included in the study.
(C) that the data come from a random sample or a randomized experiment.
(D) that the population distribution is exactly Normal.
(E) that no calculation errors are made in the confidence interval or test statistic.

9. The mean weight of a random sample of 35 athletes is found to be 165 pounds with a standard deviation of 20 pounds. It is believed that a mean weight of 160 pounds would be normal for this group. To see if there is evidence that the mean weight of the population of all athletes of this type is significantly higher than 160 pounds, the hypotheses H_0: $\mu = 160$ vs. H_a: $\mu > 160$ are tested. You obtain a *P*-value of 0.0742. Which of the following is true?

(A) At the 5% significance level, you have proved that H_0 is true.
(B) You have failed to obtain sufficient evidence against H_0.
(C) At the 5% significance level, you have failed to prove that H_0 is true, and a larger sample size is needed to do so.
(D) Only 7.42% of the athletes weigh less than 160 pounds.
(E) None of the above. A significance test is inappropriate in this setting.

10. A medical researcher wishes to investigate the effectiveness of exercise versus diet in losing weight. Two groups of 25 overweight adult subjects are used, with a subject in each group matched to a similar subject in the other group on the basis of a number of physiological variables. One of the groups is placed on a regular program of vigorous exercise but with no restriction on diet, and the other is placed on a strict diet but with no requirement to exercise. The weight losses after 20 weeks are determined for each subject, and the difference between matched pairs of subjects (weight loss of subject in exercise group - weight loss of matched subject in diet group) is computed. The mean of these differences in weight loss is found to be 2 lb with standard deviation s_x = 4 lb. Is this convincing evidence of a difference in mean weight loss for the two methods? To answer this question, you should use

(A) one-proportion *z* test.
(B) one-sample *z* interval for μ_d.
(C) one-proportion *z* interval.
(D) one-sample *t* test for μ_d.
(E) none of the above.

Check your answers below. If you got a question wrong, check to see if you made a simple mistake or if you need to study that concept more. After you check your work, identify the concepts you feel very confident about and note what you will do to learn the concepts in need of more study.

#	Answer	Concept	Right	Wrong	Simple Mistake?	Need to Study More
1	C	Writing Hypotheses				
2	A	Writing Hypotheses				
3	D	Significance Level and *P*-value				
4	B	Type I and Type II Errors				
5	B	Power				
6	D	Type I and Type II Errors				
7	B	Calculating *P*-value for proportions				
8	C	Conditions for Inference				
9	B	Interpreting *P*-value				
10	D	One-sample *t*-test				

Chapter 9 Reflection

Summarize the "Big Ideas" in Chapter 9:

My strengths in this chapter:

Concepts I need to study more and what I will do to learn them:

FRAPPY! Free Response AP® Problem, Yay!

The following problem is modeled after actual Advanced Placement Statistics free response questions. Your task is to generate a complete, concise response in 15 minutes. After you generate your response, view two example solutions and determine whether you feel they are "complete", "substantial", "developing" or "minimal". If they are not "complete", what would you suggest to the student who wrote them to increase their score? Finally, you will be provided with a rubric. Score your response and note what, if anything, you would do differently to increase your own score.

During a recent movie promotion, Fruity O's cereal placed mini action figures in some of its boxes. The advertisement on the box states 1 out of every 4 boxes contains an action figure. A group of promotional-toy collectors suspects the proportion of boxes containing the action figure may be lower than 0.25. The group purchased 70 boxes of cereal and found 12 action figures. Assuming the 70 boxes represent a random sample of all of the cereal boxes, is there evidence to support the toy collector's belief that the proportion of boxes containing the figure is less than 0.25? Provide statistical evidence to support your answer.

FRAPPY! Student Responses

Student Response 1:

We will perform a one-sample z-test for the proportion of boxes, p, that contain the action figures.

$H_o: p = 0.25$
$H_a: p < 0.25$

We are told the sample is random. Further, both np and $n(1-p)$ are greater than 10.

$z = \dfrac{0.17 - 0.25}{\sqrt{\dfrac{.25(.75)}{70}}} = -1.52$ and our p-value is 0.064.

We do not have significant evidence at the 5% level to reject the null hypothesis. The proportion of boxes that contain the action figure is 0.25.

How would you score this response? Is it substantial? Complete? Developing? Minimal? Is there anything this student could do to earn a better score?

Student Response 2:

Let p = the actual proportion of cereal boxes that contain the action figure

$H_o: p = 0.25$
$H_a: p < 0.25$

We are told the sample is random.
70(0.25) = 17.5 and 70(0.75) = 52.5 Both are greater than 10.
There are more than 700 cereal boxes in the population.

One sample z test for a proportion

$z = z = \dfrac{0.17 - 0.25}{\sqrt{\dfrac{.25(.75)}{70}}} = -1.52$ p-value = 0.064.

Since our p-value is greater than the typical significance level of 5%, we do not have significant evidence to reject the null hypothesis. There is not enough support to suggest that the actual proportion of boxes that contain the figure is less than 0.25.

How would you score this response? Is it substantial? Complete? Developing? Minimal? Is there anything this student could do to earn a better score?

FRAPPY! Scoring Rubric

Use the following rubric to score your response. Each part receives a score of "Essentially Correct," "Partially Correct," or "Incorrect." When you have scored your response, reflect on your understanding of the concepts addressed in this problem. If necessary, note what you would do differently on future questions like this to increase your score.

Intent of the Question

The goal of this question is to determine your ability to conduct a significance test for a single proportion.

Solution

The solution should contain 4 parts:

- Hypotheses must be stated appropriately. H_0: $p = 0.25$ and H_a: $p < 0.25$ where p = true proportion of boxes containing the action figure.
- Name of test and conditions: The test must be identified by name or formula as a one-sample z test for a population proportion. **Random** sample is given. **Normal:** $70(0.25) = 17.5$ and $70(0.75) = 52.5$ Both are at least 10. **Independent:** There are more than 700 cereal boxes in the population.
- Mechanics: $z = -1.52$ and P-value = 0.064
- Conclusion: Because the P-value is greater than the typical significance level of 5%, we fail to reject the null hypothesis. There is not sufficient evidence to suggest that the actual proportion of boxes with the action figure is less than 0.25.

Scoring

Each element scored as essentially correct (E), partially correct (P), or incorrect (I).

Hypotheses is essentially correct if the hypotheses are written correctly. This part is partially correct if the test isn't identified or if the hypotheses are written incorrectly.

Name & Conditions is essentially correct if the response correctly identifies the test by name or formula AND correctly checks the Normal condition and 10% condition. This part is partially correct if the test is correctly identified but only one of the conditions is checked correctly OR if the test is not identified correctly but both conditions are checked correctly.

Mechanics is essentially correct if the response correctly calculates the test statistic and p-value. This part is partially correct if one of the calculations is incorrect.

Conclusion is essentially correct if the response correctly fails to reject the null hypothesis because the P-value is greater than a significance level of 5% OR if the response is to reject the null hypothesis because the P-value is less than a significance level of 10% and provides an interpretation in context. This part is partially correct if the response fails to justify the decision by comparing the P-value to a significance level OR if the conclusion lacks an interpretation in context.

Scoring

This problem has four elements, each receiving an E, P, or I.

Assign one point to each E, 0.5 points to each P, and 0 points to each I.

Total the points to determine your score.

If a score falls between two whole values, consider the strength of the entire response to determine whether to round up or down.

My Score:
What I did well:
What I could improve:
What I should remember if I see a problem like this on the AP Exam:

Chapter 9: Testing a Claim

Across

2. the probability that a significance test will reject the null when a particular alternative value of the paramter is true
4. hypotheses always refer to the ____
6. the test ____ is a standardized value that assesses how far the estimate is from the hypothesized parameter
8. the probability that we would observe a statistic at least as extreme as the one observed, assuming the null is true (two terms)
9. we can use a _____ test to compare observed data with a hypothesis about a population
12. greek letter used to designate the significance level
14. the ____ hypothesis is a the claim for which we are seeking evidence against
15. an observed difference that is too small to have occured due to chance alone is considered statistically ____
16. the statements a statistical test is designed to compare

Down

1. if we reject the null hypothesis when it is actually true, we commit a Type I ____
3. if we calculate a very small P value, we have evidence to ____ the null
5. the ____ hypothesis is the claim about the population for which we are finding evidence for
7. reject the null hypothesis if the P value is ____ than the significance level
10. if our calculated P value is not small enough to provide convincing evidence, we ___ to reject he null
11. conclusions should always be written in ____
13. a ____ test allows us to analyze differences in responses within pairs

Chapter 10: Comparing Two Populations or Treatments

"We must be careful not to confuse data with the abstractions we use to analyze them." William James

Chapter Overview

Up to this point, our studies have focused on inference for a single parameter. However, many statistical studies involve comparing two populations or treatments. In this chapter, we will expand our collection of inference procedures to include confidence intervals and significance tests for the difference between two proportions or means. The procedures for two populations or groups follow the exact same format as those we learned for a single parameter. The only difference is that now you will need to rely on the sampling distribution of the difference between proportions or means to perform your calculations. You will learn the characteristics of each of those sampling distributions in this chapter and how to use them to construct a confidence interval or perform a significance test. As you did in the last chapter, you will follow a four-step process for carrying out the inference procedures. Your job now will be to make sure you understand *when* to use each one!

Sections in this Chapter
Section 10.1: Comparing Two Proportions
Section 10.2: Comparing Two Means
Section 10.3: Comparing Two Means: Paired Data

Plan Your Learning

Use the following *suggested* guide to help plan your reading and assignments in "The Practice of Statistics, 6th Edition." Note: your teacher may schedule a different pacing or assign different problems. Be sure to follow his or her instructions!

Read	**10.1: pp. 619 - 624**	**10.1: pp. 625 - 629**	**10.1: pp. 630 - 638**	**10.2: pp. 645 - 655**
Do	1, 3	5, 7, 9, 11, 13	15, 19, 21, 29 MC 31-33	37, 39, 41, 45, 49

Read	**10.2: pp. 655 - 665**	**10.3: pp. 673 - 682**	**10.3: pp. 683 - 685**	**Chapter Summary**
Do	51, 53, 55, 57, 67 MC 69-72	75, 79, 85	91, 93 MC 95-97	Multiple-Choice FRAPPY!

Section 10.1: Comparing Two Proportions

Before You Read: Section Summary

In this section, you will learn how to compare two population proportions using confidence intervals and significance tests. To begin, you will explore the sampling distribution of a difference between two proportions. Like you did with a single proportion, you will learn the conditions necessary for performing inference and then how to construct a confidence interval or perform a significance test. By the end of this section, you should be able to provide an estimate of the difference between two proportions and test a claim about the difference between two proportions.

Learning Targets:

____ I can describe the shape, center, and variability of the sampling distribution of $\hat{p}_1 - \hat{p}_2$.

____ I can determine whether the conditions are met for doing inference about a difference between two proportions.

____ I can construct and interpret a confidence interval for a difference between two proportions.

____ I can calculate the standardized test statistic and *P*-value for a test about a difference between two proportions.

____ I can perform a significance test about a difference between two proportions.

While You Read: Key Vocabulary and Concepts

sampling distribution of $\hat{p}_1 - \hat{p}_2$:
conditions for inference about a difference in proportions:
two-sample *z* interval for a difference between two proportions:
two-sample *z* test for the difference between two proportions:

After You Read: Check for Understanding

Concept 1: Sampling Distribution of $\hat{p}_1 - \hat{p}_2$

If we want to compare two population proportions based on data from independent random samples, we must consider what would happen in repeated sampling. That is, we must explore the sampling distribution of the difference between two proportions. If we take repeated samples of the same size n_1 from a population with proportion of interest p_1, and samples of size n_2 from a population with proportion of interest p_2, the sampling distribution of $\hat{p}_1 - \hat{p}_2$ will have the following characteristics:

1) The shape of the sampling distribution will be approximately Normal if n_1p_1, $n_1(1-p_1)$, n_2p_2, and $n_2(1-p_2)$ are all at least 10.
2) The mean of the sampling distribution is $p_1 - p_2$.
3) The standard deviation of the sampling distribution is $\sqrt{\frac{p_1(1-p_1)}{n_1} + \frac{p_2(1-p_2)}{n_2}}$

Note: The formula for the standard deviation of the sampling distribution is exactly correct if sampling is done with replacement or if the populations are infinite. It is approximately correct if sampling is done without replacement as long as the 10% condition is satisfied: the sample sizes must be less than or equal to 10% of the size of the populations of interest.

Check for Understanding:

_____ *I can describe the shape, center, and variability of the sampling distribution of $\hat{p}_1 - \hat{p}_2$.*

School officials are interested in implementing a policy that would allow students to bring their own technology to school for academic use. There are two large high schools in a town, Lakeville North and Lakeville South, each with 1700 students. At Lakeville North, 60% of students own digital tools that could be used for academic purposes. 75% of students at Lakeville South own those types of devices. The district takes an SRS of 125 students from Lakeville North and a separate SRS of 160 students at Lakeville South. The sample proportions of students who own digital tools that could be used academically are recorded and the difference, $\hat{p}_S - \hat{p}_N$, is determined to be 0.07.

a) Describe the shape, center, and variability of the sampling distribution of $\hat{p}_S - \hat{p}_N$.

b) Find the probability of getting a difference in sample proportions of 0.07 or less from the two surveys. Show your work.

c) Does the result in part (b) give you reason to doubt the study's reported value? Explain.

Concept 2: Conference Intervals for $p_1 - p_2$

To construct a confidence interval for the difference between two proportions $p_1 - p_2$, you should follow the four step process introduced in Chapter 8. The logic behind its construction is the same as the logic behind the construction of a confidence interval for a single proportion. That is, we will estimate the difference by comparing the sample proportions from two random samples and build an interval around that point estimate by using the standard error of the statistic and a critical *z* value. To construct a two-sample *z* interval for a difference between two proportions:

- **State** the parameters of interest and the confidence level you will be using to estimate the difference.
- **Plan:** Indicate the type of confidence interval you are constructing and verify that the conditions for Random, Normal, and Independent are satisfied for the two samples.
- **Do** the actual construction of the interval using the following formula:

$$(\hat{p}_1 - \hat{p}_2) \pm z^* \sqrt{\frac{\hat{p}_1(1-\hat{p}_1)}{n_1} + \frac{\hat{p}_2(1-\hat{p}_2)}{n_2}}$$

  where z^* is the critical value for the standard Normal curve with area C between $-z^*$ and z^*.
- **Conclude** by interpreting the interval in the context of the problem.

Check for Understanding:

____ *I can determine whether the conditions are met for doing inference about a difference between two proportions.* ____ *I can construct and interpret a confidence interval for a difference between two proportions.*

In 1990, 551 of 1500 randomly sampled adults indicated they smoked. In 2010, 652 of 2000 randomly sampled adults indicated they smoked. Use this information to construct and interpret a 95% confidence interval for the difference in the proportion of adults who smoke in 1990 and 2010.

Concept 3: Significance tests for $p_1 - p_2$

To test a claim about the difference between two proportions, we will follow the same four-step process learned in Chapter 9. The key difference when performing a test about $p_1 - p_2$ is that when we assume these two parameters are equal in the null hypotheses, we need to estimate just what value they are equal to. To calculate this "pooled" proportion, simply divide the sum of the successes in the two samples by the sum of the sample sizes. This pooled proportion can then be used in the formula for the standard deviation of the sampling distribution. To conduct a two-sample z test for the difference between two proportions:

1) **STATE the parameter of interest and the hypotheses you would like to test.**
   The null hypothesis assumes that the parameters are equal while the alternative hypothesis is that one proportion is greater than, less than, or not equal to the other:

   $$H_0: p_1 = p_2$$
   $$H_a: p_1 > p_2 \text{ OR } p_1 < p_2 \text{ OR } p_1 \neq p_2$$

   State a significance level.

2) **PLAN: Choose the appropriate inference method and check the conditions.**
   We must check the Random, Normal, and Independent conditions. Verify that the data come from random samples or the groups in a randomized experiment. To ensure that the sampling distribution of $\hat{p}_1 - \hat{p}_2$ is approximately Normal, check that $n_1\hat{p}_1$, $n_1(1 - \hat{p}_1)$, $n_2\hat{p}_2$, and $n_2(1 - \hat{p}_2)$ are all at least 10. If sampling is done without replacement, check that the populations are at least 10 times as large as the samples.

3) **DO: If conditions are met, calculate a test statistic and *P*-value.**
   Compute the pooled proportion $\hat{p}_C = \dfrac{X_1 + X_2}{n_1 + n_2}$ and *z* test statistic

   $$z = \frac{(\hat{p}_1 - \hat{p}_2) - 0}{\sqrt{\dfrac{\hat{p}_C(1-\hat{p}_C)}{n_1} + \dfrac{\hat{p}_C(1-\hat{p}_C)}{n_2}}}$$

   and find the *P*-value by calculating the probability of observing a *z* statistic at least this extreme in the direction of the alternative hypothesis.

4) **CONCLUDE by interpreting the results of your calculations in the context of the problem.**
   If the *P*-value is smaller than the stated significance level, you can conclude that you have significant evidence to reject the null hypothesis. If it is larger than or equal to the significance level, then you fail to reject the null hypothesis.

Check for Understanding:

____ *I can perform a significance test about a difference between two proportions.*

A school official suspects the difference in the proportion of students who own digital tools between Lakeville North and Lakeville South high schools may be a result of a difference in the socioeconomic status of the students in the two schools. The results of a random sampling of student registration records indicated 28 out of 120 students at Lakeville North came from low-income families while 30 out of 150 students at Lakeville South came from low-income families. Do these data provide convincing evidence that the proportion of low-income students at Lakeville North is higher than the proportion at Lakeville South? Use a 5% significance level.

Section 10.2: Comparing Two Means

Before You Read: Section Summary

In this section, you will learn to compare two means instead of two proportions. You will start by exploring the sampling distribution of the difference between two means $\bar{x}_1 - \bar{x}_2$ along with the conditions necessary to perform inference. Then you will learn how to estimate a difference between two means as well as how to test a claim about that difference. Like you did with a single mean, you will rely on *t* distributions when performing calculations. Because some of the calculations are complex, you may want to rely on your calculator to do most of the work. As always, though, make sure you can interpret the results your calculator gives you!

Learning Targets:

____ I can describe the shape, center, and variability of the sampling distribution of $\bar{x}_1 - \bar{x}_2$.

____ I can determine whether the conditions are met for performing inference about a difference between two means.

____ I can construct and interpret a confidence interval for a difference between two means.

____ I can calculate the standardized test statistic and *P*-value for a test about a difference between two means.

____ I can perform a significance test about a difference between two means.

While You Read: Key Vocabulary and Concepts

sampling distribution of $\bar{x}_1 - \bar{x}_2$:
conditions for inference about a difference in means:
two-sample *t* interval for $\mu_1 - \mu_2$:
two-sample *t* test for $\mu_1 - \mu_2$:

After You Read: Check for Understanding

Concept 1: Sampling Distribution of $\bar{x}_1 - \bar{x}_2$

Like proportions, if we want to compare two means based on the results of independent random samples or randomly assigned groups, we must consider what would happen in repeated randomization. That is, we must explore the sampling distribution of the difference between two means. If we take repeated samples of the same size n_1 from a population with mean μ_1 and standard deviation σ_1 and samples of size n_2 from a population with mean μ_2 and standard deviation σ_2, the sampling distribution of $\bar{x}_1 - \bar{x}_2$ will have the following characteristics:

1) The shape of the sampling distribution will be Normal if the population distributions are Normal OR approximately Normal if both sample sizes are at least 30.
2) The mean of the sampling distribution is $\mu_1 - \mu_2$.
3) The standard deviation of the sampling distribution is $\sqrt{\frac{\sigma_1^2}{n_1} + \frac{\sigma_2^2}{n_2}}$

Note: The formula for the standard deviation of the sampling distribution is exactly correct if sampling is done with replacement or if the populations are infinite. It is approximately correct if sampling is done without replacement as long as the 10% condition is satisfied: the sample sizes must be less than or equal to 10% of the size of the populations of interest.

Check for Understanding:

____ *I can describe the shape, center, and variability of the sampling distribution of $\bar{x}_1 - \bar{x}_2$.*

Researchers are interested in studying the effect of sleep on exam performance. Suppose the population of individuals who get at least 8 hours of sleep prior to an exam score an average of 96 points on the exam with a standard deviation of 18 points. The population of individuals who get less than 8 hours of sleep score an average of 72 points with a standard deviation of 9.4 points. Suppose 40 individuals are randomly sampled from each population.

a) Describe the shape, center, and variability of the sampling distribution of $\bar{x}_1 - \bar{x}_2$

b) Find the probability of observing a difference in sample means of 2 points or more from the two samples. Show your work.

Concept 2: Conference Intervals for $\mu_1 - \mu_2$

To construct a confidence interval for the difference between two population or treatment means $\mu_1 - \mu_2$, you should follow the familiar four-step process. We will estimate the difference by comparing the sample means and build an interval around that point estimate by using the standard error of the statistic and a critical *t* value. To construct a two-sample *t* interval for a difference between two means:

- **State** the parameters of interest and the confidence level you will be using to estimate the difference.
- **Plan:** Indicate the type of confidence interval you are constructing and verify that the conditions for Random, Normal, and Independent are satisfied for the two samples.
- **Do** the actual construction of the interval using the following formula:

$$(\bar{x}_1 - \bar{x}_2) \pm t^* \sqrt{\frac{s_1^2}{n_1} + \frac{s_2^2}{n_2}}$$

where t^* is the critical value for the *t* distribution curve having df = smaller of $n_1 - 1$ and $n_2 - 1$ OR given by technology with area *C* between $-t^*$ and t^*.

- **Conclude** by interpreting the interval in the context of the problem.

Check for Understanding: ____ *I can construct and interpret a confidence interval for a difference between two means.*

Researchers are interested in determining the effectiveness of a new diet for individuals with heart disease. 200 heart disease patients are selected and randomly assigned to the new diet or the current diet used in the treatment of heart disease. The 100 patients on the new diet lost an average of 9.3 pounds with standard deviation 4.7 pounds. The 100 patients continuing with their current prescribed diet lost an average of 7.4 pounds with standard deviation 4 pounds. Construct and interpret a 95% confidence interval for the difference in mean weight loss for the two diets.

Concept 3: Significance tests for $\mu_1 - \mu_2$

To test a claim about the difference between two population or treatment means, we will follow the same four-step process learned previously. To conduct a two-sample *t* test for the difference between two means:

1) **STATE the parameter of interest and the hypotheses you would like to test.**

   The null hypothesis usually states that the parameters are equal while the alternative hypothesis is that one mean is greater than, less than, or not equal to the other:

   $$H_0: \mu_1 = \mu_2$$

   $$H_a: \mu_1 > \mu_2 \text{ OR } \mu_1 < \mu_2 \text{ OR } \mu_1 \neq \mu_2$$

   State the significance level.

2) **PLAN: Choose the appropriate inference method and check the conditions.**

   We must check the Random, Normal, and Independent conditions. Verify that the data come from random samples or the groups in a randomized experiment. To ensure that the sampling distribution of $\bar{x}_1 - \bar{x}_2$ is at least approximately Normal, check that the population distributions are Normal OR n_1 and n_2 are both at least 30. If sampling is done without replacement, check that the populations are at least 10 times as large as the samples.

3) **DO: If conditions are met, calculate a test statistic and *P*-value.**

   $$t = \frac{(\bar{x}_1 - \bar{x}_2) - (\mu_1 - \mu_2)}{\sqrt{\frac{s_1^2}{n_1} + \frac{s_2^2}{n_2}}}$$

   Find the *P*-value by calculating the probability of observing a *t* statistic at least this extreme in the direction of the alternative hypothesis.

   Use the *t* distribution with df = smaller of $n_1 - 1$ and $n_2 - 1$ OR given by technology.

4) **Conclude by interpreting the results of your calculations in the context of the problem.**

   If the *P*-value is smaller than the stated significance level, you have significant evidence to reject the null hypothesis. If it is larger than or equal to the significance level, then you fail to reject the null hypothesis.

Check for Understanding: ____ *I can calculate the standardized test statistic and P-value for a test about a difference between two means.* ____ *I can perform a significance test about a difference between two means.*

Do boys have better short-term memory than girls? A random sample of 200 boys and 150 girls was administered a short-term memory test. The average score for boys was 48.9 with standard deviation 12.96. The girls had an average score of 48.4 with standard deviation 11.85. Is there significant evidence at the 5% level to suggest boys have better short-term memory than girls? Note: higher test scores indicate better short-term memory.

Section 10.3: Comparing Two Means: Paired Data

Before You Read: Section Summary

In this section, you will learn to compare two means instead of two proportions. You will start by exploring the sampling distribution of the difference between two means $\bar{x}_1 - \bar{x}_2$ along with the conditions necessary to perform inference. Then you will learn how to estimate a difference between two means as well as how to test a claim about that difference. Like you did with a single mean, you will rely on *t* distributions when performing calculations. Because some of the calculations are complex, you may wish to rely on your calculator to do most of the work. As always, though, you will want to make sure you can interpret the results that your calculator gives you!

Learning Targets:

____ I can analyze the distribution of differences in a paired data set using graphs and summary statistics.
____ I can construct and interpret a confidence interval for a mean difference.
____ I can perform a significance test about a mean difference.
____ I can determine when it is appropriate to use paired *t* procedures versus two-sample *t* procedures.

While You Read: Key Vocabulary and Concepts

paired data:
conditions for inference about a mean difference:
paired *t* interval for a mean difference:
paired *t*-test for a mean difference:

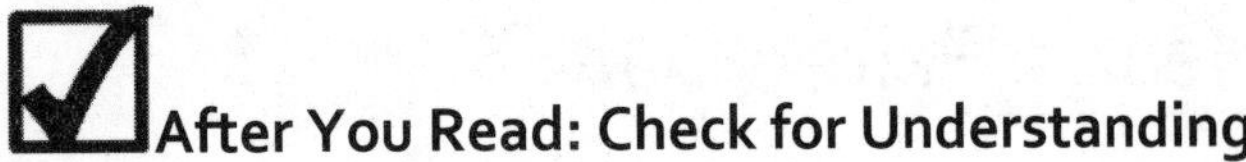

After You Read: Check for Understanding

Concept 1: Inference for Paired Data

Studies that involve making two observations on the same individual or making an observation on each of two very similar individuals result in paired data. In these types of studies, we are often interested in analyzing the differences in responses within each pair. If the conditions for inference are met, we can use one-sample *t* procedures to estimate or test a claim about the mean difference μ_d. In this case, we refer to the inference method as a paired *t* procedure.

Check for Understanding: ____ *I can perform a significance test about a mean difference.*

A study measured how fast subjects could repeatedly push a button when under the effects of caffeine. Subjects were asked to push a button as many times as possible in two minutes after consuming a typical amount of caffeine. During another test session, they were asked to push the button after taking a placebo. The subjects did not know which treatment they were administered each day and the order of the treatments was randomly assigned. The data, given in presses per two minutes for each treatment follows. Use a paired *t* procedure to determine whether or not caffeine results in a higher rate of beats, on average, per two-minute period.

Subject	Beats Caffeine	Beats Placebo	
1	251	201	
2	284	262	
3	300	283	
4	321	290	
5	240	259	
6	294	291	
7	377	354	
8	345	346	
9	303	283	
10	340	361	
11	408	411	

Chapter Summary: Comparing Two Populations or Groups

In this chapter, you learned the inference procedures that help us compare two parameters. Whether you are comparing two proportions or two means, the processes for constructing a confidence interval or performing a significance test are the same. When comparing two proportions, you use two-sample *z* procedures to reach your conclusions. If you are dealing with means, you use two-sample *t* procedures. Like we learned for one-sample procedures, it is important to identify which procedure you are using, verify that the appropriate conditions are met, perform the necessary calculations, and interpret your results in the context of the problem.

After You Read: What Have I Learned?

Complete the vocabulary puzzle, multiple-choice questions, and FRAPPY. Check your answers and performance on each of the learning targets. Be sure to get extra help on any targets that you identify as needing more work!

Learning Target	Got It!	Almost There	Needs Some Work
I can describe the shape, center, and variability of the sampling distribution of $\hat{p}_1 - \hat{p}_2$.			
I can determine whether the conditions are met for doing inference about a difference between two proportions.			
I can construct and interpret a confidence interval for a difference between two proportions.			
I can calculate the standardized test statistic and *P*-value for a test about a difference between two proportions.			
I can perform a significance test about a difference between two proportions.			
I can describe the shape, center, and variability of the sampling distribution of $\bar{x}_1 - \bar{x}_2$.			
I can determine whether the conditions are met for doing inference about a difference between two means.			
I can construct and interpret a confidence interval for a difference between two means.			
I can calculate the standardized test statistic and *P*-value for a test about a difference between two means.			
I can perform a significance test about a difference between two means.			
I can analyze the distribution of differences in a paired data set using graphs and summary statistics.			
I can construct and interpret a confidence interval for a mean difference.			
I can perform a significance test about a mean difference.			
I can determine when it is appropriate to use paired *t* procedures versus two-sample *t* procedures.			

Chapter 10 Multiple Choice Practice

Directions. *Identify the choice that best completes the statement or answers the question. Check your answers and note your performance when you are finished.*

1. Is the proportion of marshmallows in Mr. Miller's favorite breakfast cereal lower than it used to be? To determine this, you test the hypotheses H_0: $p_{old} = p_{new}$, H_a: $p_{old} > p_{new}$ at the $\alpha = 0.05$ level. You calculate a test statistic of 1.980. Which of the following is the appropriate P-value and conclusion for your test?

   (A) P-value = 0.047; fail to reject H_0; we do not have convincing evidence that the proportion of marshmallows has been reduced.
   (B) P-value = 0.047; accept H_a; there is convincing evidence that the proportion of marshmallows has been reduced.
   (C) P-value = 0.024; fail to reject H_0; we do not have convincing evidence that the proportion of marshmallows has been reduced.
   (D) P-value = 0.024; reject H_0; we have convincing evidence that the proportion of marshmallows has been reduced.
   (E) P-value = 0.024; fail to reject H_0; we have convincing evidence that the proportion of marshmallows has not changed.

2. An SRS of 100 teachers showed that 64 owned smartphones. An SRS of 100 students showed that 80 owned smartphones. Let p_T be the proportion of all teachers who own smartphones, and let p_S be the proportion of all students who own smartphones. A 95% confidence interval for the difference $p_T - p_S$ is

   (A) (0.264, 0.056)
   (B) (0.098, 0.222)
   (C) (-0.222, -0.098)
   (D) (-0.264, -0.056)
   (E) (-0.283, -0.038)

3. A school receives textbooks independently from two suppliers. An SRS of 400 textbooks from supplier 1 finds 20 that are defective. An SRS of 100 textbooks from supplier 2 finds 10 that are defective. Let p_1 and p_2 be the proportions of all textbooks from suppliers 1 and 2, respectively that are defective. Which of the following represents a 95% confidence interval for $p_1 - p_2$?

   (A)
   $$-0.05 \pm 1.96\sqrt{\frac{(0.05)(0.95)}{400} - \frac{(0.1)(0.9)}{100}}$$

   (B)
   $$-0.05 \pm 1.96\sqrt{\frac{(0.05)(0.95)}{400} + \frac{(0.1)(0.9)}{100}}$$

   (C)
   $$-0.05 \pm 1.64\sqrt{\frac{(0.05)(0.95)}{400} - \frac{(0.1)(0.9)}{100}}$$

   (D)
   $$-0.05 \pm 1.64\sqrt{\frac{(0.05)(0.95)}{400} + \frac{(0.1)(0.9)}{100}}$$

   (E)
   $$-0.05 \pm 1.64\sqrt{\frac{(0.06)(0.94)}{500}}$$

4. An agricultural researcher wishes to see if a new fertilizer helps increase the yield of tomato plants. One hundred tomato plants in individual containers are randomly assigned to two different groups. Plants in both groups are treated identically, except that the plants in group 2 are sprayed weekly with the fertilizer, while the plants in group 1 are not. After 4 weeks, 12 of the 50 plants in group 1 exhibited an increased yield, and 18 of the 50 plants in group 2 showed an increased yield. Let p_1 be the actual proportion of all tomato plants of this variety that would experience an increased yield under the fertilizer treatment, and let p_2 be the actual proportion of all tomato plants of this variety that would experience an increased yield under with no fertilizer treatment, assuming that the tomatoes are grown under conditions similar to those in the experiment. Is there evidence of an increase in the proportion of tomato plants with increased yield for those sprayed with fertilizer? To determine this, you test the hypotheses $H_0: p_1 = p_2$, $H_a: p_1 < p_2$. The *P*-value of your test is

(A) greater than 0.10.
(B) between 0.05 and 0.10.
(C) between 0.01 and 0.05.
(D) between 0.001 and 0.01.
(E) below 0.001.

5. An SRS of 45 male employees at a large company found that 36 felt that the company was supportive of female and minority employees. An independent SRS of 40 female employees found that 24 felt that the company was supportive of female and minority employees. Let p_1 represent the proportion of all male employees at the company and p_2 represent the proportion of all female employees members at the company who hold this opinion. We wish to test the hypotheses $H_0: p_1 - p_2 = 0$ vs. $H_a: p_1 - p_2 < 0$. Which of the following is the correct expression for the test statistic?

(A) $$\frac{0.8-0.6}{\sqrt{\frac{(0.8)(0.2)}{45}+\frac{(0.6)(0.4)}{40}}}$$

(B) $$\frac{0.8-0.6}{\sqrt{\frac{(0.706)(0.294)}{45}+\frac{(0.706)(0.294)}{40}}}$$

(C) $$\frac{0.8-0.6}{\sqrt{\frac{(0.706)(0.294)}{45}-\frac{(0.706)(0.294)}{40}}}$$

(D) $$\frac{0.8-0.6}{\frac{(0.8)(0.2)}{\sqrt{45}}+\frac{(0.6)(0.4)}{\sqrt{40}}}$$

(E) $$\frac{0.8}{\frac{(0.8)(0.2)}{\sqrt{45}}}+\frac{0.6}{\frac{(0.6)(0.4)}{\sqrt{40}}}$$

6. Some researchers have conjectured that stem-pitting disease in peach tree seedlings might be controlled with weed and soil treatment. An experiment was conducted to compare peach tree seedling growth with soil and weeds treated with one of two herbicides. In a field containing 20 seedlings, 10 were randomly selected from throughout the field and assigned to receive Herbicide A. The remaining 10 seedlings were to receive Herbicide B. Soil and weeds for each seedling were treated with the appropriate herbicide, and at the end of the study period, the height (in centimeters) was recorded for each seedling. A box plot of each data set showed no indication of non-Normality. The following results were obtained:

	$\overline{x}$ (cm)	s_x (cm)
Herbicide A	94.5	10
Herbicide B	109.1	9

Suppose we wished to determine if there is a significant difference in mean height for the seedlings treated with the different herbicides. Based on our data, which of the following is the value of test statistic?

(A) 14.60
(B) 7.80
(C) 3.43
(D) 2.54
(E) 1.14

7. A researcher wished to test the effect of the addition of extra calcium on the "tastiness" of yogurt. Sixty-two adult volunteers were randomly divided into two groups of 31 subjects each. Group 1 tasted yogurt containing the extra calcium. Group 2 tasted yogurt from the same batch as group 1 but without the added calcium. Both groups rated the flavor on a scale of 1 to 10, with 1 being "very unpleasant" and 10 being "very pleasant." The mean rating for group 1 was 6.5 with a standard deviation of 1.5. The mean rating for group 2 was 7.0 with a standard deviation of 2.0. Let μ_1 and μ_2 represent the true mean ratings we would observe for the entire population represented by the volunteers if all of them tasted, respectively, the yogurt with and without the added calcium. Which of the following would lead us to believe that the *t*-procedures were not safe to use in this situation?

(A) The sample medians and means for the two groups were slightly different.
(B) The distributions of the data for the two groups were both slightly skewed right.
(C) The data are integers between 1 and 10 and so cannot be normal.
(D) The standard deviations from both samples were very different from each other.
(E) None of the above.

8. A researcher wishes to compare the effect of two stepping heights (low and high) on heart rate in a step-aerobics workout. The researcher constructs a 98% confidence interval for the difference in mean heart rate between those who did the high and those who did the low stepping heights. Which of the following is a correct interpretation of this interval?

(A) 98% of the time, the true difference in the mean heart rate of subjects in the high-step *vs.* low-step groups will be in this interval.
(B) We are 98% confident that this interval captures the true difference in mean heart rate of subjects like these who receive the high-step and low-step treatments.
(C) There is a 0.98 probability that the true difference in mean heart rate of subjects in the high-step *vs.* low-step groups in this interval.
(D) 98% of the intervals constructed in this way will contain the value 0.
(E) There is a 98% probability that we have <u>not</u> made a Type I error.

9. Using the setting from problem 8. The researcher decides to test the hypotheses $H_0: \mu_1 - \mu_2 = 0$ *vs.* $H_a: \mu_1 - \mu_2 < 0$ at the $\alpha = 0.05$ level and produces a *P*-value of 0.0475. Which of the following is a correct interpretation of this result?

(A) The probability that there is a difference is 0.0475.
(B) The probability that this test resulted in a Type II error is 0.0475.
(C) If this test were repeated many times, we would make a Type I error 4.75% of the time.
(D) If the null hypothesis is true, the probability of getting a difference in sample means as far or farther from 0 as the difference in our samples is 0.0475.
(E) If the null hypothesis is false, the probability of getting a difference in sample means as far or farther from 0 as the difference in our samples is 0.0475.

10. The researcher in question 8 randomly assigned 50 adult volunteers to two groups of 25 subjects each. Group 1 did a standard step-aerobics workout at the low height. The mean heart rate at the end of the workout for the subjects in group 1 was 90 beats per minute with a standard deviation of 9 beats per minute. Group 2 did the same workout but at the high step height. The mean heart rate at the end of the workout for the subjects in group 2 was 95.2 beats per minute with a standard deviation of 12.3 beats per minute. Assuming the conditions are met, which of the following could be the 98% confidence interval for the difference in mean heart rates based on these results?

(A) (2.15, 8.25)
(B) (-0.77, 11.17)
(C) (-2.13, 12.54)
(D) (-2.16, 12.56)
(E) (-4.09, 14.49)

Check your answers below. If you got a question wrong, check to see if you made a simple mistake or if you need to study that concept more. After you check your work, identify the concepts you feel very confident about and note what you will do to learn the concepts in need of more study.

#	Answer	Concept	Right	Wrong	Simple Mistake?	Study More
1	D	Significance Test, *P*-value, and Conclusion				
2	E	Confidence Interval for Difference in Proportions				
3	B	Confidence Interval for Difference in Proportions				
4	B	P-value for Test of Significance				
5	B	Test Statistic				
6	C	Significance Test for Difference in Means				
7	E	Conditions for Inference				
8	B	Interpret Confidence Interval				
9	D	Interpret P-value				
10	D	Confidence Interval for Difference in Means				

Chapter 10 Reflection

Summarize the "Big Ideas" in Chapter 10:

My strengths in this chapter:

Concepts I need to study more and what I will do to learn them:

FRAPPY! Free Response AP® Problem, Yay!

The following problem is modeled after actual Advanced Placement Statistics free response questions. Your task is to generate a complete, concise response in 15 minutes. After you generate your response, view two example solutions and determine whether you feel they are "complete", "substantial", "developing" or "minimal". If they are not "complete", what would you suggest to the student who wrote them to increase their score? Finally, you will be provided with a rubric. Score your response and note what, if anything, you would do differently to increase your own score.

Researchers are interested in whether or not women who are part of a prenatal care program give birth to babies with a higher average birth weight than those who do not take part in the program. A random sample of hospital records indicates that the average birth weight for 75 babies born to mothers enrolled in a prenatal care program was 3100 g with standard deviation 420 g. A separate random sample of hospital records indicates that the average birth weight for 75 babies born to women who did not take part in a prenatal care program was 2750 g with standard deviation 425 g. Do these data provide convincing evidence that mothers who participate in a prenatal care program have babies with a higher average birth weight than those who don't?

FRAPPY! Student Responses

Student Response 1:

We will perform a two-sample t-test for the difference in means.
$H_0: \mu_Y = \mu_N$
$H_a: \mu_Y > \mu_N$ (Note, Y = enrolled in program, N = not enrolled)

We are told the samples are random. Both sample sizes (75) are greater than 30.

$t = \dfrac{3100 - 2750}{\sqrt{\dfrac{420^2}{75} + \dfrac{425^2}{75}}} = 5.07$ and our *P*-value is 0.000000578.

We have significant evidence at the 5% level to reject the null hypothesis. The babies born to mothers enrolled in the prenatal care program have higher average birth weights than the babies born to mothers who are not enrolled in the program.

How would you score this response? Is it substantial? Complete? Developing? Minimal? Is there anything this student could do to earn a better score?

Student Response 2:

I will construct a 95% confidence interval for the difference in mean weights. We are told we have a random samples and both sample sizes are greater than 30.

$$(3100 - 2750) \pm t^* \sqrt{\frac{420^2}{75} + \frac{425^2}{75}} = (213.65, 486.34)$$

We are 95% confident that the true difference in average? Birth weights between babies born to mothers enrolled in the program and babies born to mothers not enrolled in the program is between 213.65 g and 486.34 g. Since 0 is not contained in this interval, we have evidence to suggest babies born to mothers enrolled in the program have an average weight between 213.65g and 486.34g greater than the average weight of babies born to mothers not enrolled in the program.

How would you score this response? Is it substantial? Complete? Developing? Minimal? Is there anything this student could do to earn a better score?

FRAPPY! Scoring Rubric

Use the following rubric to score your response. Each part receives a score of "Essentially Correct," "Partially Correct," or "Incorrect." When you have scored your response, reflect on your understanding of the concepts addressed in this problem. If necessary, note what you would do differently on future questions like this to increase your score.

Intent of the Question

The goal of this question is to determine your ability to conduct a significance test for a single proportion the difference between two means.

Solution

The solution should contain 4 parts:

Hypotheses: Must be stated correctly: $H_0: \mu_Y - \mu_N = 0$ and $H_a: \mu_Y - \mu_N > 0$ where μ_Y = true mean birth weight of babies born to mothers enrolled in a prenatal care program and μ_N= true mean birth weight of babies born to mothers who were not enrolled in a prenatal care program.

Name of Test and Conditions: The test must be identified by name or formula as a two-sample *t* test for a difference in population means. Two random samples were obtained. Both sample sizes are greater than 30. Both samples are less than 10% of their respective populations since this is a large hospital.

Mechanics: $t = 5.07$ and *P*-value = 5.78×10^{-7} with 147.97 degrees of freedom.

Conclusion: Because the *P*-value is smaller than any reasonable significance level we reject the null in favor of the alternative hypothesis. There is significant evidence to suggest the mean birth weight of babies born to mothers who participate in a prenatal care program is higher than the weight of those born to mothers who do not participate.

Scoring

Each element scored as essentially correct (E), partially correct (P), or incorrect (I).

Hypotheses are essentially correct if the hypotheses are written correctly. This part is partially correct if the test isn't identified or if the hypotheses are written incorrectly.

Name and Conditions is essentially correct if the response correctly identifies the test by name or formula AND correctly checks the Random, Normal, and 10% conditions. This part is partially correct if at least one of the conditions is not checked and incorrect if only one condition is checked.

Mechanics is essentially correct if the response correctly calculates the test statistic and *p*-value. This part is partially correct if one of the calculations is incorrect.

Conclusion is essentially correct if the response correctly rejects the null hypothesis due to the very small *p*-value and provides an interpretation in context. This part is partially correct if the response fails to justify the decision by indicating that the p-value is very small OR if the conclusion lacks an interpretation in context.

Scoring

This problem has four elements, each receiving an E, P, or I.

Assign one point to each E, 0.5 points to each P, and 0 points to each I.

Total the points to determine your score.

If a score falls between two whole values, consider the strength of the entire response to determine whether to round up or down.

My Score:
What I did well:
What I could improve:
What I should remember if I see a problem like this on the AP Exam:

Chapter 10: Comparing Two Populations or Groups

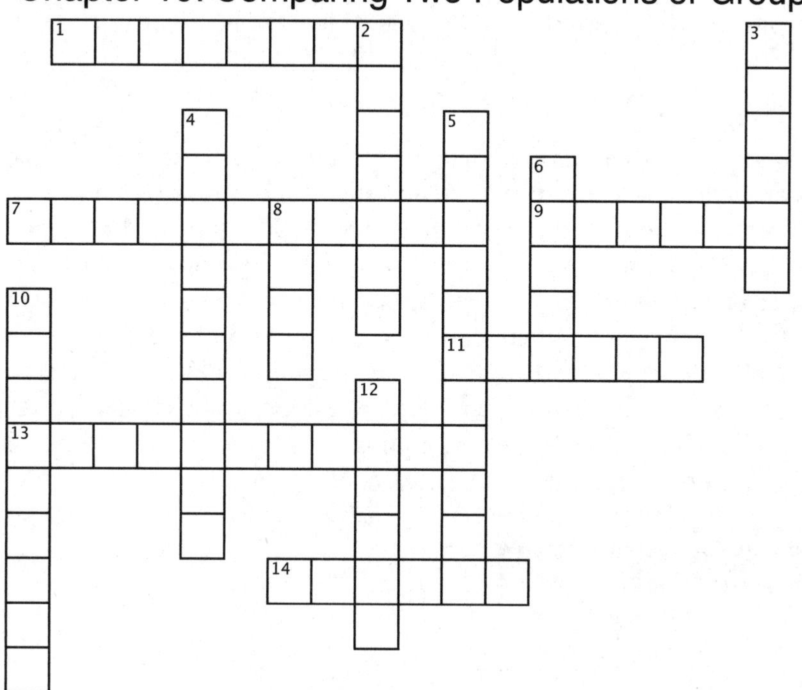

Across

1. an alternative hypothesis the seeks evidence of a difference requires the use of a ____ test (two words)
7. to perform two-sample procedures, the random samples should be____
9. procedures that yield accurate results, even when a condition is violated, are ____
11. if the number of successes and failures from both samples are greater than ten, the sampling distribution for the difference between the two proportions will be approximately ____
13. results that are too unlikely to be due to chance alone are considered statistically ____
14. in a two-proportion test, we calculate a ____ proportion

Down

2. particular t distributions are distinguished by ____ of freedom
3. the sampling distribution for the difference between means will be approximately Normal if both sample sizes are greater than
4. hypotheses should always be written in terms of the ____
5. the hypthesis we are seeking evidence for
6. the margin of error in a confidence interval consists of a critical value and standard ____ of the statistic
8. the hypothesis we are seeking evidence against
10. to compare two proportions, we can construct a ____ z interval (two words)
12. if we want to compare two population proprtions, we must have data from two ____ samples

Chapter 11: Inference for Distributions of Categorical Data

"By a small sample, we may judge the whole piece." Miguel de Cervantes (Don Quixote)

Chapter Overview

So far, our study of inference has focused on how to estimate and test claims about single means and proportions as well as about differences between means and proportions for quantitative variables. In this chapter, we will shift our focus to inference about distributions of and relationships between categorical variables. You will learn how to perform three different significance tests for distributions of categorical data, allowing you to determine (1) whether or not a sample distribution differs significantly from a hypothesized distribution, (2) whether or not the distribution of a categorical variable differs between multiple populations, or (3) whether or not an association exists between two categorical variables. As before, each test will consist of hypotheses, conditions to be checked, a test statistic and *P*-value, and a conclusion in context. It is easy to confuse the three tests, so be sure to study the differences between them so you know when to use each one!

Sections in this Chapter
Section 11.1: Chi-Square Tests for Goodness-of-Fit **Section 11.2**: Inference for Two-Way Tables

Plan Your Learning

Use the following *suggested* guide to help plan your reading and assignments in "The Practice of Statistics, 6th Edition." Note: your teacher may schedule a different pacing or assign different problems. Be sure to follow his or her instructions!

Read	**11.1: pp. 708 - 717**	**11.1: pp. 717 - 722**
Do	1, 3, 5, 7	9, 13 MC 19-21

Read	**11.2: pp. 726 - 740**	**11.2: pp. 740 - 752**	**Chapter Summary**
Do	27, 29, 31, 33, 35	41, 43, 47, 49, 51, 55-60	Multiple-Choice FRAPPY!

Section 11.1: Chi-Square Tests for Goodness-of-Fit

Before You Read: Section Summary

In this section, you will be introduced to the chi-square distributions and the chi-square test statistic. The chi-square statistic provides us with a way to measure the difference between an observed and a hypothesized distribution of categorical data. When certain conditions are satisfied, we can model the sampling distribution of this statistic with a chi-square distribution and calculate *P*-values. We can use the chi-square goodness-of-fit test to determine whether or not a significant difference between an observed and hypothesized distribution of a categorical variable exists. Finally, we can use a follow-up analysis to determine which categories contributed the most to the difference.

Learning Targets:

____ I can state appropriate hypotheses and compute the expected counts and chi-square test statistic for a chi-square test for goodness of fit.
____ I can state and check the Random, 10%, and Large Counts conditions for performing a chi-square test for goodness of fit.
____ I can calculate the degrees of freedom and *P*-value for a chi-square test for goodness of fit.
____ I can perform a chi-square test for goodness of fit.
____ I can conduct a follow-up analysis when the results of a chi-square test are statistically significant.

While You Read: Key Vocabulary and Concepts

expected count:
chi-square statistic:
chi-square distribution:
conditions for performing a chi-square test for goodness of fit:
chi-square test for goodness-of-fit:

☑ After You Read: Check for Understanding

Concept 1: The Chi-Square Test Statistic and Distributions

When testing a claim about a distribution of categorical data, we are interested in knowing how the observed counts from the sample compare to the counts that would be expected if the null hypothesis were true. Like we did with means and proportions, we will standardize the difference between observed and expected values using a test statistic. To determine the chi-square test statistic for a table of observed counts, use the following formula:

$$\chi^2 = \sum \frac{(\text{observed count-expected count})^2}{\text{expected count}}$$

This test statistic can then be used with the appropriate chi-square distribution to determine a *P*-value. If the observed values are far from the expected values, our test statistic will be large, giving evidence against the null hypothesis. The chi-square distributions are a family of right-skewed distributions that are defined by degrees of freedom based on the number of categories the variable takes on. We can use a table or technology to find a *P*-value for a particular χ^2 value.

Check for Understanding: _______ *I can state appropriate hypotheses and compute the expected counts and chi-square test statistic for a chi-square test for goodness of fit.*

After playing a dice game with a friend, you suspect the 6-sided die may not be fair. That is, you suspect some numbers may be rolled more often than you would expect. To test your suspicion, you roll the die 300 times and record the results.

Value:	1	2	3	4	5	6
Frequency:	42	55	38	57	64	44

Use these observed counts to determine the chi-square statistic for this example.

Concept 2: Chi-Square Test for Goodness-of-Fit

To perform a chi-square test for goodness-of-fit for a claim about a population distribution of categorical data, we will follow the same basic process as we did for means and proportions. That is, to test a claim about the population distribution of a categorical variable:

1) **STATE the hypotheses you would like to test.**

   When testing a claim about a distribution of a categorical variable, we start by defining hypotheses. The null hypothesis assumes that the hypothesized distribution is correct as stated while the alternative hypothesis is that the specified distribution is different:

   H_0: *The specified distribution of the categorical variable is correct.*
   H_a: *The specified distribution of the categorical variable is not correct.*

   State a significance level.

2) **PLAN: Choose the appropriate inference method and check the conditions.**

   We must check the conditions to ensure that we can use a chi-square distribution to determine our *P*-value. The data must come from a random sample from the population. When sampling without replacement, check that the population is at least 10 times as large as the sample. All expected counts must be at least 5.

3) **DO: If conditions are met, calculate a test statistic and *P*-value.**

   Compute the chi-square test statistic $\chi^2 = \sum \frac{(\text{observed count-expected count})^2}{\text{expected count}}$

   Find the *P*-value by using a chi-square distribution with *k* - 1 degrees of freedom (where *k* is the number of categories).

4) **CONCLUDE by interpreting the results of your calculations in the context of the problem.**

   If the *P*-value is smaller than the stated significance level, you can conclude that you have significant evidence to reject the null hypothesis. If the *P*-value is larger than or equal to the significance level, then you fail to reject the null hypothesis.

If you find evidence to reject the null hypothesis, you should perform a follow up analysis to determine the components that contributed the most to the chi-square statistic. That is, determine which observed categories differed the most from the expected values.

Check for Understanding: ____ *I can state and check the Random, 10%, and Large Counts conditions for performing a chi-square test for goodness of fit.* ____ *I can calculate the degrees of freedom and P-value for a chi-square test for goodness of fit.* ____ *I can perform a chi-square test for goodness of fit.* ____ *I can conduct a follow-up analysis when the results of a chi-square test are statistically significant.*

Use the observed counts from the previous Check for Understanding to determine whether or not you have convincing evidence that the die is not fair. If you have significant evidence, perform a follow-up analysis. Recall, to test your suspicion, you rolled the die 300 times.

Value:	1	2	3	4	5	6
Frequency:	42	55	38	57	64	44

Section 11.2: Inference for Two-Way Tables

Before You Read: Section Summary

Chi-square goodness-of-fit tests allow us to compare the distribution of one categorical variable to a hypothesized distribution. However, sometimes we are interested in comparing the distribution of a categorical variable across several populations or treatments. Chi-square tests for homogeneity allow us to do this. In this section you will learn how to conduct the chi-square test for homogeneity and how to perform a follow-up analysis, just like you did for the goodness-of-fit test. Also, you will learn how to conduct a chi-square test for association/independence to determine whether or not there is convincing evidence that two categorical variables are related. While this test is the same as the test for homogeneity in its mechanics, the hypotheses are different. Be sure to note the distinction!

Learning Targets:

____ I can state appropriate hypotheses and compute the expected counts and chi-square statistic for a chi-square test based on data in a two-way table.

____ I can state and check the Random, 10%, and Large Counts conditions for a chi-square test based on data in a two-way table.

____ I can calculate the degrees of freedom and *P*-value for a chi-square test based on data in a two-way table.

____ I can perform a chi-square test for homogeneity.

____ I can perform a chi-square test for independence.

____ I can choose the appropriate chi-square test in a given setting.

While You Read: Key Vocabulary and Concepts

expected counts:
conditions for a chi-square test based on data in a two-way table:
chi-square test for homogeneity:
chi-square test for independence:

After You Read: Check for Understanding

Concept 1: Expected Counts and the Chi-Square Statistic

While a one-way table can summarize data for a single categorical variable, a two-way table can be used to summarize data on the relationship between two categorical variables. When these data are produced using independent random samples from several populations or from a randomized comparative experiment, we can test whether or not the actual distribution of the categorical variable is the same for each population or treatment. If the data are produced using a single random sample from a population and classified according to two categorical variables, we can test whether or not there is a relationship between those variables. In each case, we must compare the observed counts to expected counts.

To determine an expected count, we use the general formula

$$\text{expected count} = \frac{\text{row total} \bullet \text{column total}}{\text{table total}}.$$

Then, just like we did with the goodness-of-fit test, we calculate the chi-square statistic using

$$\chi^2 = \sum \frac{(\text{observed count - expected count})^2}{\text{expected count}}.$$

Concept 2: Chi-Square Test for Homogeneity

To determine whether or not a distribution of a categorical variable differs for two or more populations or treatments, we use a chi-square test for homogeneity. This test is similar to the test for goodness-of-fit in the sense that we compare observed counts to expected counts using a chi-square test statistic. However, the hypotheses and degrees of freedom differ slightly.

1) **STATE the hypotheses you would like to test.**
   When testing a claim about the distribution of a single categorical variable for two or more populations or treatments, we start by defining hypotheses. The null hypothesis says that the variable has the same distribution for all of the populations or treatments. The alternative hypothesis is that the distribution of that variable is not the same for all of the populations or treatments:
   H_0: There is no difference in the distribution of a categorical variable for several populations or treatments.
   H_a: There is a difference in the distribution of a categorical variable for several populations or treatments.

2) **PLAN: Choose the appropriate inference method and check the conditions.**
   We must check the conditions to ensure that we can use a chi-square distribution to determine our *P*-value. The data must come from independent random samples or from groups in a randomized experiment. When sampling without replacement, check that the population is at least 10 times as large as each sample. All expected counts must be at least 5.

3) **DO: If conditions are met, calculate a test statistic and *P*-value.**
   Compute the chi-square test statistic $\chi^2 = \sum \frac{(\text{observed count - expected count})^2}{\text{expected count}}$ and find the *P*-value by using a chi-square distribution with *(number of rows - 1)(number of columns – 1)* degrees of freedom.

4) **CONCLUDE by interpreting the results of your calculations in the context of the problem.**
   If the *P*-value is smaller than the stated significance level, you can conclude that you have significant evidence to reject the null hypothesis. If *the* P-value is larger than or equal to the significance level, then you fail to reject the null hypothesis.

If you find sufficient evidence to reject the null hypothesis, you should perform a follow up analysis to determine the components that contributed the most to the chi-square statistic. That is, determine which observed categories differed the most from the expected values.

Check for Understanding: ____ *I can state appropriate hypotheses and compute the expected counts and chi-square statistic for a chi-square test based on data in a two-way table.* ____ *I can calculate the degrees of freedom and P-value for a chi-square test based on data in a two-way table.* ____ *I can perform a chi-square test for homogeneity.*

A recent study tracked the television viewing habits of 100 randomly selected first-grade boys and 200 randomly selected first-grade girls. Each child was asked to identify their favorite TV show. The following table summarizes the results:

	Zooboomafoo	iCarly	Phineas and Ferb
Boys	20	30	50
Girls	70	80	50

Do these data provide convincing evidence that television preferences differ significantly for boys and girls?

Concept 3: Chi-Square Test for Independence

To determine whether or not two categorical variables are related in a population, we can use a chi-square test for independence. The mechanics of this test are almost the same as those for the test for homogeneity. The only difference is that the data come from a single random sample from the population of interest. The hypotheses are defined in terms of an association between the two categorical variables.

H_o: There is no association between two categorical variables in the population of interest.

H_a: There is an association between two categorical variables in the population of interest.

Check for Understanding: ____ *I can perform a chi-square test for independence.*

A recent study looked into the relationship between political views and opinions about nuclear energy. A survey administered to 100 randomly selected adults asked their political leanings as well as their approval of nuclear energy. The results are below:

	Liberal	**Conservative**	**Independent**
Approve	10	15	20
Disapprove	9	2	16
No Opinion	8	2	18

Do these data provide convincing evidence that political leanings and views on nuclear energy are associated in the larger population of adults from which the sample was selected?

Chapter Summary: Inference for Distributions of Categorical Data

In this chapter, you learned the inference procedures for distributions of categorical data. A chi-square goodness-of-fit test can be used to determine whether or not an observed distribution of a categorical variable differs from a hypothesized distribution. When examining the distribution of a single categorical variable in multiple populations or treatments, a test for homogeneity can be used to determine whether the distributions differ. Finally, we can use a test for independence to determine whether or not two categorical variables are related in a population. In each test, we use the chi-square test statistic to measure how much the observed counts differ from the expected counts. When our test provides significant evidence against the null hypothesis, we can use a follow-up analysis to determine which component(s) contributed the most to the test statistic.

After You Read: What Have I Learned?

Complete the vocabulary puzzle, multiple-choice questions, and FRAPPY. Check your answers and performance on each of the learning targets. Be sure to get extra help on any targets that you identify as needing more work!

Learning Target	**Got It!**	**Almost There**	**Needs Some Work**
I can state appropriate hypotheses and compute the expected counts and chi-square test statistic for a chi-square test for goodness of fit.			
I can state and check the Random, 10%, and Large Counts conditions for performing a chi-square test for goodness of fit.			
I can calculate the degrees of freedom and *P*-value for a chi-square test for goodness of fit.			
I can perform a chi-square test for goodness of fit.			
I can conduct a follow-up analysis when the results of a chi-square test are statistically significant.			
I can state appropriate hypotheses and compute the expected counts and chi-square statistic for a chi-square test based on data in a two-way table.			
I can state and check the Random, 10%, and Large Counts conditions for a chi-square test based on data in a two-way table.			
I can calculate the degrees of freedom and *P*-value for a chi-square test based on data in a two-way table.			
I can perform a chi-square test for homogeneity.			
I can perform a chi-square test for independence.			
I can choose the appropriate chi-square test in a given setting.			

Chapter 11 Multiple Choice Practice

Directions. *Identify the choice that best completes the statement or answers the question. Check your answers and note your performance when you are finished.*

1. To test the effectiveness of your calculator's random number generator, you randomly select 1000 numbers from a standard Normal distribution. You classify these 1000 numbers according to whether their values are at most −2, between −2 and 0, between 0 and 2, or at least 2. The results are given in the following table. The expected counts, based on the 68-95-99.7 rule, are given as well.

	At most -2	Between -2 and 0	Between 0 and 2	At least 2
Observed Count	18	492	468	22
Expected Count	25	475	475	25

To test to see if the distribution of observed counts differs significantly from the distribution of expected counts, we can use a χ^2 goodness of fit test. For this test, the test statistic has approximately a χ^2 distribution. How many degrees of freedom does this distribution have?

(A) 3
(B) 4
(C) 7
(D) 999
(E) 1000

2. Refer to problem 1. Which of the following is the component of the χ^2 statistic corresponding to the category "at most −2"?

(A) (43)(1000)/2000
(B) (43)(25)/1000
(C) 18/1000
(D) $(18-25)^2/25$
(E) $(18-25)^2/18$

3. Which of the following statements is true of chi-square distributions?

(A) As the number of degrees of freedom increases, their density curves look less and less like a normal curve.
(B) As the number of degrees of freedom increases, their density curves look more and more like a uniform distribution.
(C) Their density curves are skewed to the left.
(D) They take on only positive values.
(E) All of the above are true.

4. A student at a large high school suspects that Mr. Andreasen is grading his students too harshly. Over the past 10 years the proportions of students in *all* sections of statistics (taught by many different teachers) received grades of A, B, C, D, or F in the following proportions: A: 0.20; B: 0.30; C: 0.30; D: 0.10; and F: 0.10. An SRS of 90 students who took statistics with Mr. Andreasen in the past 10 years produces the following information:

Grade	A	B	C	D	F
Number of students	12	26	28	15	9

Which of the following conditions must be met before the student can use the χ^2 procedure in this situation?

(A) The distribution of grades in all introductory statistics courses must be approximately Normal.
(B) The number of categories is small relative to the number of observations.
(C) All the observed counts are greater than 5.
(D) Each observation was randomly selected from the population of all students.
(E) All expected counts are approximately equal.

5. Which of the following expressions represents the expected count of the grade category D?

(A) 90/5
(B) (0.10)(90)
(C) (0.1)(15)
(D) $15^2/90$
(E) $(15-9)^2/9$

6. Anne wants to know if males and females prefer different brands of frozen pizzas. She bakes four dozen pizzas made by each of four manufacturers, which she labels brands A, B, C, and D. She then selects a simple random sample of 48 students, records their gender, gives them one slice of each brand and asks which brand they like best. Here are her results:

	A	B	C	D	Total
Males	2	4	6	7	**19**
Females	11	5	6	7	**29**
Total	**13**	**9**	**12**	**14**	**48**

If we want to compare the conditional distributions for preferred pizza brand for males to the same distribution for females, which of the following is an appropriate graph to use?

(A) Parallel dotplots
(B) Back-to-back stemplots
(C) Segmented bar graphs
(D) Side-by-side bar graphs
(E) Scatterplot

7. The appropriate null hypothesis for Anne's question in this problem is:

(A) There is an association between gender and preferred frozen pizza.
(B) Gender and pizza preference are independent.
(C) The distribution of preferred pizza for each gender is different.
(D) The observed count in each cell is equal to the expected count.
(E) The males and females subjects in this experiment have the same distribution of pizza brand preference.

8. Are the conditions for a chi-square test of association/independence met?

(A) Yes, because the sample size is greater than 30.
(B) Yes, because a simple random sample was selected.
(C) No, because the distribution for each gender is different.
(D) No, because not all observed counts are greater than 5.
(E) No, because 25% of expected counts are less than 5.

9. Below is a table of individual components of the chi-square test of association/independence for a study done on amount of time spent at a computer and whether or not a person wears glasses:

		Wear Glasses?	
		Yes	**No**
Amount of Computer Screen Time	**Above Average**	8.7	6.3
	Average	0.5	0.3
	Below Average	3.1	2.2

Which of the following statements is supported by the information in this table?

(A) Above-average screen time individuals wore glasses much less often than expected.
(B) Average screen time individuals wore glasses much less often than expected.
(C) Below-average screen time individuals wore glasses about as often as expected.
(D) You can't determine this without the original observed counts.
(E) The chi-square statistic for this test is about 3.5.

10. A random sample of 200 Canadian students were asked about their hand dominance and whether they suffer from allergies. Here are the results:

		Allergies?	
		Yes	**No**
Hand Dominance	**Ambidextrous**	12	7
	Left-handed	11	9
	Right-handed	95	66

What can you conclude about the relationship between hand dominance and allergies?

(A) Using a test for association/independence, there is not enough evidence ($P = 0.13$) to conclude that there is a relationship between hand dominance and allergies.
(B) Using a test for association/independence, there is enough evidence ($P = 0.87$) to conclude that there is a relationship between hand dominance and allergies.
(C) Using a test for association/independence, there is not enough evidence ($P = 0.87$) to conclude that there is a relationship between hand dominance and allergies.
(D) Using a test for association/independence, there is not enough evidence ($P = 0.13$) to conclude that there is a relationship between hand dominance and allergies.
(E) We cannot perform a chi-square test on these data.

Check your answers below. If you got a question wrong, check to see if you made a simple mistake or if you need to study that concept more. After you check your work, identify the concepts you feel very confident about and note what you will do to learn the concepts in need of more study.

#	Answer	Concept	Right	Wrong	Simple Mistake?	Need to Study More
1	A	Degrees of Freedom				
2	D	Chi-square Components				
3	D	Chi-square Distribution Characteristics				
4	D	Conditions for Chi-square Procedures				
5	B	Expected Counts				
6	D	Conditional Distributions				
7	B	Chi-square test of Independence				
8	E	Conditions for Chi-Square Procedures				
9	D	Follow-up Analysis				
10	C	Chi-square Conclusions				

Chapter 11 Reflection

Summarize the "Big Ideas" in Chapter 11:

My strengths in this chapter:

Concepts I need to study more and what I will do to learn them:

FRAPPY! Free Response AP® Problem, Yay!

The following problem is modeled after actual Advanced Placement Statistics free response questions. Your task is to generate a complete, concise response in 15 minutes. After you generate your response, view two example solutions and determine whether you feel they are "complete", "substantial", "developing" or "minimal". If they are not "complete", what would you suggest to the student who wrote them to increase their score? Finally, you will be provided with a rubric. Score your response and note what, if anything, you would do differently to increase your own score.

A study was performed to determine whether or not the name of a course had an effect on student registrations. A statistics course in a large school district was given 4 different names in a course catalog. Each name corresponded to the exact same statistics course. A random sample of student registrations was recorded and the results are given below:

Course Name	Number of Registrations
Statistical Applications	25
Statistical Reasoning	22
Statistical Analysis	30
The Practice of Statistics	40
Total	117

Do these data suggest the name of the course has an effect on student registrations? Conduct an appropriate statistical test to support your conclusion.

FRAPPY! Student Responses

Student Response 1:

It is obvious there is a difference. If there wasn't, all of the course names would have received about 30 registrations. However, "The Practice of Statistics" was by far the most popular course name, earning 10 more registrations than expected. "Statistical Reasoning" is the least popular, earning 8 fewer registrations than expected. It appears course name does have an effect on student registrations.

How would you score this response? Is it substantial? Complete? Developing? Minimal? Is there anything this student could do to earn a better score?

Student Response 2:

We will perform a chi-square goodness-of-fit test.

H_0: the distribution of registrations is uniform.
H_a: the distribution of registrations is not uniform.

Conditions. We are told the data comes from a random sample. All observed registration counts are greater than 5. We will assume the sample is less than 10% of all registrations.

$$x^2 = \frac{(25-29.25)^2}{29.25} + \frac{(22-29.25)^2}{29.25} + \frac{(30-29.25)^2}{29.95} + \frac{(40-29.95)^2}{29.95} = 6.38$$

df = 4 − 1 =3 *P*-value = 0.09

Since the *P*-value is greater than the 5% rule of thumb, we fail to reject the null hypothesis. We do not have evidence that the name of the course has an effect on student registrations.

How would you score this response? Is it substantial? Complete? Developing? Minimal? Is there anything this student could do to earn a better score?

FRAPPY! Scoring Rubric

Use the following rubric to score your response. Each part receives a score of "Essentially Correct," "Partially Correct," or "Incorrect." When you have scored your response, reflect on your understanding of the concepts addressed in this problem. If necessary, note what you would do differently on future questions like this to increase your score.

Intent of the Question

The primary goals of this question are to assess your ability to (1) state the appropriate hypotheses; (2) identify and compute the appropriate test statistic; (3) make a conclusion in the context of the problem;

Solution

The solution should contain 4 parts:
Test and Hypotheses: The test must be identified by name or formula as a chi-square goodness-of-fit test and hypotheses must be stated appropriately. H_0: *student registrations do not differ by course name* and H_a: *student registrations do differ by course name.*
Conditions: Random sample is given. All expected counts are 29.25, which is at least 5.
Independence: we can assume that the sample is less than 10% of all registrations for the course.
Mechanics: $\chi^2 = 6.38$ and *P*-value = 0.094 with 3 degrees of freedom. Conclusion: Because the *P*-value is larger than a significance level of 5%, we do not have significant evidence to suggest the course name has an effect on the number of registrations.

Scoring:

Each element scored as essentially correct (E), partially correct (P), or incorrect (I).

Name and Hypotheses is essentially correct if the response correctly identifies the test by name or formula and writes hypotheses correctly. This part is partially correct if the test isn't identified or if the hypotheses are written incorrectly.

Conditions is essentially correct if the response correctly checks the Random, Large Sample Size, and 10% conditions. This part is partially correct if one of the conditions is not checked and incorrect if only one condition is checked.

Mechanics is essentially correct if the response correctly calculates the test statistic and *P*-value. This part is partially correct if one of the calculations is incorrect.

Conclusion is essentially correct if the response correctly fails to reject the null hypothesis because the *P*-value is greater than a significance level of 5% and provides an interpretation in context. This part is partially correct if the response fails to justify the decision by comparing the *P*-value to a significance level OR if the conclusion lacks an interpretation in context.

Scoring

This problem has four elements, each receiving an E, P, or I.

Assign one point to each E, 0.5 points to each P, and 0 points to each I.

Total the points to determine your score.

If a score falls between two whole values, consider the strength of the entire response to determine whether to round up or down.

My Score:
What I did well:
What I could improve:
What I should remember if I see a problem like this on the AP Exam:

Chapter 11: Inference for Distributions of Categorical Data

Across

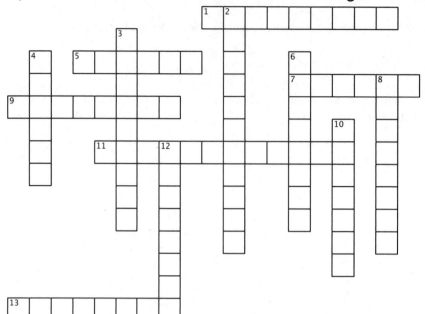

1. test statistic used to test hypotheses about distributions of categorical data (two words)
5. when comparing two or more categorical variables, or one categorical variable over multiple groups, we arrange our data in a _____ table (two words)
7. a ____ table summarizes the distribution of a single categorical variable (two words)
9. the problem of doing many comparisons at once with an overall measure of confidence is the problem of ____ comparisons
11. to test whether there is an association between two categorical variables, use a chi-square test of association / ____
13. type of categorical count gathered from a sample of data

Down

2. to test whether there is a difference in the distribution of a categorical variable over several populations or treatments, use a chi-square test of ____
3. a follow up analysis involves identifying the ____ that contribued the most to the chi-square statistic
4. when calculating chi-square statistics, observations should be expressed in ____, not percents
6. ____ of fit test: used to determine whether a population has a certain hypothesized distribution of proportions for a categorical variable
8. after conducting a chi-square test, be sure to carry out a follow-up ____
10. specific chi-square distributions are distinguished by ____ of freedom
12. type of categorical count that we would see if the null hypothesis were true

Chapter 12: More About Regression

"A judicious man looks on statistics not to get knowledge, but to save himself from having ignorance foisted on him." Thomas Carlyle

Chapter Overview

In the last chapter, you learned how to determine whether or not there was convincing evidence of a relationship between two categorical variables. In this chapter, you will learn how to determine whether or not there is convincing evidence of a relationship between two quantitative variables. You will first learn how to perform inference about the slope of a least-squares regression line. By the end of the first section, you will know how to construct and interpret a confidence interval for the slope as well as how to perform a significance test about it. In the second section, you will learn how to transform data to achieve linearity when a scatterplot shows a curved relationship between two quantitative variables. It has been a while since you studied least-squares regression. This chapter will refresh your memory on some of the key concepts of regression while introducing you to some new inference techniques. As you wrap up your studies and begin your preparations for the AP exam, it might be helpful to look back at the learning targets throughout this guide to determine what topics are in need of a little extra review!

Sections in this Chapter

Section 12.1: Inference for Linear Regression
Section 12.2: Transforming to Achieve Linearity

Plan Your Learning

Use the following *suggested* guide to help plan your reading and assignments in "The Practice of Statistics, 6th Edition." Note: your teacher may schedule a different pacing or assign different problems. Be sure to follow his or her instructions!

Read	**12.1: pp. 768 - 776**	**12.1: pp. 776 – 782**	**12.1: pp. 782 - 787**
Do	1, 3, 5	7, 9, 11	15 MC 23-28

Read	**12.2: pp. 795 – 802**	**12.2: pp 803 - 811**	**Chapter Summary**
Do	33, 35, 37, 39	43, 45, 47 MC 51-54	Multiple-Choice FRAPPY!

Section 12.1: Inference for Regression

Before You Read: Section Summary

When you construct a least-squares regression line based on sample data, your line is an approximation of the population (true) regression line. If you took a different sample and constructed a least-squares regression line, it is very likely you would end up with a slightly different line. In this section, you will learn about the sampling distribution of the sample slope *b* so that you can perform inference about the true slope of the regression line. Like you did with other inference procedures, you will start by learning how to check the conditions for performing inference. Then you will learn how to construct and interpret a confidence interval for the true slope, and how to perform a significance test about the true slope. You will also be re-introduced to computer output for a linear regression analysis. It is important that you be able to interpret this output as a number of AP exam questions involve standard computer output for regression.

Learning Targets:

____ I can check the conditions for performing inference about the slope β_1 of the population (true) regression line.

____ I can interpret the values of b_0, b_1, s, and SE_{b1} in context, and determine these values from computer output.

____ I can construct and interpret a confidence interval for the slope β_1 of the population (true) regression line.

____ I can perform a significance test about the slope β_1 of the population (true) regression line.

While You Read: Key Vocabulary and Concepts

sample regression line:
population (true) regression line:
sampling distribution of a slope:
conditions for regression inference:
t interval for the slope:
t test for the slope:

After You Read: Check for Understanding

Concept 1: The Sampling Distribution of b_1

If the conditions for performing inference are met, we can use the slope b_1 of the sample regression line $\hat{y} = b_0 + b_1x$ to estimate or test a claim about the slope β_1 of the population (true) regression line $\mu_y = \beta_0 + \beta_1x$. The required conditions are:

- **Linear**: There is a true linear relationship between the variables given by $\mu_y = \beta_0 + \beta_1x$.
  Is a scatterplot of the sample data linear?

- **Independent**: Individual observations are independent.
  If sampling without replacement, is the sample size less than 10% of the population?

- **Normal**: For each x-value, the responses *y* follow a Normal distribution.
  Does a graph of the residuals exhibit strong skewness or other signs of non-Normality?

- **Equal SD**: At each *x*-value, the standard deviation σ of the responses *y* is the same.
  Is the amount of scatter above and below 0 on the residual plot roughly the same from the smallest to largest x-values?

- **Random**: The data are produced by a random sample or randomized experiment.

If we take repeated samples of the same size *n* from a population with a true regression line $\mu_y = \beta_0 + \beta_1x$ and determine the sample regression line $\hat{y} = b_0 + b_1x$, the sampling distribution of the slope b_1 will have the following characteristics:

1) The shape of the sampling distribution will be approximately Normal.

2) The mean of the sampling distribution is $\mu_{b_1} = \beta_1$.

3) The standard deviation of the sampling distribution of b_1 is $\sigma_{b1} = \dfrac{\sigma}{\sigma_x\sqrt{n}}$

   where σ is the standard deviation of the residuals and σ_x is the standard deviation of the explanatory variable.

To perform inference about the slope, we note that the sampling distribution of the standardized slope values has a *t* distribution with $n - 2$ degrees of freedom.

Also, since we don't know σ for the true regression line, we estimate the variability of the sampling distribution of *b* with the *standard error of the slope*

$$SE_{b1} = \frac{s}{s_x\sqrt{n-1}}$$

Concept 2: Confidence Intervals for β_1

When we construct a least-squares regression line $\hat{y} = b_0 + b_1x$ and the conditions noted above are met, b_1 is our estimate of the true slope β_1. A level C confidence interval for the true slope has the form $b_1 \pm t^*SE_{b1}$ where t^* is the critical value for the t distribution with $n - 2$ degrees of freedom that has area C between $-t^*$ and t^*.

Check for Understanding: ____ *I can construct and interpret a confidence interval for the slope β_1 of the population (true) regression line.*

A study by *Consumer Reports* rated 10 randomly selected cereals on a 100 point scale (higher numbers are better) and recorded the number of grams of sugar in each serving. The data from the study are below:

Sugar	6.0	8.0	5.0	0.0	8.0	10.0	14.0	8.0	6.0	5.0
Rating	68.40	33.98	59.43	93.70	34.38	29.51	33.17	37.04	49.12	53.31

The LSRL for this data is $\hat{r} = 82.62 - 4.77(s)$ where r = rating and s = sugar.

Construct and interpret a 95% confidence interval for the true slope of the regression line. Assume the conditions for performing inference are met.

Concept 3: Significance Test for β_1

When the conditions for inference are met, not only can we estimate the true slope from b_1, we can also test whether or not a specified value for β_1 is plausible. The process for testing a claim about a population (true) slope follows the same format used for other significance tests.

1) **STATE the parameter of interest and the hypotheses you would like to test.**

   The null hypothesis states that the population (true) slope is equal to a particular value (usually zero) while the alternative hypothesis is that the population (true) slope is greater than, less than, or not equal to that value:

   $$H_0: \beta_1 = 0$$
   $$H_a: \beta_1 > 0 \text{ OR } \beta_1 < 0 \text{ OR } \beta_1 \neq 0$$

   State a significance level.

2) **PLAN: Choose the appropriate inference method and check the conditions.**

   Check the linear, independent, Normal, equal SD, and random conditions.

3) **DO: If conditions are met, calculate a test statistic and *P*-value.**

   Compute the t test statistic: $t = \dfrac{b_1 - \textbf{hypothesized slope}}{SE_{b_1}}$

   Find the *P*-value by calculating the probability of observing a t statistic at least this extreme in the direction of the alternative hypothesis in a t distribution with $n - 2$ degrees of freedom.

4) **CONCLUDE by interpreting the results of your calculations in the context of the problem.**

   If the *P*-value is smaller than the stated significance level, you can conclude that you have sufficient evidence to reject the null hypothesis. If the *P*-value is larger than the significance level, then you fail to reject the null hypothesis.

Generally, we test whether or not the true slope is zero, which would indicate no linear relationship between x and y. If you want to test a null hypothesis other than zero, get the slope and standard error from computer output and use the formula to calculate the t statistic.

Like other inference situations, a significance test can tell us whether or not a claim about the parameter is plausible, while using a confidence interval can give us additional information about its true value.

Check for Understanding: ____ *I can interpret the values of b_0, b_1, s, and SE_{b_1} in context, and determine these values from computer output.* ____ *I can perform a significance test about the slope β_1 of the population (true) regression line.*

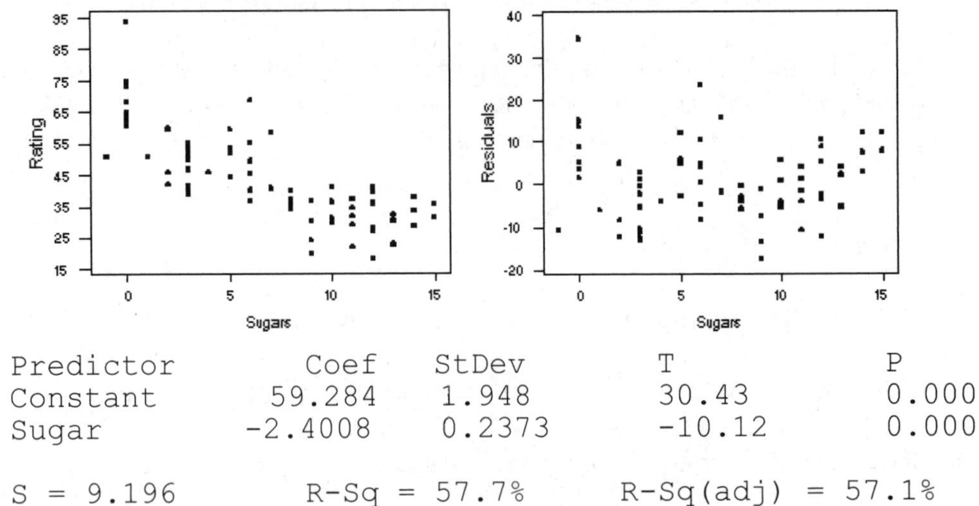

Predictor	Coef	StDev	T	P
Constant	59.284	1.948	30.43	0.000
Sugar	-2.4008	0.2373	-10.12	0.000

S = 9.196 R-Sq = 57.7% R-Sq(adj) = 57.1%

The study in the previous Check for Understanding was expanded to include a total of 77 randomly selected cereals. The scatterplot, residual plot, and computer output of the regression analysis are noted above.

Use this output to determine the LSRL for the sample data.

Interpret the slope in the context of the situation.

Is there convincing evidence that the slope of the true regression line is less than zero?

Section 12.2: Transforming to Achieve Linearity

Before You Read: Section Summary

You learned how to analyze linear relationships between two quantitative variables back in Chapter 3. In this section, you will learn how to deal with curved relationships. Since you already know how to model a linear relationship, you will learn how to transform data that show a curved relationship so that a linear model would be appropriate. That is, you will apply mathematical transformations to one or both variables to "straighten" out the scatterplot. By finding the linear model for the transformed data, you can make predictions involving the original data. The better the fit of your model, the better your prediction!

Learning Targets:

____ I can use transformations involving powers and roots to find a power model that describes the relationship between two quantitative variables, and use the model to make predictions.

____ I can use transformations involving logarithms to find a power model that describes the relationship between two quantitative variables, and use the model to make predictions.

____ I can use transformations involving logarithms to find an exponential model that describes the relationship between two quantitative variables, and use the model to make predictions.

____ I can determine which of several transformations does a better job of producing a linear relationship.

While You Read: Key Vocabulary and Concepts

transforming data:
power model:
logarithm:
exponential model:

After You Read: Check for Understanding

Concept 1: Transforming with Powers and Roots

When we know or suspect that a nonlinear relationship between two variables can be described by a model of the form $y = ax^p$, we have two strategies to transform the data to achieve linearity:

1) Raise all of the x values to the p power and plot (x^p, y)
2) Take the pth root of the y values and plot $(x, \sqrt[p]{y})$.

We can then determine the least-squares regression line for the transformed data and use this equation to make predictions about the original data.

Check for Understanding: ____ *I can use transformations involving powers and roots to find a power model that describes the relationship between two quantitative variables, and use the model to make predictions.*

The following data represent the lengths (mm) and diameters (mm) of the humerus bones of the *Moleskius Primateum* species of monkeys once thought to inhabit Northern Minnesota.

Diameter	17.6	26	31.9	38.9	45.8	51.2	58.1	64.7	66.7	80.8	82.9
Length	159.9	206.9	236.8	269.9	300.6	323.6	351.7	377.6	384.1	437.2	444.7

Previous studies suggest the diameter and length are related by a power model of the form *length* = *a*(*diameter*)$^{0.7}$. Transform the original data and use least-squares regression to find an appropriate model for the transformed data.

You discover a portion of a *Moleskius Primateum* humerus bone with a diameter of 47mm. Use your model to predict how long the entire bone was.

Concept 2: Transforming with Logarithms

In general, we don't know whether or not a power model is appropriate for describing the relationship between two quantitative variables. Some curved relationships are better summarized using an exponential model. A more efficient method for linearizing curved scatterplots involves using logarithms. To determine an appropriate model, use the following process:

1) Use logarithms to transform your data.
   - Plot (x, log y)
   - Plot (log x, log y)
2) Determine which transformation is most linear.
   - If (x, log y) is most linear, an exponential model may best describe (x, y)
   - If (log x, log y) is most linear, a power model may best describe (x, y)
3) Find the appropriate linear model.
   - If (x, log y) is most linear, find the LSRL of the transformed data $\widehat{\log y} = a + bx$
   - If (log x, log y) is most linear, find the LSRL of the transformed data

     $\widehat{\log y} = a + b(\log x)$
4) Use your model to make predictions for the original data.

Check for Understanding: ____ *I can use transformations involving logarithms to find a power model or an exponential model that describes the relationship between two quantitative variables, and use the model to make predictions.* ____ *I can determine which of several transformations does a better job of producing a linear relationship.*

The following data describe the number of police officers (thousands) and the violent crime rate (per 100,000 population) in a sample of states. Use these data to determine a model for predicting violent crime rate based on number of police officers employed. Show all appropriate plots and work.

Police	86.2	9.2	45	39.9	6	11.8	2.9	14.6	30.5	12.3	46.2	15.2	10.9
Crime	1090	559	1184	1039	303	951	132	763	635	726	840	373	523

Use your model to predict the violent crime rate for a state with 25,400 police officers.

Chapter Summary: More About Regression

In this chapter, you learned how to apply your knowledge of inference to linear relationships. When you find a least-squares regression line for a set of sample data, you are constructing a model that approximates the true relationship between x and y. By considering the sampling distribution of b, you can construct a confidence interval for the slope of the true regression line as well as test claims about the slope. In the event the relationship between two variables is curved, you can use transformations to "straighten" the scatterplot. You can find the least-squares regression line for the transformed data and make better predictions for the original relationship. The most common methods for transforming data involve taking powers, roots, or logarithms of one or both variables. To determine which transformation does a better job of "straightening" the relationship, examine residual plots.

After You Read: What Have I Learned?

Complete the vocabulary puzzle, multiple-choice questions, and FRAPPY. Check your answers and performance on each of the learning targets. Be sure to get extra help on any targets that you identify as needing more work!

Learning Target	Got It!	Almost There	Needs Some Work
I can check the conditions for performing inference about the slope β_1 of the population (true) regression line.			
I can interpret the values of b_0, b_1, s, and SE_{b_1} in context, and determine these values from computer output.			
I can construct and interpret a confidence interval for the slope β_1 of the population (true) regression line.			
I can perform a significance test about the slope β_1 of the population (true) regression line.			
I can use transformations involving powers and roots to find a power model that describes the relationship between two quantitative variables, and use the model to make predictions.			
I can use transformations involving logarithms to find a power model that describes the relationship between two quantitative variables, and use the model to make predictions.			
I can use transformations involving logarithms to find an exponential model that describes the relationship between two quantitative variables, and use the model to make predictions.			
I can determine which of several transformations does a better job of producing a linear relationship.			

Chapter 12 Multiple Choice Practice

Directions. *Identify the choice that best completes the statement or answers the question. Check your answers and note your performance when you are finished.*

1. Is it possible to predict a student's GPA in their senior year from their GPA in the first marking period of their freshman year? A random sample of 15 seniors from the graduating class of 468 students is selected and both full-year GPA in their senior year ('Senior") and first-marking-period GPA in their freshman year ("Fresh") is recorded. A computer regression analysis and a residual plot for these data are given below.

Predictor	Coef	SE Coef	T	P
Constant	1.6310	0.5328	3.06	0.009
Fresh	0.5304	0.1789	2.96	0.011

S = 0.3558 R-Sq = 40.3% R-Sq(adj) = 35.7%

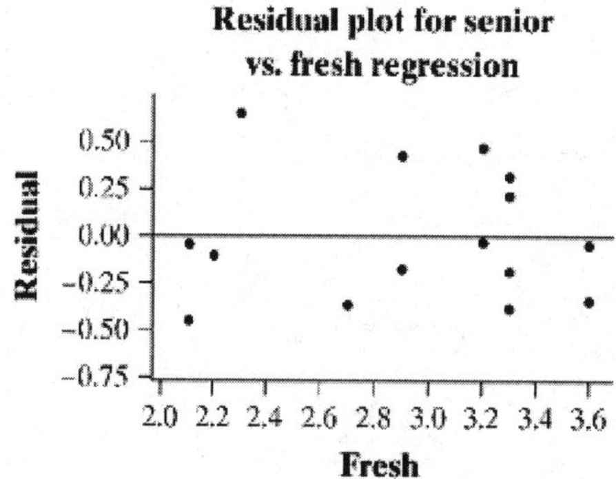

Which of the following is the estimate for the standard deviation of the sampling distribution of slopes?

(A) 0.1789
(B) 0.3558
(C) 0.5304
(D) 0.5328
(E) 1.6310

2. The equation of the least-squares regression line is

(A) $\widehat{Fresh} = 1.6310(Senior) + 0.5304$
(B) $\widehat{Fresh} = 1.6310 + 0.5304(Senior)$
(C) $\widehat{Senior} = 1.6310(Fresh) + 0.5304$
(D) $\widehat{Senior} = 1.6310 + 0.5304(Fresh)$
(E) $\widehat{Senior} = 0.5304(Fresh) + 0.1789$

3. Can we predict annual household electricity costs in a specific region from the number of rooms in the house? Below is a scatterplot of annual electricity costs (in dollars) *versus* number of rooms for 30 randomly-selected houses in Michigan, along with computer output for linear regression of electricity costs on number of rooms.

```
Predictor    Coef   SE Coef      T       P
Constant    406.9     164.8   2.47   0.020
Rooms       58.45     24.77   2.36   0.026

S = 246.735 R-Sq = 16.6% R-Sq(adj) = 13.6%
```

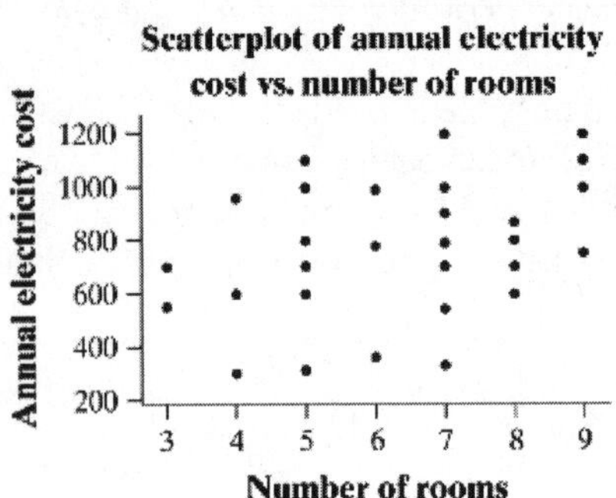

Assume the conditions for inference have been met. If we test the hypotheses H_0: β = 0 vs. H_a: β > 0 at the α = 0.05 level. Which of the following is the appropriate conclusion?

(A) Since the *P*-value of 0.020 is less than α, we reject H_0. There is convincing evidence of a linear relationship between annual electricity costs and number of rooms in the population of Michigan homes.
(B) Since the *P*-value of 0.020 is greater than α we fail to reject H_0. We do not have enough evidence to conclude that there is a linear relationship between annual electricity costs and number of rooms in the population of Michigan homes.
(C) Since the *P*-value of 0.026 is greater than α, we accept H_0. We have convincing evidence that there is not a linear relationship between annual electricity costs and number of rooms in the population of Michigan homes.
(D) Since the *P*-value of 0.026 is less than α, we accept H_0. We have convincing evidence that there is not a linear relationship between annual electricity costs and number of rooms in the population of Michigan homes.
(E) Since the *P*-value of 0.026 is less than α, we reject H_0. We have convincing evidence of a linear relationship between annual electricity costs and number of rooms in the population of Michigan homes.

4. Are high school students who like their English class more likely to enjoy their history class as well? Here is a residual plot for 30 randomly-selected students who were asked to rate how much they liked both English and history on a 0 to 5 scale (a higher rating means the student liked the subject more). [Data from 2004-5 Census at Schools survey in Canada.]

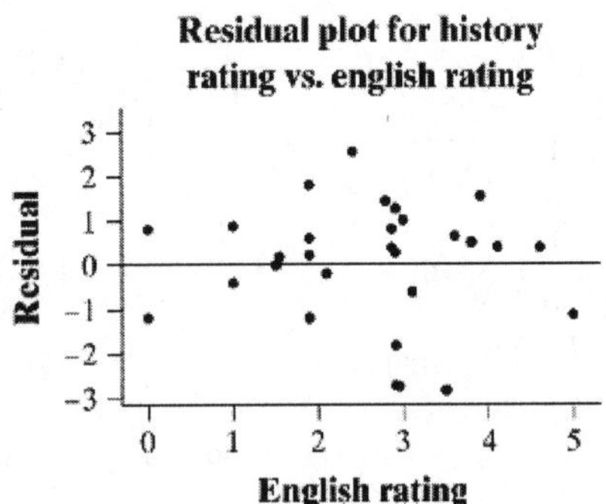

Which of the following conditions for inference does the residual plot suggest has not been satisfied?

(A) The data come from a random sample.
(B) Observations for each student are independent.
(C) The variance of residuals is roughly equal for each value of English rating.
(D) For each value of English rating, the distribution of history rating is roughly Normal.
(E) Mean History rating is a linear function of English rating.

5. Consider the output for the regression analysis on the situation from question 4.

```
Predictor     Coef   SE Coef       T      P
Constant    1.1867    0.5574    2.13  0.042
English     0.5254    0.1995    2.63  0.014

S = 1.37707   R-Sq = 19.8%   R-Sq(adj) = 17.0%
```

Assume the conditions for regression inference have been satisfied. What does the quantity R-Sq = 19.8% represent?

(A) The correlation of history rating and English rating—a measure of the strength of the linear relationship between the two variables.
(B) The average deviation of observed history ratings from the predicted history ratings, expressed as a percentage of the predicted history rating.
(C) The average deviation of observed English ratings from the predicted English ratings, expressed as a percentage of the predicted English rating.
(D) The percentage of variation in history rating that can be explained by the regression of history rating on English rating.
(E) The LSRL is accurately predicts history rating 19.8% of the time.

6. Which of the following is the 95% confidence interval for the population slope?

(A) 1.1867 ± 1.960(0.5574)
(B) 1.1867 ± 2.048(0.5574)
(C) 0.5254 ± 1.960(0.1995)
(D) 0.5254 ± 2.048(0.1995)
(E) 0.5254 ± 2.630(1.37707)

7. Suppose we measure a response variable *Y* for several values of an explanatory variable *X*. A scatterplot of log *Y* versus log *X* looks approximately like a negatively-sloping straight line. We may conclude that

(A) the rate of growth of *Y* is positive, but slowing down over time.
(B) an exponential growth model would approximately describe the relationship between *Y* and *X*.
(C) a power model would approximately describe the relationship between *Y* and *X*.
(D) the relationship between *Y* and *X* is a positively-sloping straight line.
(E) the residual plot of the regression of log *Y* on log *X* would have a "U-shaped" pattern suggesting a non-linear relationship.

8. Suppose the relationship between a response variable *y* and an explanatory variable *x* is modeled well by the equation $y = 3.6(0.32)^x$. Which of the following plots is most likely to be roughly linear?

(A) A plot of *y* against *x*
(B) A plot of *y* against log *x*
(C) A plot of log *y* against *x*
(D) A plot of 10^y against *x*
(E) A plot of log *y* against log *x*

9. Use of the Internet worldwide increased steadily from 1990 to 2002. A scatterplot of this growth shows a strongly non-linear pattern. However, a scatterplot of *ln* Internet Users *versus* Year is much closer to linear. Below is a computer regression analysis of the transformed data (note that natural logarithms are used).

```
Predictor      Coef   SE Coef        T      P
Constant    -951.10     43.45   -21.89  0.000
Year         0.4785   0.02176    21.99  0.000

S = 0.2516   R-Sq = 98.2%   R-Sq(adj) = 98.0%
```

Which of the following best describe the model that is given by this computer printout?

(A) The linear model: $\widehat{users} = -951.10 + 0.4785(year)$

(B) The power model: $\widehat{users} = e^{-951.10}(year)^{0.4785}$

(C) The power model: $\widehat{users} = 10^{-951.10}(year)^{0.4785}$

(D) The exponential model: $\widehat{users} = e^{-951.10}(e^{0.4785})^{year}$

(E) The exponential model: $\widehat{users} = 10^{-951.10}(10^{0.4785})^{year}$

10. Like most animals, small marine crustaceans are not able to digest all the food they eat. Moreover, the percentage of food eaten that is assimilated (that is, digested) decreases as the amount of food eaten increases. A residual plot for the regression of Assimilation rate (as a percentage of food intake) on Food Intake (in g/day) is shown below.

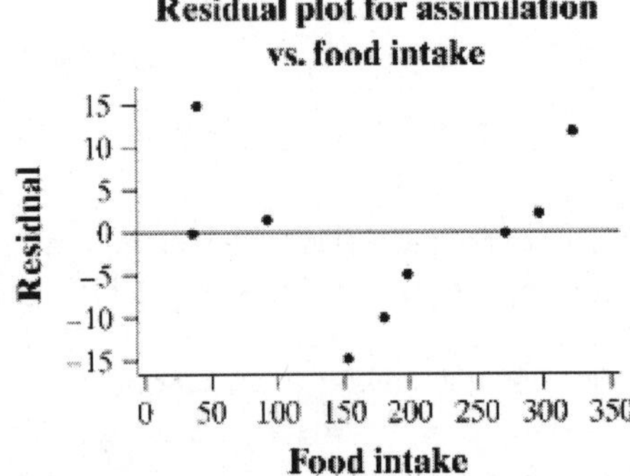

A scatterplot of ln Assimilation *versus* ln Food Intake is strongly linear, suggesting that a linear regression of these transformed variables may be more appropriate. Below is a computer regression analysis of the transformed data (note that natural logarithms are used).

```
Predictor          Coef   SE Coef       T      P
Constant         6.3324    0.5218   12.14  0.000
ln Food Intake  -0.6513    0.1047   -6.22  0.000

S = 0.247460   R-Sq = 84.7%   R-Sq(adj) = 82.5%
```

When food intake is 250 g/day, what is the predicted assimilation rate from this model?

(A) 2.7%
(B) 15.4%
(C) 27.4%
(D) 34.3%
(E) 54.4%

Check your answers below. If you got a question wrong, check to see if you made a simple mistake or if you need to study that concept more. After you check your work, identify the concepts you feel very confident about and note what you will do to learn the concepts in need of more study.

#	Answer	Concept	Right	Wrong	Simple Mistake?	Need to Study More
1	A	SE of Slope from Output				
2	D	Regression Equation from Output				
3	E	Significance Test from Output				
4	C	Conditions for Inference				
5	D	Confidence Interval for Slope				
6	D	Conclusion from Significance Test				
7	C	Interpreting log-log Scatterplot				
8	C	Exponential Functions/Transformations				
9	D	Semi-log Transformation				
10	B	Prediction from log-log Transformation				

Chapter 12 Reflection

Summarize the "Big Ideas" in Chapter 12:

My strengths in this chapter:

Concepts I need to study more and what I will do to learn them:

FRAPPY! Free Response AP® Problem, Yay!

The following problem is modeled after actual Advanced Placement Statistics free response questions. Your task is to generate a complete, concise response in 15 minutes. After you generate your response, view two example solutions and determine whether you feel they are "complete", "substantial", "developing" or "minimal". If they are not "complete", what would you suggest to the student who wrote them to increase their score? Finally, you will be provided with a rubric. Score your response and note what, if anything, you would do differently to increase your own score.

Paul is interested in purchasing a digital camera and notices that as each model's image quality (in megapixels) increases, the cost appears to increase linearly. A scatterplot and regression output of the megapixels vs. cost for seven randomly chosen camera models is below:

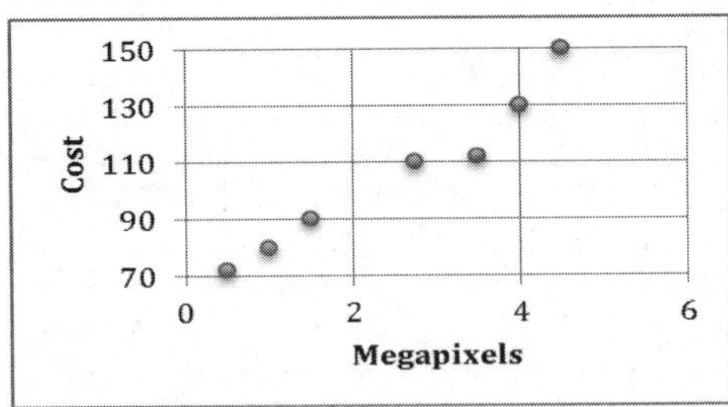


```
Regression Analysis: Cost versus Megapixels

Predictor      Coef       SE Coef     T       P
Constant      63.457      2.387      26.58   0.000
Speed         16.2809     0.8192     19.88   0.000

S = 3.087    R-Sq = 98.7%        R-Sq (adj) = 98.5%
```

a) Using the regression output, write the equation of the fitted regression line in context.

b) Interpret the slope and y-intercept in the context of the problem.

c) Construct and interpret a 98% confidence interval for the slope. Assume the conditions for inference have been met.

FRAPPY! Student Responses

Student Response 1:

a) y = 63.457 + 16.2809x

b) The intercept of 63.457 means a camera with 0 megapixels will cost about 63 dollars. This doesn't really make any sense. The slope of 16.28 means for each increase of one megapixel, the cost goes up $16.28.

c) 98% CI for slope: $b \pm t^* SE_b$
16.2809 ± 3.365 (0.8192) = (13.52, 19.03)

We are 98% confident the true slope falls between 13.52 and 19.03.

How would you score this response? Is it substantial? Complete? Developing? Minimal? Is there anything this student could do to earn a better score?

Student Response 2:

a) $\widehat{cost} = 63.453 + 16.2809(megapixels)$

b) The intercept doesn't make sense in terms of the problem as it states a camera with NO megapixels would be predicted to cost $63.46. The slope tells us that for each increase of 1 megapixel in picture quality, we predict the cost will increase by approximately $16.28 on average.

c) Since the conditions are met, we can construct a 98% t-interval for the true slope.

16.2809 ± 3.365 (0.8192) = (13.52, 19.03)

We are 98% confident the interval from 13.52 to 19.03 captures the true slope of the relationship between megapixels and cost. That is, we are 98% confident for each increase of one megapixel, the predicted cost of a camera will increase between $13.52 and $19.03.

How would you score this response? Is it substantial? Complete? Developing? Minimal? Is there anything this student could do to earn a better score?

FRAPPY! Scoring Rubric

Use the following rubric to score your response. Each part receives a score of "Essentially Correct," "Partially Correct," or "Incorrect." When you have scored your response, reflect on your understanding of the concepts addressed in this problem. If necessary, note what you would do differently on future questions like this to increase your score.

Intent of the Question

The primary goals of this question are to assess your ability to (1) interpret standard computer output; (2) interpret a linear model in context; (3) construct and interpret a confidence interval for the slope of a regression line;

Solution

a) $\widehat{cost} = 63.453 + 16.2809(megapixels)$

b) Intercept = 63.457. This is the predicted cost when megapixels = 0.
Slope = 16.2809. We predict a cost increase of $16.28 for each increase of 1 megapixel.

c) 98% Confidence Interval: 16.2809 ± 3.365(0.8192) = (13.52, 19.03). We are 98% confident the interval from 13.52 to 19.03 captures the true slope of the regression line for megapixels and cost.

Scoring:

Each element scored as essentially correct (E), partially correct (P), or incorrect (I).

a) Essentially correct if the equation is properly stated and variables are defined in the context of the problem. Partially correct if the equation is properly stated, but the variables are not defined or left as x and y.

b) Essentially correct if both the intercept and slope are identified and interpreted correctly. Partially correct if only one is properly identified and interpreted. Note, the slope must indicate a *predicted* increase.

c) Essentially correct if a correct interval is constructed (with formula shown) and interpreted in the context of the problem. Partially correct if supporting work is not provided or if the interpretation is incorrect or lacks context.

4 Complete Response
All three parts essentially correct

3 Substantial Response
Two parts essentially correct and one part partially correct

2 Developing Response
Two parts essentially correct and no parts partially correct
One part essentially correct and two parts partially correct
Three parts partially correct

1 Minimal Response
One part essentially correct and one part partially correct
One part essentially correct and no parts partially correct
No parts essentially correct and two parts partially correct

My Score:
What I did well:
What I could improve:
What I should remember if I see a problem like this on the AP Exam:

Chapter 12: More About Regression

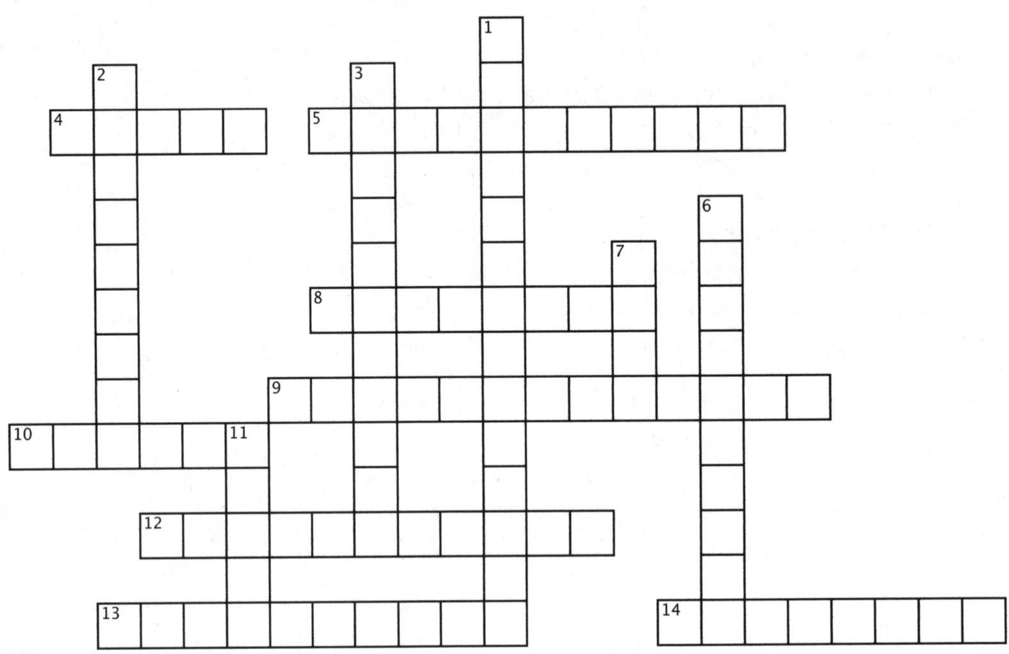

Across

4. if (ln(x), ln(y)) is linear, a/an ____ model may be most appropriate for (x,y)
5. another name for the x-variable
8. another name for the y variable
9. the coefficient of ____ indicates the fraction of variability in predicted y values that can be explained by the LSRL of y on x
10. when we construct a LSRL for a set of observed data, we construct a ____ regression line
12. the _____ coefficient is a measure of the strength of the linear relationship between two quantitative variables
13. the line that describes the relationship between an explanatory and response variable
14. observed y - predicted y

Down

1. when data displays a curved relationship, we can perform a ____ to "straighten" it out
2. the common mathematical transformation used in this chapter to achieve linearity for a relationship between two quantitative variables
3. if (x, ln(y)) is linear, a/an ____ model may be most appropriate for (x,y)
6. to estimate the population (true) slope, we can construct a ____ interval
7. the Greek letter we use to represent the slope of the population (true) regression line
11. to calculate a confidence interval, we must use the standard ____ of the slope

Solutions

Chapter 1: Solutions

Introduction Concept 2:

____ *I can identify individuals and variables in a set of data.*

1) This data set describes the students in Mr. Buckley's class (James, Jen, ..., Sharon)

2) Quantitative: ACT Score, GPA Categorical: Gender, Favorite Subject

3) The ACT scores range from 28 to 35 with a center around 32. Four of the scores are "bunched up" in the 32-35 range, while Jonathan's score of 28 doesn't quite fit with the rest of the scores.

4) It appears the students who prefer math and science had higher scores on the ACT. However, we can not infer there is a large difference between the scores of these students and those who do not prefer math and science.

Section 1.1 Concept 1:

____ *I can make and interpret bar graphs for categorical data.*

1) It appears the majority of individuals prefer Goodbye Blue Monday and One Mean Bean. Very few people prefer the national chain.

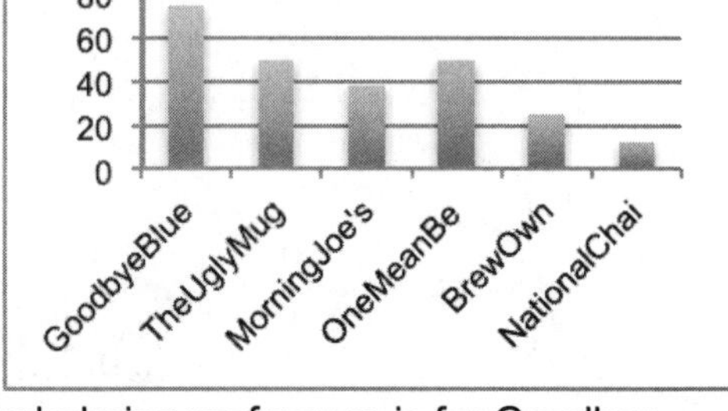

2) The pie chart does not just show actual counts. Rather, it displays the percent of the total that preferred each coffee-shop. It is clear, again, from this chart that the overwhelming preference is for Goodbye Blue Monday, One Mean Bean, and The Ugly Mug with 70% of residents polled preferring them.

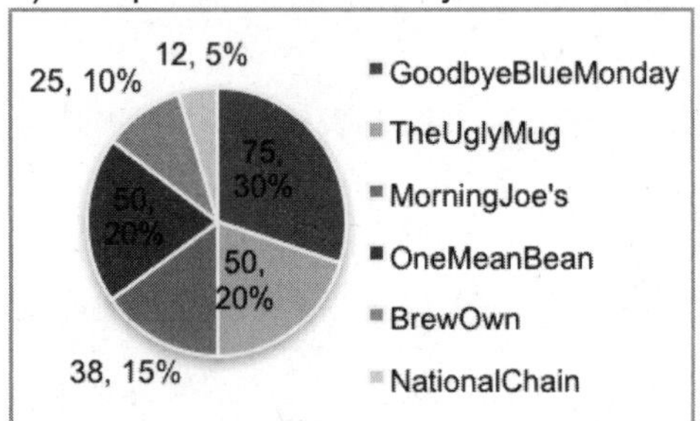

Section 1.1 Concept 2:

____ *I can describe the nature of the association between two categorical variables.*

We suspect gender might influence coffee preference, so we'll compare the conditional distributions of coffee preference for men alone and women alone.

Preference	Male	Female
National Chain	17.8%	14.0%
One Mean Bean	2.8%	18.3%
The Ugly Mug	27.1%	5.4%
Goodbye Blue Monday	31.8%	19.4%
Home-brewed	18.7%	34.4%
Don't drink coffee	1.9%	8.6%

We'll make a side-by-side bar graph to compare the preferences of males and females.

Based on the sample, it appears men prefer to get their coffee from The Ugly Mug and Goodbye Blue Monday while women are more likely (34.4% vs 18.7%) to home brew their coffee.

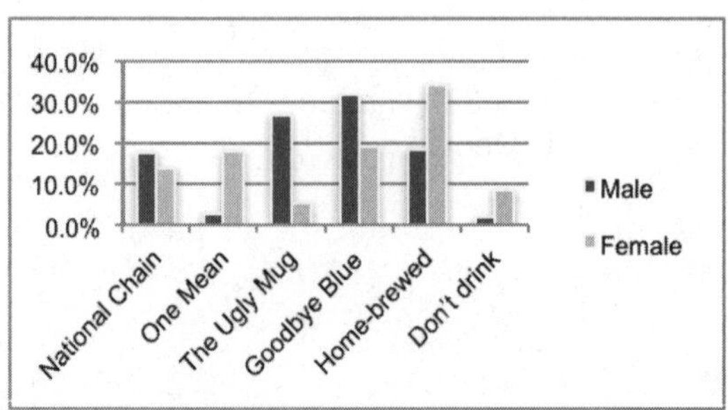

Section 1.2 Concept 1:

____ *I can make and interpret dotplots and stemplots.*

1) The distribution of gas mileage appears to be fairly symmetric, centered at about 37 mpg, and ranging from 31.8 to 41 mpg. There do not appear to be any extreme values.

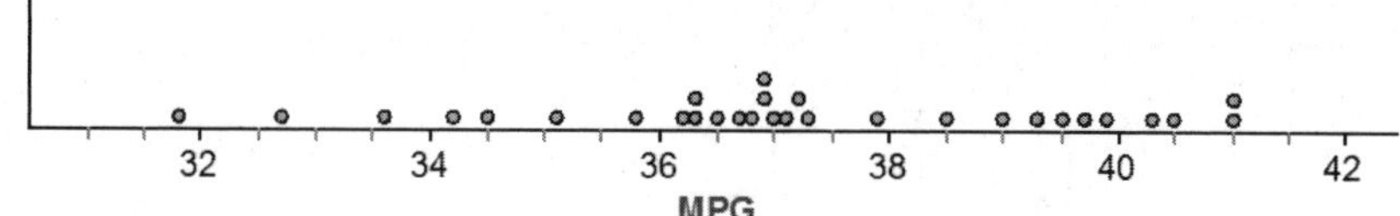

2) Stemplot:

```
31 | 8
32 | 7
33 | 6
34 | 2 5
35 | 1
36 | 2 3 3 5 7 8 9 9
37 | 0 1 2 3 9
38 | 5
39 | 0 3 5 7 9
40 | 3 5
41 | 0 0
```

Section 1.2 Concept 2:

____ *I can make and interpret a histogram.*

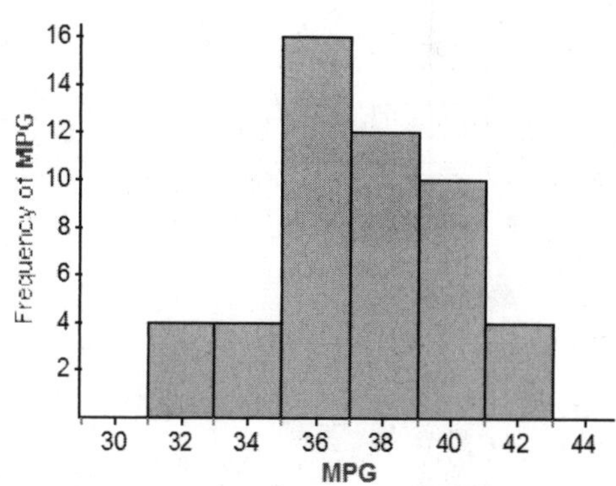

(a) Histograms may vary depending on "bin widths" and starting values. The distribution of mileage for the 50 cars is mound shaped, with a slight skew to the right. The center of the distribution is approximately 37mpg with a range from 31 to 42. There do not appear to be any extreme values.

(b) To get your calculator to match the histogram you constructed by hand you will need to adjust the Ymin and Yscl in window settings.

Section 1.3 Concept 2:

____ *I can calculate and interpret measures of center.*

1) The plot suggests the distribution is slightly skewed to the right. Therefore, the mean will probably be greater than the median since it tends to get pulled in the direction of the tail.

2) The average amount of time necessary to complete the logic puzzle was 32.67 seconds.

3) Median = 29. This is the "middle" time in the distribution. Half of the students took longer than 29 seconds to complete the puzzle and half took less than 29 seconds.

4) Since there is some skewness, the median would be a better measure as it is not affected by the tail or extreme values. It is a more accurate measure of center for this distribution.

Section 1.3 Concept 3:

____ *I can calculate and interpret measures of variability.*

(a) The distribution of book lengths (in pages) is fairly symmetric centered at about 330 with outliers at the minimum of 170 and 242, Q1 = 314, Median = 330, Q3 = 344, and maximum of 374.

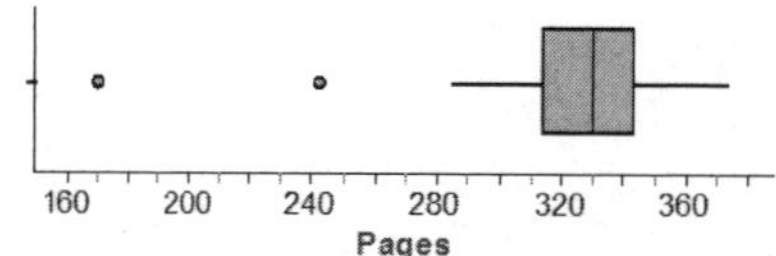

(b) These data have a mean of 316.27 pages and standard deviation 51.63 pages. Anne's favorite books tend to be within about 52 pages of 316, on average.

Chapter 1: Exploring Data

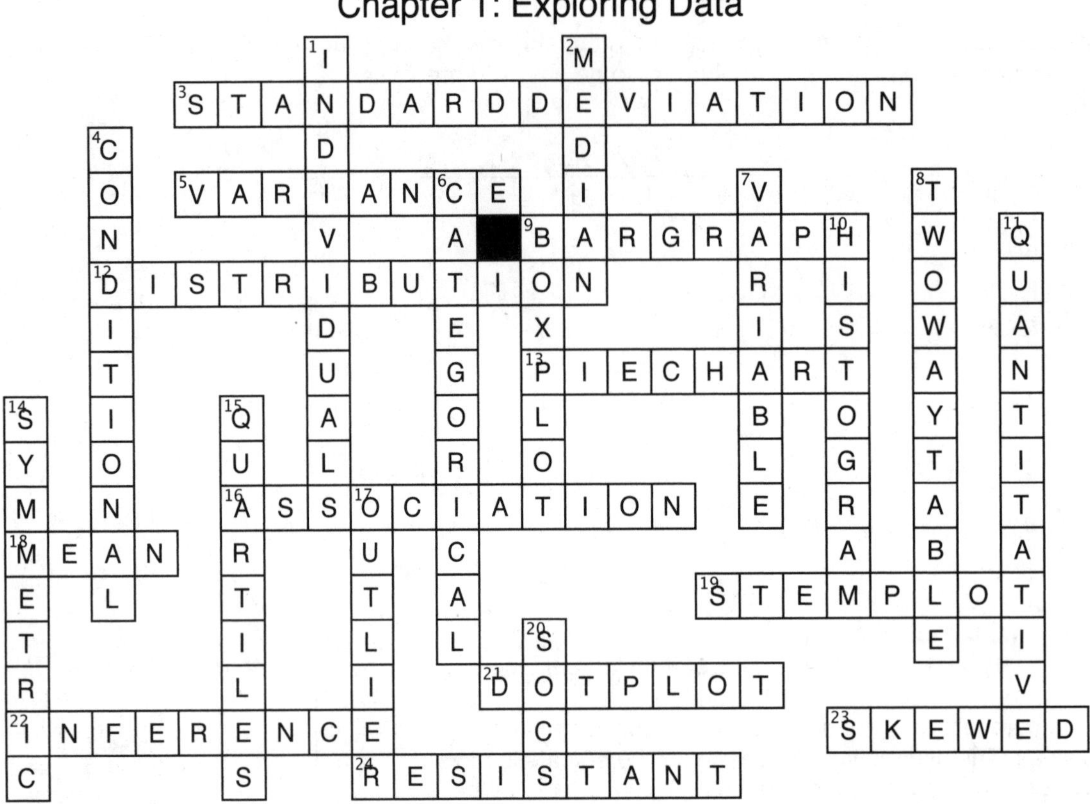

Across

3. The average distance of observations from their mean (two words) [STANDARDDEVIATION]
5. The average squared distance of the observations from their mean [VARIANCE]
9. Displays the counts or percents of categories in a categorical variable through differing heights of bars [BARGRAPH]
12. Tells you what values a variable takes and how often it takes these values [DISTRIBUTION]
13. Displays a categorical variable using slices sized by the counts or percents for the categories [PIECHART]
16. When specific values of one variable tend to occur in common with specific values of another [ASSOCIATION]
18. A measure of center, also called the average [MEAN]
19. A graphical display of quantitative data that involves splitting the individual values into two components [STEMPLOT]
21. One of the simplest graphs to construct when dealing with a small set of quantitative data [DOTPLOT]
22. Drawing conclusions beyond the data at hand [INFERENCE]
23. The shape of a distribution if one side of the graph is much longer than the other [SKEWED]
24. What we call a measure that is relatively unaffected by extreme observations [RESISTANT]

Down

1. The objects described by a set of data [INDIVIDUALS]
2. The midpoint of a distribution of quantitative data [MEDIAN]
4. A _____ distribution describes the distribution of values of a categorical variable among individuals who have a specific value of another variable. [CONDITIONAL]
6. A variable that places an individual into one of several groups or categories [CATEGORICAL]
7. A characteristic of an individual that can take different values for different individuals [VARIABLE]
8. When comparing two categorical variables, we can orgainze the data in a ___-___ _____. [TWOWAYTABLE]
9. A graphical display of the five-number summary [BOXPLOT]
10. A graphical display of quantitative data that shows the frequency of values in intervals by using bars [HISTOGRAM]
11. A variable that takes numerical values for which it makes sense to find an average [QUANTITATIVE]
14. The shape of a distribution whose right and left sides are approximate mirror images of each other [SYMMETRIC]
15. These values lie one-quarter, one-half, and three-quarters of the way up the list of quantitative data [QUARTILES]
17. A value that is at least 1.5 IQRs above the third quartile or below the first quartile [OUTLIER]
20. When exploring data, don't forget your ____ [SOCS]

Chapter 2: Solutions

Section 2.1 Concept 1:

____ *I can find and interpret the percentile of an individual value in a distribution.*

____ *I can estimate percentiles and individual values using a cumulative relative frequency graph.*

1) James earned a score of 33. His score falls above 17 of the 23 students. Approximately 78% of students scored at or below 33.

2) Heather earned a score of 12. Her score falls above 3 of the 23 students. Approximately 17% of students scored at or below 12.

3) Using the ogive, a score of 38 is at about the 80th percentile.

4) Using the ogive, Q1 is about 14, the median is about 25, and Q3 is about 30.

Section 2.1 Concept 2:

____ *I can find and interpret the standardized score (z-score) of an individual value in a distribution of data.*

1) Mean = 25.04 Standard Deviation = 14.20

2) $z = \frac{45 - 25.04}{14.2} = 1.41$ Paul's score falls 1.41 standard deviations above the mean score.

3) $z = \frac{2 - 25.04}{14.2} = -1.62$ Carl's score falls 1.62 standard deviations below the mean score.

4) First quiz: $z = \frac{45 - 25.04}{14.2} = 1.41$ Next quiz: $z = \frac{47 - 32}{6.5} = 2.31$

Paul's z-score is greater for the next quiz. He scored better relative to the mean on this quiz.

Section 2.1 Concept 3:

____ *I can describe the effect of adding, subtracting, multiplying by, or dividing by a constant on the shape, center, and variability of a distribution of data.*

1) The distribution of scores on Ms. Blockhus' quiz is roughly symmetric, centered at a mean of 25.04. The scores have a minimum of 0 and a maximum of 50 (range = 50), with a standard deviation of 14.2.

2) If Ms. Blockhus added 5 points to each score, the new distribution would still be symmetric. The mean score would increase by 5 to 30.04, and the variability would remain the same, with a standard deviation of 14.2.

3) If Ms. Blockhus doubled the scores, the new distribution would still be symmetric. The mean score would double to 50.08, and the variability would increase. The standard deviation would double to 28.4 and the range would double to 100.

Section 2.2 Concept 1:

____ *I can use a density curve to model distributions of quantitative data.*

1) The total area under the curve is 1.

2) 12% of scores fell between 28 and 40?

3) Since this curve is skewed left, the mean will be less than the median. The median probably falls around 50, while the mean is likely to be between 40 and 50. It is difficult to tell without the actual values, though.

Section 2.2 Concept 3:

____ *I can find the proportion of values in a specified interval in a Normal distribution using Table A or technology.*

1)

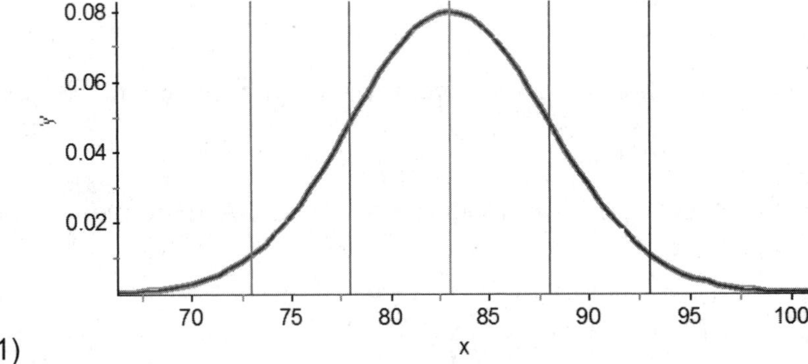

2) $P(78<x<83) = 0.34$ $P(83<x<93) = .475$ $P(78<x<93) = 0.34 + 0.475 = 81.5\%$

3) $z=\frac{90-83}{5}=1.4$ $P(z>1.40) = 1 - 0.9192 = 0.0808$. 8.08% of quizzes earn an “A”.

4) $z=\frac{71-83}{5}=-2.4$ $z=\frac{95-83}{5}=2.4$ -> 0.9918 – 0.0082 = 0.9836. 98.36% of scores would fall between 71 and 95.

5) The 20th percentile corresponds to a *z*-score of about -0.84. x = 83 - 0.84(5) = 78.8.

Section 2.2 Concept 4:

____ *I can determine whether a distribution of data is approximately Normal from graphical and numerical evidence.*

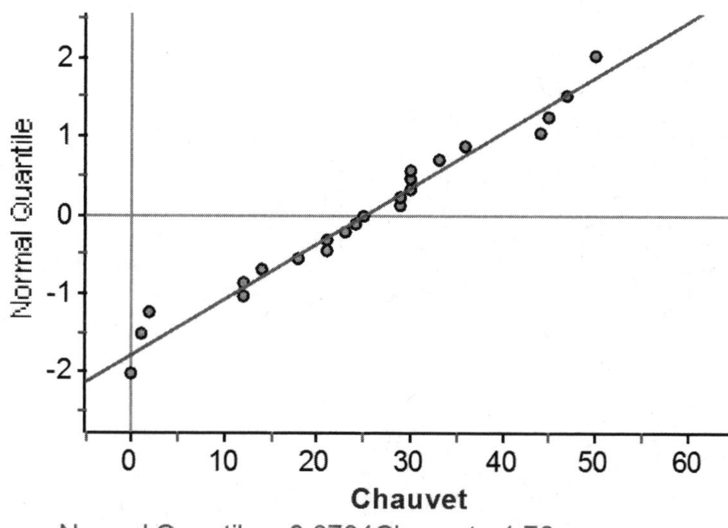

The quickest way to assess Normality is to construct a Normal probability plot. If the plot displays a linear pattern, the data can be assumed to be approximately Normal. It appears the quiz scores for Ms. Chauvet's class are approximately Normally distributed. You could also justify this by counting how many scores fall within one, two, and three standard deviations of the mean and determining whether or not that reflects the 68-95-99.7 Rule. Keep in mind that you will rarely find a distribution that is perfectly Normal!

Chapter 2: Modeling Distributions of Data

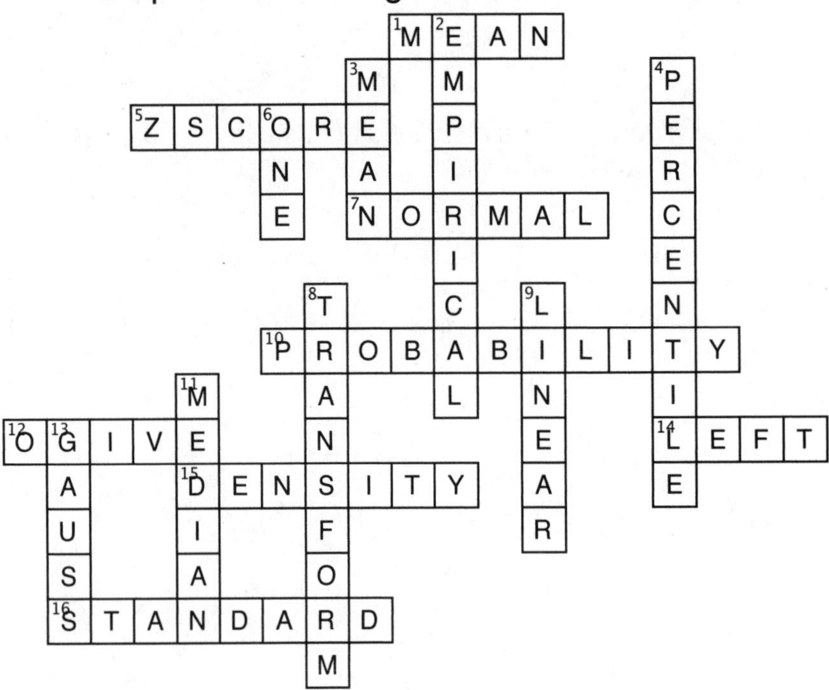

Across

1. The balance point of a density curve, if it were made of solid material [MEAN]
5. The standardized value of an observation [ZSCORE]
7. These common density curves are symmetric and bell-shaped [NORMAL]
10. A Normal ______ plot provides a good assessment of whether a data set is approximately Normally distributed [PROBABILITY]
12. Another name for a cumulative relative frequency graph [OGIVE]
14. The standard Normal table tells us the area under the standard Normal curve to the ___ of z [LEFT]
15. A ___ curve is a smooth curve that can be used to model a distribution [DENSITY]
16. This Normal distribution has mean 0 and standard deviation 1 [STANDARD]

Down

2. The ____ rule is also known as the 68-95-99.7 rule for Normal distributions [EMPIRICAL]
3. To standardize a value, subtract the ___ and divide by the standard deviation [MEAN]
4. The value with p percent of the observations less than it [PERCENTILE]
6. The area under any density curve is always equal to [ONE]
8. We ___ data when we change each value by adding a constant and/or multiplying by a constant. [TRANSFORM]
9. If a Normal probability plot shows a _____ pattern, the data are approximately Normal [LINEAR]
11. The point that divides the area under a density curve in half [MEDIAN]
13. This mathematician first applied Normal curves to data to errors made by astronomers and surveyors [GAUSS]

Chapter 3: Solutions

Section 3.1 Concept 1:

_____ *I can distinguish between explanatory and response variables for quantitative data.*

Explanatory: test anxiety Response: performance on the test
Explanatory: volume of hippocampus Response: verbal retention

Section 3.1 Concept 2:

_____ *I can make a scatterplot to display the relationship between two quantitative variables.*

_____ *I can describe the direction, form, and strength of a relationship displayed in a scatterplot and identify unusual features..*

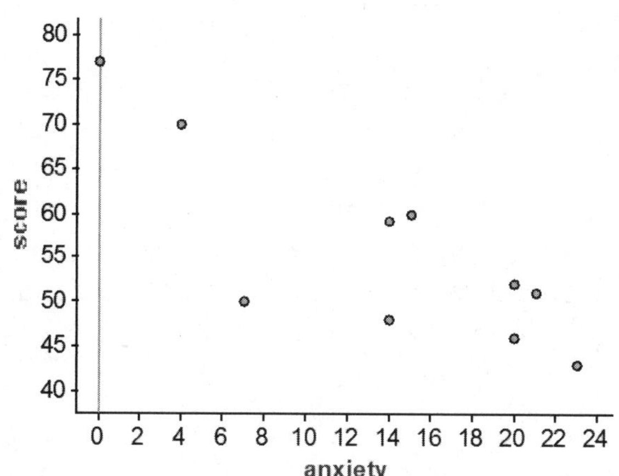

The scatterplot suggests a moderate, negative, linear relationship between anxiety and exam performance. As anxiety increases, exam performance decreases.

Section 3.1 Concept 3:

_____ *I can interpret the correlation.*

r = -0.8185 There is a strong, negative, linear relationship between anxiety and test score.

Section 3.2 Concept 3:

_____ *I can make predictions using regression lines, keeping in mind the dangers of extrapolation.*

_____ *I can calculate and interpret a residual*

_____ *I can interpret the slope and y intercept of a regression line.*

_____ *I can determine the equation of a least-squares regression line using technology or computer output.*

a) predicted score = 71.3 -1.14(anxiety)

b) slope = -1.14. For each increase of one unit in anxiety scores, we predict about a 1.14 point decrease in exam performance.

c) predicted score = 71.3 -1.14(15) = 54.2.

d) residual = actual – predicted = 60 – 54.2 = 5.8.

e) No, 35 is far greater than the maximum observed anxiety score. Extrapolation such as this could result in a poor prediction as we do not know if the linear model applies beyond the observed data. (For all we know, the exam scores could actually start to increase!)

Section 3.2 Concept 4:

____ *I can construct and interpret residual plots to assess whether a regression model is appropriate.*

____ *I can interpret the standard deviation of the residuals and r^2 and use these values to assess how well a least-squares regression line models the relationship between two variables.*

1) The residual plot does not show any obvious pattern. It appears the linear model may be appropriate for making predictions of exam scores.

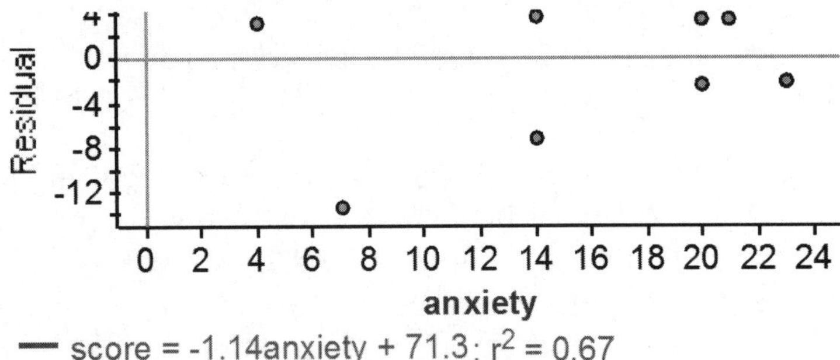

2) $r = 0.8185$ r^2=0.67 There is a strong, negative, linear relationship between anxiety score and exam score. Approximately 67% of the variability in predicted exam scores can be explained by the least squares regression model of anxiety score on exam score.

Section 3.2 Concept 5:

____ *I can determine the equation of a least-squares regression line using technology or computer output.*

1) predicted algae level = 42.8477 + 0.4762(temperature)

2) Slope = 0.4762. For each increase of 1 degree F, we predict an increase of 0.4762 parts per million in algae level.

3) $r = \sqrt{0.917} = 0.957$. There is a strong, positive, linear relationship between the temperature and algae level.

4) s = 0.4224. The predicted algae levels differ from the observed levels by an average of about 0.4224 units.

Chapter 3: Describing Relationships

Across

2. the difference between an observed value of the response and the value predicted by a regression line [RESIDUAL]
7. Important note: Association does not imply ______. [CAUSATION]
10. graphical display of the relationship between two quantitative variables [SCATTERPLOT]
11. line that describes the relationship between two quantitative variables [REGRESSION]
14. the coefficient of _______ describes the fraction of variability in y values that is explained by least squares regression on x. [DETERMINATION]
15. A _____ association is defined when above average values of one variable are accompanied by below average values of the other. [NEGATIVE]
16. individual points that substantially change the correlation or slope of the regression line [INFLUENTIAL]

Down

1. the use of a regression line to make a prediction far outside the observed x values [EXTRAPOLATION]
3. the amount by which y is predicted to change when x increases by one unit [SLOPE]
4. The ______ of a relationship in a scatterplot is determined by how closely the point follow a clear form. [STRENGTH]
5. the ____-_____ regression line is also known as the line of best fit (2 words) [LEASTSQUARES]
6. an individual value that falls outside the overall pattern of the relationship [OUTLIER]
7. value that measures the strength of the linear relationship between two quantitative variables [CORRELATION]
8. A ______ association is defined when above average values of the explanatory are accompanied by above average values of the response [POSITIVE]
9. y-hat is the _______ value of the y-variable for a given x [PREDICTED]
11. variable that measures the outcome of a study [RESPONSE]
12. variable that may help explain or influence changes in another variable [EXPLANATORY]
13. The _____ of a scatterplot indicates a positive or negative association between the variables. [DIRECTION]
17. The ____ of a scatterplot is usually linear or nonlinear. [FORM]

Chapter 4: Solutions

Section 4.1 Concept 1:

____ *I can identify the population and sample in a statistical study.*

The population is all sentences and words used in the popular Algebra 1 textbooks. The sample consists of the sentences and words in the 10 randomly selected paragraphs.

Section 4.1 Concept 2:

____ *I can describe how to select a simple random sample with technology or a table of random digits.*

Assign each student a two-digit label 01-23. Read two-digit blocks across the random number table until 4 of the labels are selected, ignoring repeats and labels from 24-00.
The four selected are 19: Rohnkol, 22: Wilcock, 05: Buckley, 13: Lacey

Section 4.1 Concept 3:

____ *I can explain how undercoverage, nonresponse, question wording, and other aspects of a sample survey can lead to bias*

A practical problem with this survey is that recent graduates may not give truthful answers about whether or not they cheated on an exam their senior year. The actual proportion of students who cheated on at least one exam their senior year is most likely higher than the 64 percent observed in the study.

Section 4.2 Concept 1:

____ *I can explain the concept of confounding and how it limits the ability to make cause-and effect conclusions.*

1. This is an observational study since no treatment was imposed on the subjects. Test scores and shoe size were observed. No effort was made to influence either variable.
2. The explanatory variable would be the shoe size and response would be test score.
3. One possible confounding variable could be age of the students. Older students would have bigger feet, on average, and would most likely have higher test scores.

Section 4.2 Concept 2:

____ *I can identify experimental units and treatments in an experiment.*
____ *I can describe a completely randomized design for an experiment*

1. Experimental units: Mr. Tyson's students
   Explanatory variable: listening to classical music (yes or no)
   Response variable: test scores
2. A potential confounding variable could be an existing musical preference based on test score. Perhaps higher performing students prefer listening classical music. Maybe students who listen to classical music have more resources/opportunities to score well than those who don't.
3. The 150 students should be randomly assigned to two groups. This could be accomplished by drawing names from a hat until 75 are in one group and 75 are in another. Both groups will receive the same instruction from Mr. Tyson. However, one group will be assigned to listen to classical music while studying and the other group will study in silence. All students will take the same assessment and their average results will be compared.

Section 4.2 Concept 3:

____ *I can describe a randomized block design and a matched pairs design for an experiment and explain the purpose of blocking in an experiment.*

(example) Suppose Mr. Tyson suspects students enrolled in a music class may have higher scores than those who don't. In order to ensure not all students enrolled in a music class are assigned to listen to classical music (which could happen through random assignment), he should block by enrollment. He should block all students enrolled in a music class together and all students who are not enrolled in music should be blocked separately. Then, he should randomly assign half of the students in each block to listen to classical music while studying and the other half should study in silence. Then all students should take the same assessment and the results within each block should be compared.

Section 4.3 Concept 1:

____ *I can explain the concept of sampling variability when making an inference about a population and how sample size affects sampling variability.*

1. No. It is unlikely the true average amount of change carried by all AP Statistics students is \$0.42. Different samples of size 50 would produce different average amounts.

2. A random sample of 200 students would be more likely to give an estimate closer to the truth, because larger random samples tend to produce more precise estimates than smaller random samples.

Chapter 4: Designing Studies

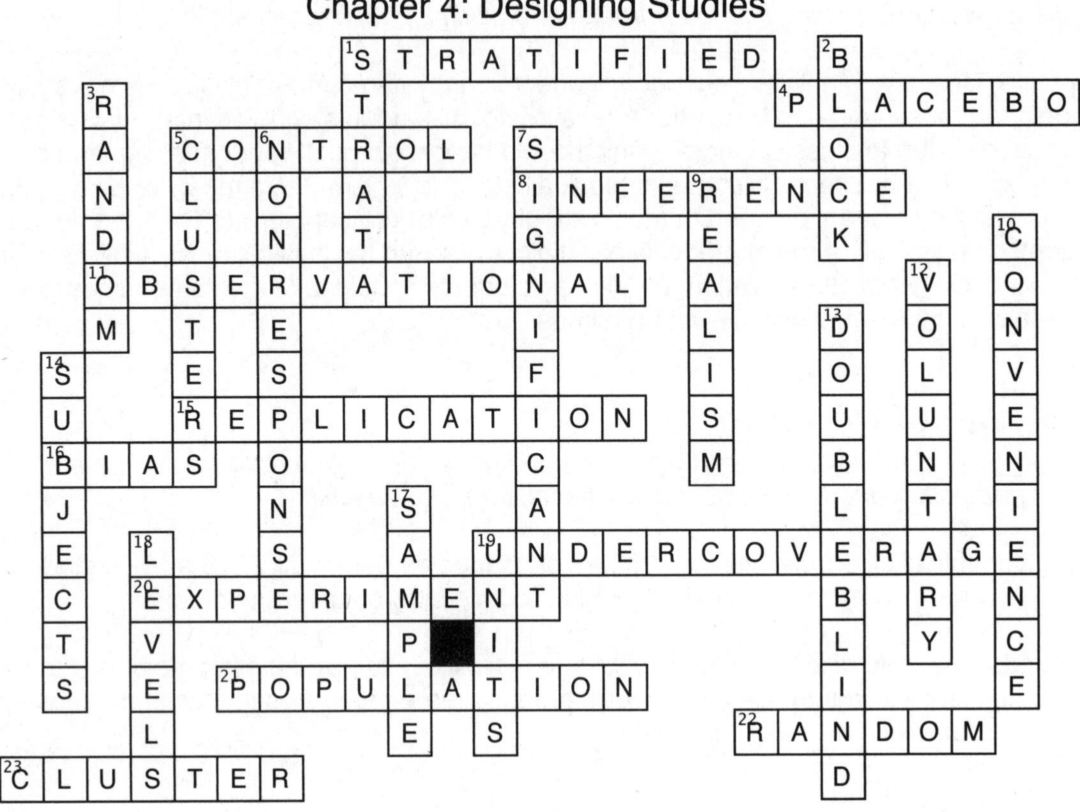

Across

1. a ______ random sample consists of separate simple random samples drawn from groups of similar individuals [STRATIFIED]
4. a "fake" treatment that is sometimes used in experiments [PLACEBO]
5. the effort to minimize variability in the way experimental units are obtained and treated [CONTROL]
8. the process of drawing a conclusion about the population based on a sample [INFERENCE]
11. this type of student can not be used to establish cause-effect relationships [OBSERVATIONAL]
15. the practice of using enough subjects in an experiment to reduce chance variation [REPLICATION]
16. a study that systematically favors certain outcomes shows this [BIAS]
19. this occurs when some groups in the population are left out of the process of choosing the sample [UNDERCOVERAGE]
20. a study in which a treatment is imposed in order to observe a reasponse [EXPERIMENT]
21. the entire group of individuals about which we want information [POPULATION]
22. a simple _____ sample consists of individuals from the population, each of which has an equally likely chance of being chosen [RANDOM]
23. a _____ sample consists of a simple random sample of small groups from a population [CLUSTER]

Down

1. groups of similar individuals in a population [STRATA]
2. a group of experimental units that are similar in some way that may affect the response to the treatments [BLOCK]
3. the rule used to assign experimental units to treatments is ____ assignment [RANDOM]
5. smaller groups of individuals who mirror the population [CLUSTERS]
6. this occurs when an individual chosen for the sample can't be contacted or refuses to participate [NONRESPONSE]
7. an observed effect that is too large to have occurred by chance alone [SIGNIFICANT]
9. a lack of ____ in an experiment can prevent us from generalizing the results [REALISM]
10. a sample in which we choose individuals who are easiest to reach [CONVENIENCE]
12. a ____ response sample consists of people who choose themselves by responding to a general appeal [VOLUNTARY]
13. neither the subjects nor those measuring the response know which treatment a subject received (two words) [DOUBLEBLIND]
14. when units are humans, they are called [SUBJECTS]
17. the part of the population from which we actually collect information [SAMPLE]
18. another name for treatments [LEVELS]
19. the individuals on which an experiement is done are experimental ____ [UNITS]

Chapter 5: Solutions

Section 5.1 Concept 1:

____ *I can interpret probability as a long-run relative frequency.*

a) If you were to repeatedly draw cards, with replacement, from a shuffled deck, you would draw a jack, queen, or king 23% of the time.

b) No, we'd expect to draw a jack, queen, or king 23 times out of 100, but we are not guaranteed *exactly* 23.

Section 5.1 Concept 2:

____*I can use simulation to model chance behavior.*

Let the digits 01-95 represent a passenger showing up for the flight.
Let 96-00 represent a "no show".

Select 12 two-digit numbers from a line of the random number table or a random number generator. Ignore repeats until you have 12 numbers selected. Count how many times 96-00 occurs. If 96-00 occurs less than two times, the flight is overbooked. Repeat this procedure 20 times. Determine how many of the 20 trials result in an overbooked flight.

Section 5.1 Concept 1:

____ *I can give a probability model for a chance process with equally likely outcomes and use it to find the probability of an event.*

____ *I can use basic probability rules, including the complement rule and addition rule for mutually exclusive events.*

1. There are 52 possible outcomes. Each has a probability of 1/52.

2. $P(A) = 4/52 = 0.0769$. $P(B) = 13/52 = 0.25$.

3. $P(A^C) = 1 - P(A) = 48/52 = 0.9231$.

4. No, A and B are not mutually exclusive because a card can be both an Ace and a heart.

Section 5.2 Concept 2:

____ *I can use a two-way table or Venn diagram to model a chance process and calculate probabilities involving two events.*

____ *I can apply the general addition rule to calculate probabilities.*

1. Use a two-way table to display the sample space.

	A	A^C
B	1	12
B^C	3	36

2.

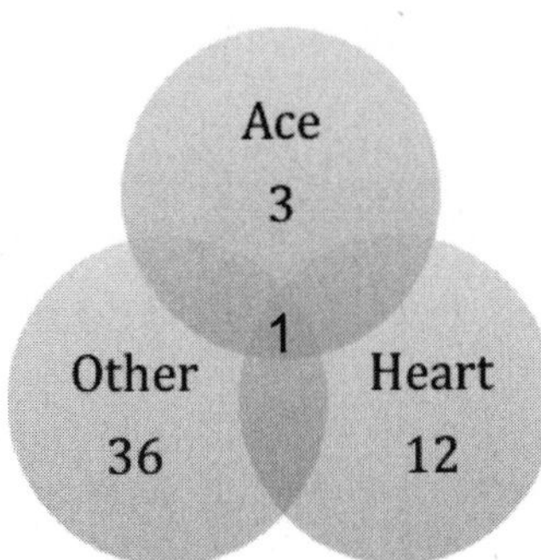

3. $P(A \cup B)$ = P(A) + P(B) – P(A and B) = 4/52 + 13/52 – 1/52 = 16/52 = 0.3077

Section 5.3 Concept 1:

____ *I can calculate and interpret conditional probabilities.*
____ *I can determine if two events are independent.*

If A and B are independent, P(A) = P(A|B). P(A) = 120/200 = 0.60. P(A|B) = 80/140 = 0.5714. Since these probabilities are not the same, A and B are not independent.

Section 5.3 Concept 2:

____ *I can use a tree diagram to describe chance behavior*
____ *I can use the general multiplication rule to solve probability questions*

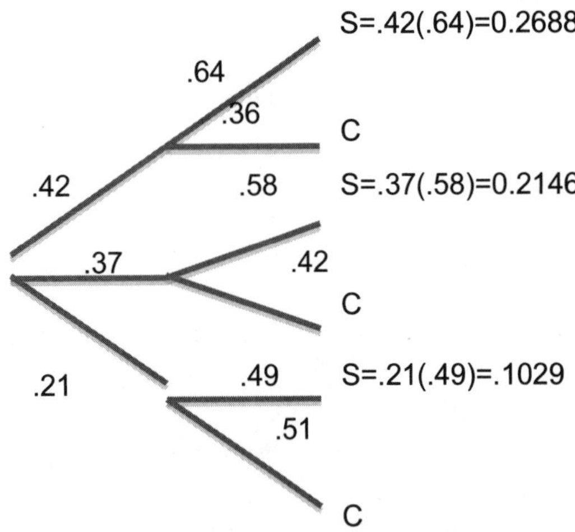

P(Statistics) = 0.2688+0.2146+0.1029 = 0.5863

Section 5.3 Concept 3:

____ *I can use a tree diagram to model a chance process involving a sequence of outcomes and to calculate probabilities.*

P(Lakeville|Statistics) = P(Lakeville and Statistics) / P(Statistics)
P(Lakeville|Statistics) = 0.2688 / 0.5863 = 0.4585

Chapter 5: Probability

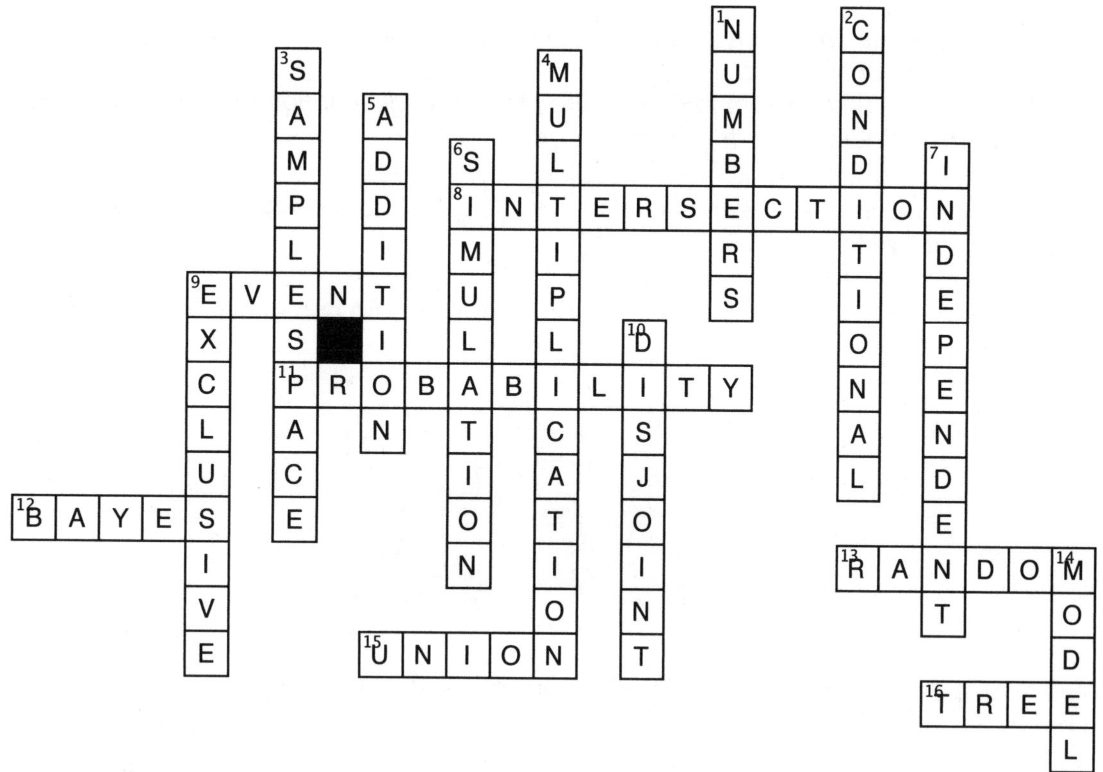

Across

8. The collection of outcomes that occur in both of two events. [INTERSECTION]
9. A collection of outcomes from a chance process. [EVENT]
11. The proportion of times an outcome would occur in a very long series of repetitions. [PROBABILITY]
12. _____ Theorem can be used to find probabilities that require going "backward" in a tree diagram. [BAYES]
13. In statistics, this doesn't mean "haphazard." It means "by chance." [RANDOM]
15. The collection of outcomes that occur in either of two events. [UNION]
16. A ____ diagram can help model chance behavior that involves a sequence of outcomes. [TREE]

Down

1. The law of large _____ states that the proportion of times an outcome occurs in many repetitions will approach a single value. [NUMBERS]
2. The probability that one event happens given another event is known to have happened. [CONDITIONAL]
3. The set of all possible outcomes for a chance process (two words). [SAMPLESPACE]
4. The probability that two events both occur can be found using the general _____ rule. [MULTIPLICATION]
5. P(A or B) can be found using the general _____ rule. [ADDITION]
6. The imitation of chance behavior, based on a model that reflects the situation. [SIMULATION]
7. The occurrence of one event has no effect on the chance that another event will happen. [INDEPENDENT]
9. Another term for disjoint: Mutually _____. [EXCLUSIVE]
10. Two events that have no outcomes in common and can never occur together. [DISJOINT]

Chapter 6: Solutions

Section 6.1 Concept 1:

____ *I can use the probability distribution of a discrete random variable to calculate the probability of an event.*

____ *I can make a histogram to display the probability distribution of a discrete random variable and describe its shape.*

Consider two 4-sided dice, each having sides labeled 1, 2, 3, 4. Let X = the sum of the numbers that appear after a roll of the dice.

a) X is a discrete random variable. We are most likely to roll a sum of 5 and least likely to roll a sum of 2 or 8.

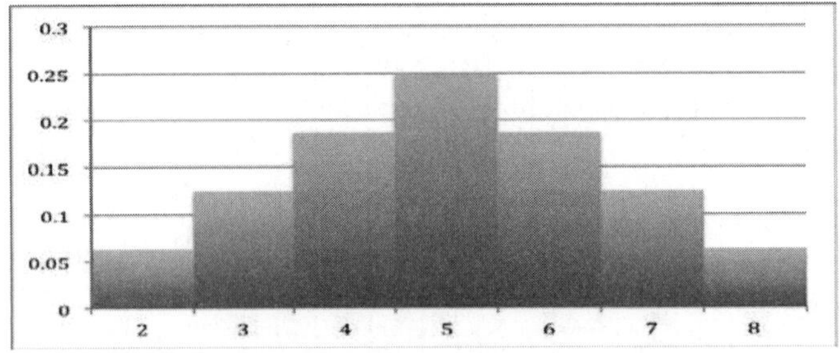

b) Yes, we should be surprised. In 10 rolls we would expect to see a sum of 3 or less about once or twice.

Section 6.1 Concept 2:

____ *I can calculate and interpret the mean (expected value) of a discrete random variable.*

____ *I can calculate and interpret the standard deviation of a discrete random variable.*

____ *I can use the probability distribution of a continuous random variable (uniform or Normal) to calculate the probability of an event..*

a)

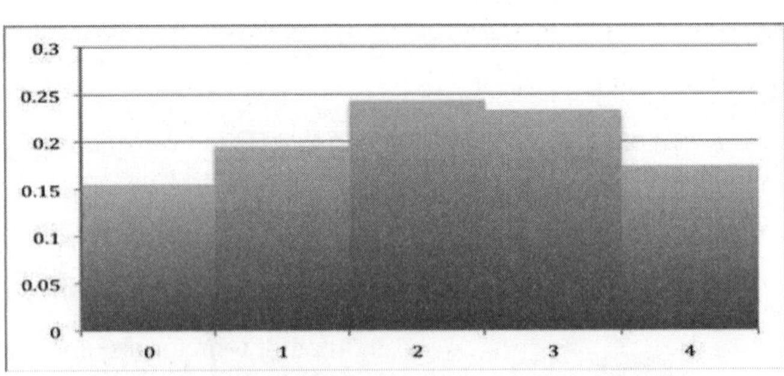

$E(Y) = 0(0.155) + 1(0.195) + 2(0.243) + 3(0.233) + 4(0.174) = 2.076$.
In the long run, we'd expect to see an average of 2.076 goals per game for many, many games.

$\sigma^2_Y = (0-2.076)^2(0.155) + \ldots + (4-2.076)^2(0.174) = 1.7382$
$\sigma_Y = \sqrt{(1.7382)} = 1.3184$.
We would expect the number of goals per game to vary by about 1.3184 from 2.076 in the long run.

b) The weights of toddler boys follow an approximately Normal distribution with mean 34 pounds and standard deviation 3.5 pounds. Suppose you randomly choose one toddler boy and record his weight. What is the probability that the randomly selected boy weighs less than 31 pounds?

$z = (31-34)/3.5 = -0.8571$ $P(z < -0.8571) = 0.1956$.

Section 6.2 Concept 1:

____ *I can describe the effect of adding or subtracting a constant or multiplying or dividing by a constant on the probability distribution of a random variable.*

The shape will be slightly skewed to the right.

$\mu_Y = 1.5(\mu_X) - 2 = -0.35$.
In the long run, we would expect to lose \$0.35 each time we play the game, on average.

$\sigma_Y = 1.5(0.943) = 1.4145$.
On average, we would expect our profit to vary by about \$1.42 around a loss of \$0.35.

Section 6.2 Concept 2:

____ *I can calculate the mean and standard deviation of the sum or difference of random variables.*

a) $\mu_{total} = 1.4+1.2+0.9+1 = 3.3$ min
$\sigma_{total} = \sqrt{(0.1^2+0.4^2+0.8^2+0.7^2)} = \sqrt{1.3} = 1.14$ min

b) $\mu_{(Doug - Alan)} = 1 - 1.4 = -0.4$ min
On average, Doug is faster by 0.4 min.

$\sigma_{(Doug - Alan)} = \sqrt{(0.7^2+0.1^2)} = 0.7071$
The difference between Doug and Allan's times will vary by 0.7071 min around 0.4 min on average.

Section 6.2 Concept 3:

____ *I can find probabilities involving the sum or difference of independent Normal random variables.*

a) $\mu_{(Molesky - Liberty)} = 110-100 = 10$
$\sigma_{(Molesky - Liberty)} = \sqrt{(10^2+8^2)}=\sqrt{164}=12.81$

b) Find P(Molesky - Liberty < 0): $z = (0-10)/12.81 = -0.78$. $P(z < -0.78) = 0.2177$.
There is about a 21.77% chance Mr. Molesky will finish before Mr. Liberty on any given day.

Section 6.3 Concept 1:

____ *I can determine whether the conditions for a binomial setting have been met.*

____ *I can calculate and interpret probabilities involving binomial distributions.*

a) Show that *X* is a binomial random variable.
B: A card is either a heart or it isn't
I: Each draw is independent since cards are replaced and the deck is shuffled
N: There are 10 observations in each game
S: *P*(heart) = 0.25 in each draw.

b) $P(X < 4) = P(X=0) + P(X=1) + P(X=2) + P(X=3) = 0.7759$

Section 6.3 Concept 2:

____ *I can calculate and interpret the mean and standard deviation of a binomial random variable.*

____ *I can use the Normal approximation to the binomial distribution to calculate probabilities, when appropriate.*

a) Show that *X* is approximately a binomial random variable.
X is approximately binomial with $n = 500$ and $p = 0.72$. Because $np = 500(0.72) = 360 \geq 10$ and $n(1-p) = 500(0.28) = 140 \geq 10$, we can approximate *X* with a Normal distribution.

b) Use a Normal approximation to find the probability that 400 or more students would give their teacher a positive rating in this sample.

$\mu_X = np = 500(.72) = 360$
$\sigma_X = \sqrt{(np(1-p))} = 10.04$
$z = (400-360)/10.04 = 3.98$
$P(z \geq 3.98) = 0.000034$

Section 6.3 Concept 3:

____ *I can find probabilities involving geometric random variables.*

a) Show that *X* is a geometric random variable.
There are two outcomes (ring or no ring). Each box is independent. The probability of a ring in any given box is 0.2. We are interested in how long it will take to find a ring.

b) $P(X=7) = 0.8^6(0.2) = 0.0524$

c) $P(X<4) = 0.488$

d) $E(X) = 1/.2 = 5$ boxes.

Chapter 6: Random Variables

Across

2. The average of the squared deviations of the values of a variable from its mean. [VARIANCE]
5. Random variables are _____ if knowing whether an event in X has occurred tells us nothing about the occurrence of an event involving Y. [INDEPENDENT]
7. The probability _____ of a random variable gives its possible values and their probabilities. [DISTRIBUTION]
10. The number of ways of arranging k successes among n observations is the binomial ____. [COEFFICIENT]
11. The sum or difference of independent Normal random variables follows a _____ distribution. [NORMAL]
12. When you combine independent random variable, you always add these. [VARIANCES]
14. A linear _____ occurs when we add/subtract and multiply/divide by a constant. [TRANSFORMATION]
16. An easy way to remember the requirements for a geometric setting. [BITS]

Down

1. A ____ variable takes numerical values that describe the outcomes of some chance process. [RANDOM]
3. When n is large, we can use a Normal _____ to determine probabilities for binomial settings. [APPROXIMATION]
4. A random variable that takes on all values in an interval of numbers. [CONTINUOUS]
6. A random variable that takes a fixed set of possible values with gaps between. [DISCRETE]
8. A ____ setting arises when we perform independed trials of the same chance process and record the number of trials until a particular outcome occurs. [GEOMETRIC]
9. An easy way to remember the requirements for a binmial setting. [BINS]
13. Adding a constant to each value of a random variable has no effect on the shape or ____ of the distribution. [SPREAD]
15. Multiplying each value of a random variable by a constant has no effect on the ____ of the distribution. [SHAPE]

Chapter 7: Solutions

Section 7.1 Concept 1:

____ *I can distinguish between a parameter and a statistic.*

a) Population: All males with high blood pressure
Parameter: Mean arterial pressure (μ) for all males with high blood pressure
Statistic: Sample mean arterial pressure for the 500 males in the study

b) Population: All 16- to 24-year old drivers
Parameter: Proportion of all 16- to 24-year old drivers who text while driving
Statistic: Sample proportion 0.12.

Section 7.1 Concept 2:

____ *I can distinguish among the distribution of a population, the distribution of a sample and the sampling distribution of a statistic.*

____ *I can create a sampling distribution using all possible samples from a small population.*

A breakfast cereal includes marshmallow shapes in the following distribution: 10% stars, 10% crescent moons, 20% rockets, 40% astronauts, 20% planets. We are interested in examining the proportion of rockets in a random sample of 2000 marshmallows from the cereal.

a) The population distribution of marshmallow shapes.

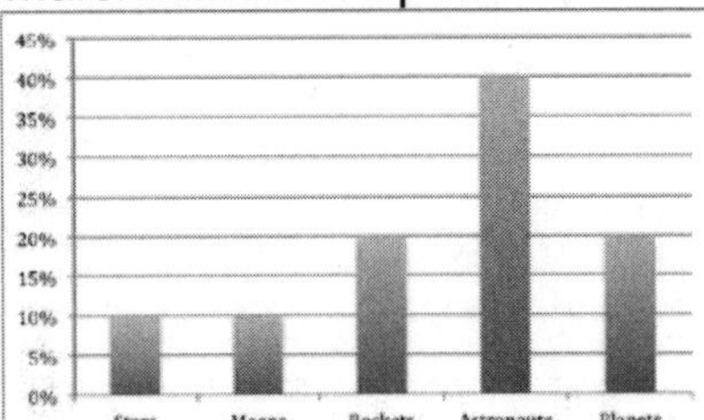

b) The distribution of sample data we would expect to see (n=2000).

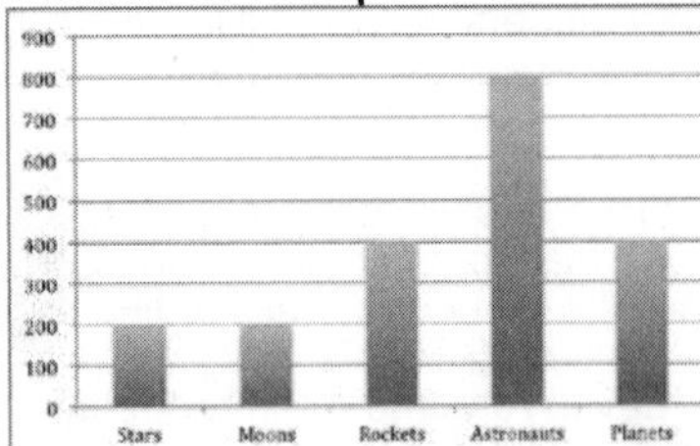

We would expect to see about 400 rockets.

c) We would expect a symmetric distribution with mean 0.2 and standard deviation .009.

Section 7.2 Concept 1:

____ *I can calculate the mean and standard deviation of the sampling distribution of a sample proportion $\hat{p}$ and interpret the standard deviation.*

____ *I can determine if the sampling distribution of $\hat{p}$ is approximately Normal.*

Suppose your job at a potato chip factory is to check each shipment of potatoes for quality assurance. Further, suppose that a truckload of potatoes contains 95% that are acceptable for processing. If more than 10% are found to be unacceptable in a random sample, you must reject the shipment. To check, you randomly select and test 250 potatoes. Let $\hat{p}$ be the sample proportion of unacceptable potatoes.

a) The mean of the sampling distribution of $\hat{p}$ is 0.05.
b) Both np=12.5 and $n(1-p)$=237.5 are greater than 10. The standard deviation of the sampling distribution of $\hat{p}$ is $\sigma_{\hat{p}} = \sqrt{\frac{.05(1-.05)}{250}} = .0137.$

c) Since both np and n(1-p) are greater than 10, we can assume the sampling distribution is approximately Normal. Since 0.10 is more than 3 standard deviations above the mean, it is unlikely we would observe a sample proportion greater than 10%. It is not likely we would reject the truckload based on a sample of 250 potatoes if 95% were truly acceptable.

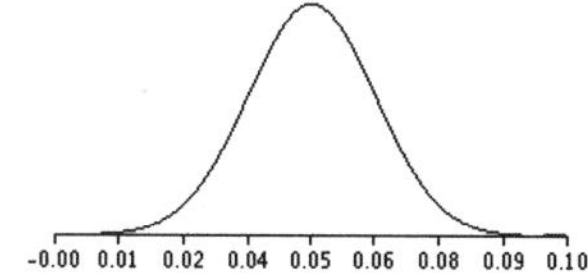

Section 7.2 Concept 2:

____ *I can use a Normal distribution to calculate probabilities involving $\hat{p}$, if appropriate.*

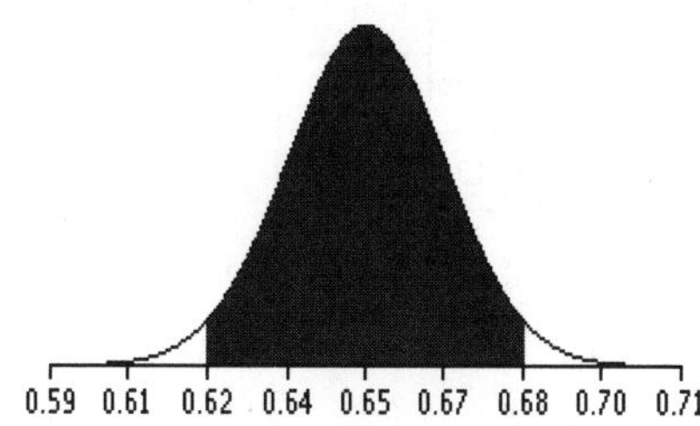

Since both np=650 and $n(1-p)$=350 are greater than 10, we can assume the sampling distribution of $\hat{p}$ is approximately Normal with mean 0.65 and standard deviation 0.015. 0.62 and 0.68 are 2 standard deviations above and below the mean. By the 68-95-99.7 rule, the probability a random sample of 1000 students will result in a $\hat{p}$ within 3-percentage points of the true proportion is approximately 95%.

Section 7.3 Concept 1:

____ *I can calculate the mean and standard deviation of the sampling distribution of a sample mean $\bar{x}$ and interpret the standard deviation.*

____ *I can use a Normal distribution to calculate probabilities involving $\bar{x}$, if appropriate.*

P(a randomly chosen 5th grader will take more than 2.5 minutes)

$z = \frac{2.5-2}{0.8} = 0.625$ $P(z>0.625)=0.2659$

The sampling distribution of $\bar{x}$ will be N(2.5, 0.8/√20).

$z = \frac{2.5-2}{0.1789} = 2.795$ $P(\bar{x}>2.5)=0.002596$

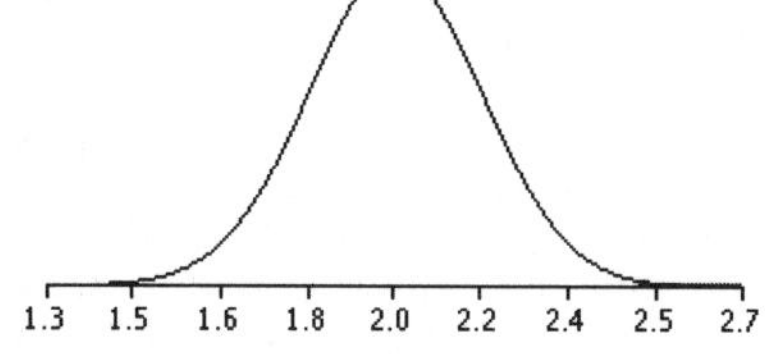

Section 7.3 Concept 2:

____ *I can use a Normal distribution to calculate probabilities involving $\bar{x}$, if appropriate.*

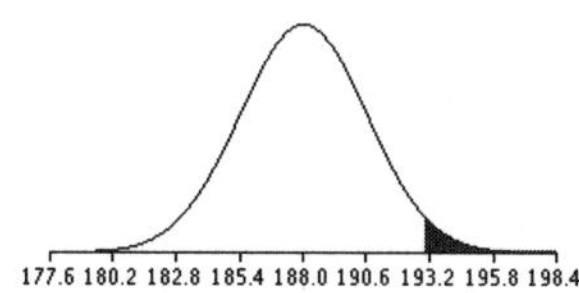

The sampling distribution of $\bar{x}$ will be Normal with mean =188 and standard deviation = 41/√250. $z = \frac{193-188}{2.593} = 1.928$

$P(\bar{x} > 193) = 0.0269.$

Chapter 7: Sampling Distributions

Across

2. _____ distribution: the distribution of values taken by the statistic in all possible samples of the same size from the population [SAMPLING]
4. _____ distribution: the distribution of all values of a variable in the population [POPULATION]
8. _____ of a statistic is described by the spread of the sampling distribution [VARIABILITY]
12. Greek letter used for the population standard deviation [SIGMA]
14. the Normal approximation for the sampling distribution of a sample proportion can be used when both the number of successes and failures are greater than _____ [TEN]
15. sampling distributions and sampling variability provide the foundation for performing _____ [INFERENCE]
16. central _____ theorem tells us if the sample size is large, the sampling distribution of the sample mean is approximately Normal, regardless of the shape of the population [LIMIT]

Down

1. a statistic is an _____ estimator if the mean of the sampling distribution is equal to the true value of the parameter being estimated. [UNBIASED]
2. a number, computed from sample data, that estimates a parameter [STATISTIC]
3. Greek letter used for the population mean [MU]
5. standard _____ : measure of spread of a sampling distribution [DEVIATION]
6. sampling _____ notes the value of a statistic may be different from sample to sample [VARIABILITY]
7. a number that describes a population [PARAMATER]
9. the rule of thumb for using the central limit theorem - the sample size should be greater than _____ [THIRTY]
10. when the sample size is large, the sampling distribution of a sample proportion is approximately _____ [NORMAL]
11. to draw a conclusion about a population parameter, we can look at information from a _____ sample [RANDOM]
13. center of a sampling distribution [MEAN]

Chapter 8: Solutions

Section 8.1 Concept 1:

____ *I can interpret a confidence level in context.*
____ *I can interpret a confidence interval in context.*

a) We are 90% confident the interval from 19.10 to 20.74 captures the true mean contents of a "20 oz." bottle of water.

b) If we were to collect many samples of 50 "20 oz." water bottles and construct a confidence interval for the mean contents in the same manner, 90% of the intervals would capture the true mean contents.

c) The average contents for the sample was 19.92 oz. Since 20 oz. is in the interval, we do not have evidence to suggest the population mean is less or greater than 20 oz.

Section 8.1 Concept 2:

____ *I can describe how the sample size and confidence level affect the margin of error.*
____ *I can explain how practical issues like nonresponse, undercoverage, and response bias can affect the interpretation of a confidence interval.*

a) The confidence interval would be narrower because decreasing the confidence level decreases the margin of error.

b) The confidence interval would be narrower because increasing the sample size decreases the margin of error.

c) The margin of error does not account for the fact that adults who are at home in the late-morning may more likely to be homemakers and, therefore more likely to bake at least twice a week. The true proportion of adults who bake at home may be less than that observed in the sample.

Section 8.2 Concept 2:

____ *I can state and check the Random, 10%, and Large Counts conditions for constructing a confidence interval for a population proportion.*
____ *I can construct and interpret a confidence interval for a population proportion.*

State: We wish to construct a 90% confidence interval for the true proportion of adults who can roll their tongue.

Plan: We have a random sample, the number of successes and failures are both greater than 10 (68>10 and 232>10), and there are more than 3000 adults. We can construct a 90% confidence interval for the true proportion.

Do: $95\%CI = 0.23 \pm 1.645\sqrt{\frac{0.23(0.77)}{300}} = (0.18691, 0.26643)$

Conclude: We are 90% confident the interval from 0.19 to 0.27 captures the true proportion of adults who can roll their tongue.

Section 8.2 Concept 3:

____ *I can determine the sample size required to obtain a C% confidence interval for a population proportion with a specified margin of error.*

$$1.96*\sqrt{\frac{.5(.5)}{n}}\leq 0.02\Rightarrow\sqrt{\frac{.5(.5)}{n}}\leq 0.0102\Rightarrow\frac{0.25}{n}\leq 0.000104$$

$$n\geq 2401$$

We would need at least 2401 adults to estimate the proportion who can roll their tongues to their satisfaction.

Section 8.3 Concept 2:

____ I can determine the critical value for calculating a C% confidence interval for a population mean using a table or technology.

a) $t^*=1.330$ b) $t^*=1.984$ c) $t^*=2.756$

Section 8.3 Concept 3:

____ *I can construct and interpret a confidence interval for a population mean.*

This is a random sample, however the sample size is less than 30. A boxplot of the sample data does not suggest strong skewness or outliers, so we can construct a 95% confidence interval for the true mean.

$$95\%CI: 22.5\pm 2.365\frac{7.191}{\sqrt{8}}=(16.488, 28.512)$$

We are 95% confident the interval from 16.488 to 28.512 captures the true mean amount of sugar for this manufacturer's soft drinks.

$$z*\frac{\sigma}{\sqrt{n}}\leq ME$$

Section 8.3 Concept 4:

____ *I can determine the sample size required to obtain a level C confidence interval for a population mean with a specified margin of error.*

$$1.96*\frac{4}{\sqrt{n}}\leq 0.5\Rightarrow\frac{4}{\sqrt{n}}\leq 0.2551\Rightarrow 15.68\leq\sqrt{n}$$

$$n\geq 245.86$$

We would need at least 246 students to estimate the mean amount of time to their satisfaction.

Chapter 8: Estimating with Confidence

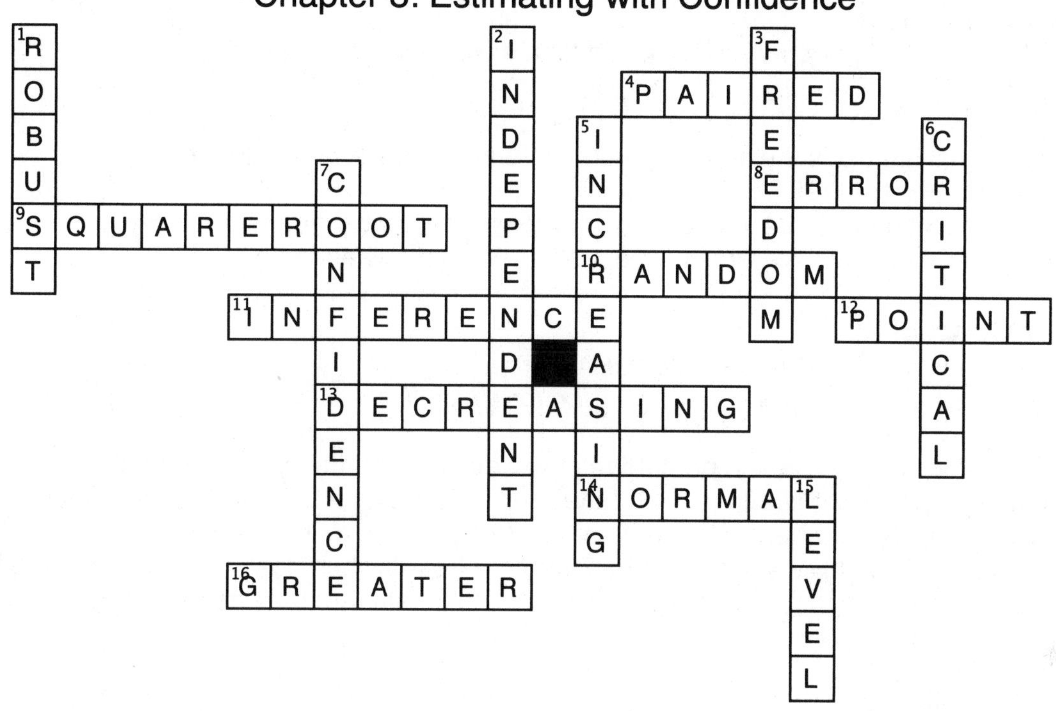

Across

4. ____ t procedures allow us to compare the responses to two treatments in a matched pairs design [PAIRED]
8. a confidence interval consists of an estimate ± margin of ____ [ERROR]
9. to find the standard error of the sample mean, divide the sample standard deviation by the ____ of the sample size (two-words) [SQUAREROOT]
10. to estimate with confidence, our estimate should be calculated from a ___ sample [RANDOM]
11. methods for drawing conclusions about a population from sample data [INFERENCE]
12. a single value used to estimate a parameter is a ____ estimator [POINT]
13. we can construct a narrow interval by ____ our confidence [DECREASING]
14. as degrees of freedom increase, the t distribution approaches the ____ distribution [NORMAL]
16. the spread of the t distributions is ____ than the spread of the standard Normal distribution [GREATER]

Down

1. inference procedures that remain fairly accurate even when a condition is violated [ROBUST]
2. another condition for confidence intervals is that observations should be ____ [INDEPENDENT]
3. particular t distributions are specified by degrees of ____ [FREEDOM]
5. we can construct a narrow confidence interval by ____ our sample size [INCREASING]
6. the margin of error consists of a ____ value and the standard error of the sampling distribution [CRITICAL]
7. a ____ interval provides an estimate for a population parameter [CONFIDENCE]
15. confidence ____: the success rate of the method in repeated sampling [LEVEL]

Chapter 9: Solutions

Section 9.1 Concept 1:

____ *I can state appropriate hypotheses for a significance test about a population parameter..*

a) The parameter of interest is the true proportion of hearts in the "chute" of cards.
b) H_0: $p = 0.25$ vs. H_a: $p > 0.25$
c) The sample proportion is 7/12 = .0583. It is possible to be dealt 7 hearts in 12 cards, however it is very unlikely. We would expect about 3 hearts when dealt 12 cards.

Section 9.1 Concept 2:

____ *I can interpret a P-value in context.*
____ *I can make an appropriate conclusion for a significance test.*

a) If the null hypothesis was true, the proportion of hearts in the "chute" would be 0.25.
b) Assuming the proportion of hearts in the "chute" is 0.25, there is a 0.4% chance we would observe 7 or more hearts when 12 cards were randomly selected. This is highly unlikely.
c) Since it is so unlikely that we would observe 7 or more hearts when dealt 12 cards from a fair chute (less than a 5% chance), we have evidence to suggest there may be more hearts than usual in the chute.

Section 9.1 Concept 3:

____ *I can interpret a Type I error and a Type II error in context, and give a consequences of each error in a given setting.*

a) A Type I error would occur if we assumed the chute had more hearts than usual (based on our sample) when, in reality, the proportion of hearts was 0.25. We were just dealt an unusual hand.
b) A Type II error would occur if we assumed the chute was fair when, in reality, the proportion of hearts was actually greater than 0.25.

Section 9.2 Concept 1:

____ *I can state and check the Random, 10%, and Large Counts conditions for performing a significance test about a population proportion.*
____ *I can calculate the standardized test statistic and P-value for a test about a population proportion.*
____ *I can perform a significance test about a population proportion.*

p = the true proportion of passages that follow the speech pattern in the work in question

$$H_0: p = 0.214$$
$$H_a: p > 0.214$$

Conditions: We have a random sample of passages from the work in question. The number of successes and failures (136 and 303) are both greater than 10.

Assuming the null hypothesis is true, the sampling distribution of the proportion of passages following the speech pattern will be Normal with mean 0.214 and standard deviation 0.0196.

$$z = \frac{0.3098 - 0.214}{0.0196} = 4.89 \quad P\text{-value} = 0.000001$$

Since the *P*-value is less than 5%, we have significant evidence to reject the null hypothesis. It appears the work in question may have too high a proportion of passages following Plato's speech pattern to be one of his actual works.

Section 9.2 Concept 2:

____ *I can use a confidence interval to make a conclusion for a two-sided test about a population parameter.*

We wish to construct a 99% confidence interval for the true proportion of teens who text while driving. We have a random sample, there are more than 270 teenage drivers and the number of successes and failures (15 and 12) are both greater than 10.

$$99\%CI: 0.56 \pm 2.576\sqrt{\frac{0.56(0.44)}{27}} = (0.30923, 0.80188)$$

We are 99% confident the interval from 0.31 to 0.80 captures the true proportion of teens who text while driving. Since 0.77 is in this interval, we do not have evidence to suggest the true proportion is different than 77%.

Section 9.3 Concept 1:

____ *I can check conditions for carrying out a test about a population mean.*
____ *I can conduct a one-sample t test about a population mean μ.*

We will conduct a one-sample *t*-test for the true mean ratio.

$H_0: \mu = 8.9$

$H_a: \mu \neq 8.9$

Conditions: We have a random sample of 41 bones and $n > 30$.

Assuming the null hypothesis is true, the sampling distribution will be *t* with 40 degrees of freedom. The mean of the sampling distribution is 8.9 and the standard error is 0.187.

$$t = \frac{9.27 - 8.9}{1.198/\sqrt{41}} = 1.97 \quad P\text{-value} = 0.056$$

Since the *P*-value is greater than 5%, we do not have significant evidence to reject the null hypothesis. There is no evidence at the 5% level to suggest these bones have a ratio different than 8.9.

Section 9.3 Concept 2:

____ *I can interpret the power of a significance test and describe what factors affect the power of a test.*

a) If the true mean power usage of the batch of processors is $\mu = 71$ watts, there is a 0.572 probability that the quality control engineer will find convincing evidence for H_a: $\mu > 70$.

b) Decrease. A smaller significance level makes it harder to reject H_0 when H_a is true.

c) Increase. A larger sample size gives more information about the true mean μ.

d) Increase. It is easier to detect a bigger difference between the null and alternative parameter value.

Chapter 9: Testing a Claim

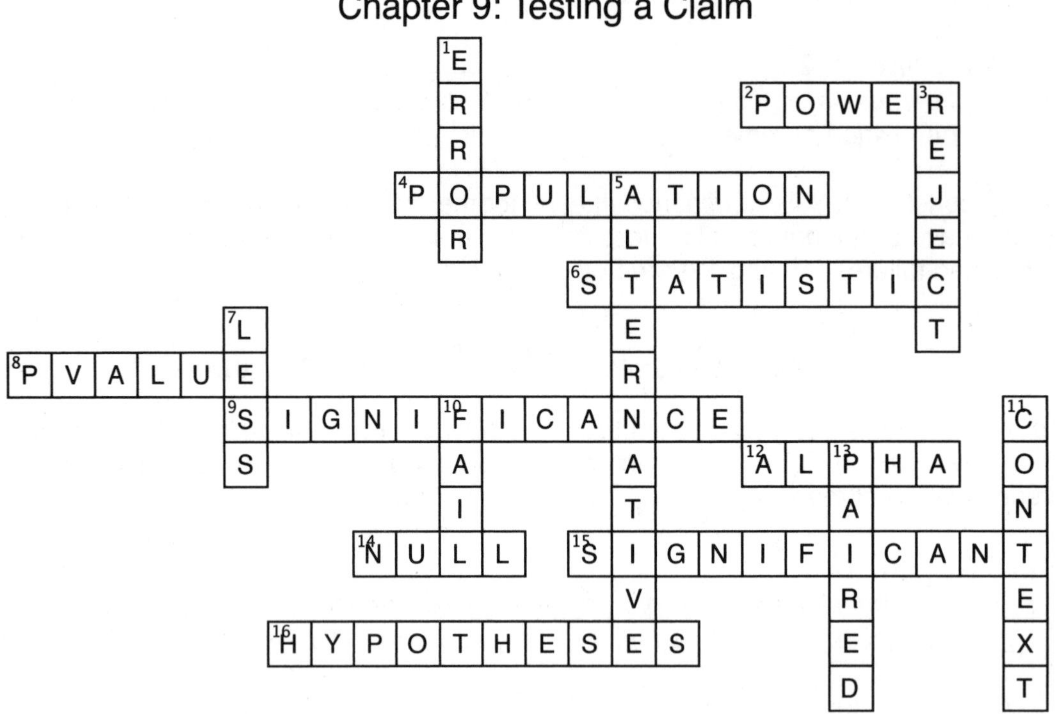

Across

2. the probability that a significance test will reject the null when a particular alternative value of the paramter is true [POWER]
4. hypotheses always refer to the ____ [POPULATION]
6. the test ____ is a standardized value that assesses how far the estimate is from the hypothesized parameter [STATISTIC]
8. the probability that we would observe a statistic at least as extreme as the one observed, assuming the null is true (two terms) [PVALUE]
9. we can use a _____ test to compare observed data with a hypothesis about a population [SIGNIFICANCE]
12. greek letter used to designate the significance level [ALPHA]
14. the ____ hypothesis is a the claim for which we are seeking evidence against [NULL]
15. an observed difference that is too small to have occured due to chance alone is considered statistically ____ [SIGNIFICANT]
16. the statements a statistical test is designed to compare [HYPOTHESES]

Down

1. if we reject the null hypothesis when it is actually true, we commit a Type I ____ [ERROR]
3. if we calculate a very small P value, we have evidence to ____ the null [REJECT]
5. the ____ hypothesis is the claim about the population for which we are finding evidence for [ALTERNATIVE]
7. reject the null hypothesis if the P value is ____ than the significance level [LESS]
10. if our calculated P value is not small enough to provide convincing evidence, we ___ to reject he null [FAIL]
11. conclusions should always be written in ____ [CONTEXT]
13. a ____ test allows us to analyze differences in responses within pairs [PAIRED]

Chapter 10: Solutions

Section 10.1 Concept 1:

____ *I can describe the shape, center, and variability of the sampling distribution of $\hat{p}_1 - \hat{p}_2$.*

a) The sampling distribution of $\hat{p}_S - \hat{p}_N$ will be approximately Normal with a mean of 0.15 and a standard deviation of 0.0556.

b) $z = \dfrac{0.07 - 0.15}{\sqrt{\dfrac{0.25(0.75)}{160} + \dfrac{0.6(0.4)}{125}}} = -1.439$ $\quad$ $P(z<-1.439) = 0.0749$.

c) Assuming the stated proportions for each high school are true, there is a 7.5% chance we'd observe differences between the sample proportions at least as extreme as those observed. We do not have evidence to doubt the study's sample proportions.

Section 10.1 Concept 2:

____ *I can determine whether the conditions are met for doing inference about a difference between two proportions.*

____ *I can construct and interpret a confidence interval for a difference between two proportions.*

Since we have two random samples, more than 10 successes in each sample (551 in 1990 and 652 in 2010), more than 10 failures in each sample (949 in 1990 and 1348 in 2010), and we sampled less than 10% of the population of interest each year, we can construct a 95% confidence interval for the difference in proportions.

$$95\%CI: (0.367 - 0.326) \pm 1.96\sqrt{\frac{0.367(1-0.367)}{1500} + \frac{0.326(1-0.326)}{2000}} = (0.0094, 0.0732)$$

We are 95% confident the interval from 0.0094 t0 0.0732 captures the true difference in the proportions of adults who smoked in 1990 and in 2010. Since 0 is not contained in this interval, we can conclude the proportion of adults who smoked in 1990 was higher than in 2010.

Section 10.1 Concept 3:

____ *I can perform a significance test to compare two proportions.*

We will perform a two-sample test for proportions.

$H_0: p_N = p_S$

$H_a: p_N > p_S$

Conditions: We have random sample from each school. The number of successes and failures at North (28 and 92) are both greater than 10. The number of successes and failures at South (30 and 120) are both greater than 10. There are at least 1200 students at North and at least 1500 students at South.

$$z = \frac{0.233 - 0.2}{\sqrt{\dfrac{.2148(1-.2148)}{120} + \dfrac{.2148(1-.2148)}{150}}} = 0.6626$$

P-Value=0.2537

Since the P-value is greater than 5%, we do not have evidence to reject the null hypothesis. We cannot conclude that the proportion of low-income students at North is higher than the proportion of low income students at South.

Section 10.2 Concept 1:

____ *I can describe the shape, center, and variability of the sampling distribution of* $\bar{x}_1 - \bar{x}_2$.

a) The sampling distribution of $\bar{x}_1 - \bar{x}_2$ will be a *t* distribution (df = 58.79). The mean will be 24 and the standard deviation will be $\sqrt{\frac{18^2}{40}+\frac{9.4^2}{40}}=3.21$.

b)

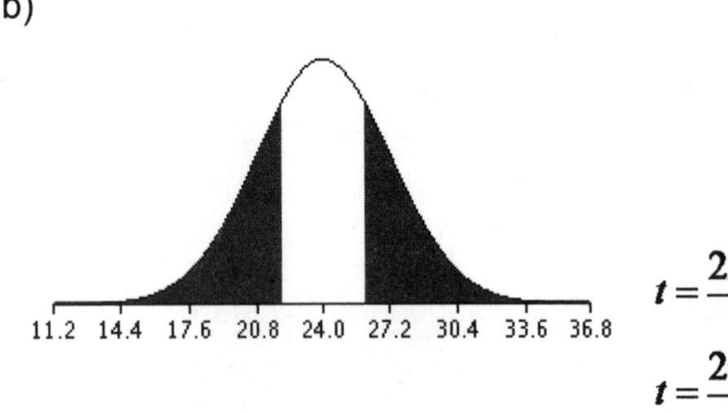

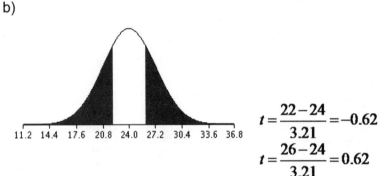

$$t=\frac{22-24}{3.21}=-0.62$$
$$t=\frac{26-24}{3.21}=0.62$$

P(difference in means is 2 points or more) = 2(0.269)=0.538 (approximate, based on df=58.79).

Section 10.2 Concept 2:

____ *I can construct and interpret a confidence interval for a difference between two means.*

We will construct a 95% confidence interval for the difference in mean weight loss. Conditions: Patients were randomly assigned to each treatment. We have at least 30 patients on each treatment.

$$95\%CI:(9.3-7.4)\pm t^*\sqrt{\frac{4.7^2}{100}+\frac{4^2}{100}}=(0.6827,3.117)$$

We are 95% confident the interval from 0.68 to 3.12 captures the true difference in mean weight loss for the two diets. Since 0 is contained in this interval, we do not have evidence to suggest the new diet is more effective than the current diet.

Section 10.2 Concept 3:

____ *I can calculate the standardized test statistic and P-value for a test about a difference between two means.*

____ *I can perform a significance test about a difference between two means.*

We will perform a two-sample *t* test for the difference in mean memory scores.

$$H_0: \mu_B = \mu_G$$
$$H_a: \mu_B > \mu_G$$

Conditions: We have two independent random samples of boys and girls. Both samples have at least 30 individuals.

$$t = \frac{48.9 - 48.4}{\sqrt{\frac{12.96}{200} + \frac{11.85}{150}}} = 0.375$$ *P*-value = 0.354

Since the *P*-value is greater than 5%, we do not have sufficient evidence to reject the null hypothesis. We cannot conclude that the boys have better short-term memory than the girls.

Section 10.3 Concept 1:

____ *I can perform a significance test about a mean difference.*

Subject	Beats Caffeine	Beats Placebo	Difference Caffeine - Placebo
1	251	201	50
2	284	262	22
3	300	283	17
4	321	290	31
5	240	259	-19
6	294	291	3
7	377	354	23
8	345	346	-1
9	303	283	20
10	340	361	-21
11	408	411	-3

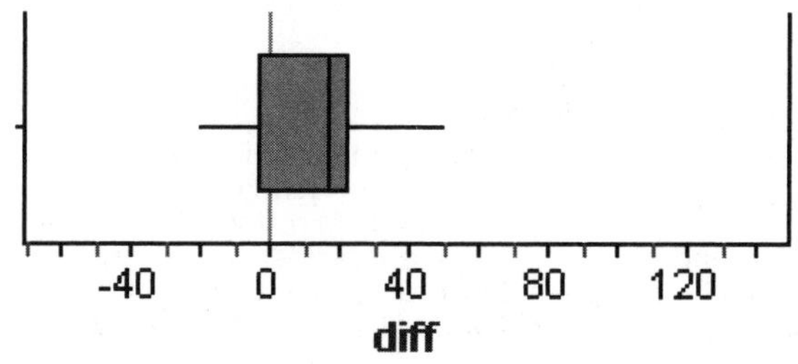

The mean difference is 11.09. We will conduct a paired *t*-test for the true mean difference between the caffeine and placebo beats per minute.

$$H_0: \mu_d = 0$$
$$H_a: \mu_d > 0$$

Conditions: The treatments were randomly assigned. We do not have more than 30 subjects. A boxplot of the differences suggests no outliers.

$$t = \frac{11.09 - 0}{21.52/\sqrt{11}} = 1.709$$ *P*-value = 0.059.

Since the *P*-value is greater than 5%, we do not have sufficient evidence to reject the null hypothesis. We cannot conclude that caffeine results in a higher beats per minute than the placebo

Chapter 10: Comparing Two Populations or Groups

Across

1. an alternative hypothesis the seeks evidence of a difference requires the use of a ____ test (two words) [TWOSIDED]
7. to perform two-sample procedures, the random samples should be____ [INDEPENDENT]
9. procedures that yield accurate results, even when a condition is violated, are ____ [ROBUST]
11. if the number of successes and failures from both samples are greater than ten, the sampling distribution for the difference between the two proportions will be approximately ____ [NORMAL]
13. results that are too unlikely to be due to chance alone are considered statistically ____ [SIGNIFICANT]
14. in a two-proportion test, we calculate a ____ proportion [POOLED]

Down

2. particular t distributions are distinguished by ____ of freedom [DEGREES]
3. the sampling distribution for the difference between means will be approximately Normal if both sample sizes are greater than [THIRTY]
4. hypotheses should always be written in terms of the ____ [POPULATION]
5. the hypthesis we are seeking evidence for [ALTERNATIVE]
6. the margin of error in a confidence interval consists of a critical value and standard ____ of the statistic [ERROR]
8. the hypothesis we are seeking evidence against [NULL]
10. to compare two proportions, we can construct a ____ z interval (two words) [TWOSAMPLE]
12. if we want to compare two population proprtions, we must have data from two ____ samples [RANDOM]

Chapter 11: Solutions

Section 11.1 Concept 1:

____ *I can state appropriate hypotheses and compute the expected counts and chi-square test statistic for a chi-square test for goodness of fit.*

H_0: Each number on the die is equally likely to be rolled.
H_a: Each number on the die is not equally likely to be rolled.

If the die was fair, we would expect each number to be rolled 300(1/6) = 50 times.

$$\chi^2 = \frac{(42-50)^2}{50} + \frac{(55-50)^2}{50} + \frac{(38-50)^2}{50} + \ldots + \frac{(44-50)^2}{50}$$

$$\chi^2 = 10.28$$

Section 11.1 Concept 2:

____ *I can state and check the Random, 10%, and Large Counts conditions for performing a chi-square test for goodness of fit.*
____ *I can calculate the degrees of freedom and P-value for a chi-square test for goodness of fit.*
____ *I can perform a chi-square test for goodness of fit.*
____ *I can conduct a follow-up analysis when the results of a chi-square test are statistically significant.*

We will perform a chi-square goodness-of-fit test.

H_0: Each value has an equal probability (1/6) of occurring
H_a: At least one value does not have a 1/6 chance of occurring

We have a random sample since the die was rolled 300 times. All expected counts (50) are greater than 5 and each roll is independent.

χ^2= 10.28. With df=5, the *P*-value will be between .05 and 0.10.

Since the *P*-value is greater than 5%, we do not have sufficient evidence to reject the null hypothesis. We cannot conclude that the die is not fair.

Section 11.2 Concept 2:

____ *I can state appropriate hypotheses and compute the expected counts and chi-square statistic for a chi-square test based on data in a two-way table.*
____ *I can calculate the degrees of freedom and P-value for a chi-square test based on data in a two-way table.*
____ *I can perform a chi-square test for homogeneity.*

	Zooboomafoo	**iCarly**	**Phineas and Ferb**
Boys	20 (30)	30 (36)	50 (33.3)
Girls	70 (60)	80 (73.3)	50 (66.6)

Do these data provide convincing evidence that television preferences differ significantly for boys and girls?

We will perform a chi-square test for homogeneity.

H_0: the distribution of preferences is the same for boys and girls.
H_a: the distributions of preference are not the same.

Conditions: We have two independent random samples. All expected counts are greater than 5.

$\chi^2 = 19.32$. With df=2, the *P*-value = 6.38×10^{-5}

Since the *P*-value is less than 5%, we have significant evidence to reject the null hypothesis. It appears the distribution of preferences for boys and girls are not the same. A follow-up analysis suggests the biggest difference occurs in the preferences for Phineas and Ferb, with more boys and fewer girls preferring that show than expected.

Section 11.2 Concept 3:

____ *I can perform a chi-square test for independence*

A recent study looked into the relationship between political views and opinions about nuclear energy. A survey administered to 100 randomly selected adults asked their political leanings as well as their approval of nuclear energy. The results are below:

	Liberal	**Conservative**	**Independent**
Approve	10 (12.15)	15 (8.55)	20 (24.3)
Disapprove	9 (7.29)	2 (5.13)	16 (14.58)
No Opinion	8 (7.56)	2 (5.32)	18 (15.12)

We will perform a chi-square test for association/independence.

H_0: there is no association between political leanings and views on nuclear energy
H_a: there is an association between political leanings and views on nuclear energy

Conditions: We have a single random sample of adults. All expected counts are greater than 5 and there are at least 1000 adults in the population.

$\chi^2 = 11.10$. With df=4, the *P*-value = 0.025.

Since the *P*-value is less than 5%, we have significant evidence to reject the null hypothesis. It appears there is an association between political leanings and views on nuclear energy. A follow-up analysis suggests the biggest difference occurs with more Conservatives approving of nuclear energy than expected.

Chapter 11: Inference for Distributions of Categorical Data

									1 C	2 H	I	S	Q	U	A	R	E	
					3 C					O								
	4 C		5 T	W	O	W	A	Y		M			6 G					
	O				M					O			7 O	N	E	W	8 A	Y
9 M	U	L	T	I	P	L	E			G			O				N	
	N				O					E			D		10 D		A	
	T			11 I	N	D	12 E	P	E	N	D	E	N	C	E		L	
	S				E		X			E			E		G		Y	
					N		P			I			S		R		S	
					T		E			T			S		E		I	
							C			Y					E		S	
							T								S			
							E											
13 O	B	S	E	R	V	E	D											

Across

1. test statistic used to test hypotheses about distributions of categorical data (two words) [CHISQUARE]
5. when comparing two or more categorical variables, or one categorical variable over multiple groups, we arrange our data in a _____ table (two words) [TWOWAY]
7. a ____ table summarizes the distribution of a single categorical variable (two words) [ONEWAY]
9. the problem of doing many comparisons at once with an overall measure of confidence is the problem of ____ comparisons [MULTIPLE]
11. to test whether there is an association between two categorical variables, use a chi-square test of association / ____ [INDEPENDENCE]
13. type of categorical count gathered from a sample of data [OBSERVED]

Down

2. to test whether there is a difference in the distribution of a categorical variable over several populations or treatments, use a chi-square test of ____ [HOMOGENEITY]
3. a follow up analysis involves identifying the ____ that contribued the most to the chi-square statistic [COMPONENT]
4. when calculating chi-square statistics, observations should be expressed in ____, not percents [COUNTS]
6. ____ of fit test: used to determine whether a population has a certain hypothesized distribution of proportions for a categorical variable [GOODNESS]
8. after conducting a chi-square test, be sure to carry out a follow-up ____ [ANALYSIS]
10. specific chi-square distributions are distinguished by ____ of freedom [DEGREES]
12. type of categorical count that we would see if the null hypothesis were true [EXPECTED]

Chapter 12: Solutions

Section 12.1 Concept 2:

____ *I can construct and interpret a confidence interval for the slope β_1 of the population (true) regression line.*

$$95\%CI: -4.77 \pm 2.306(1.0105) = (-7.10, -2.44)$$

We are 95% confident the interval from -7.10 to -2.44 captures the true slope of the regression line.

Section 12.1 Concept 3:

____ *I can interpret the values of b_0, b_1, s, and SE_{b1} in context, and determine these values from computer output.*

____ *I can perform a significance test about the slope β_1 of the population (true) regression line.*

The study in the previous Check for Understanding was expanded to include a total of 77 randomly selected cereals. The scatterplot, residual plot, and computer output of the regression analysis are noted above. Use this output to determine the LSRL for the sample data.

LSRL: predicted rating score = 59.284 – 2.4008(sugar)

The slope = -2.400. For each increase of 1g of sugar, we estimate approximately a 2.4 point decrease in rating score.

Is there convincing evidence that the slope of the true regression line is less than zero?
We will perform a *t*-test for the slope of the regression line.

$$H_0: \beta_1 = 0$$
$$H_a: \beta_1 < 0$$

Conditions: The cereals are randomly selected and independent. The scatterplot suggests a negative linear relationship. The residual plot does not display a distinct pattern, although the residuals do appear to be slightly smaller for greater sugar content.

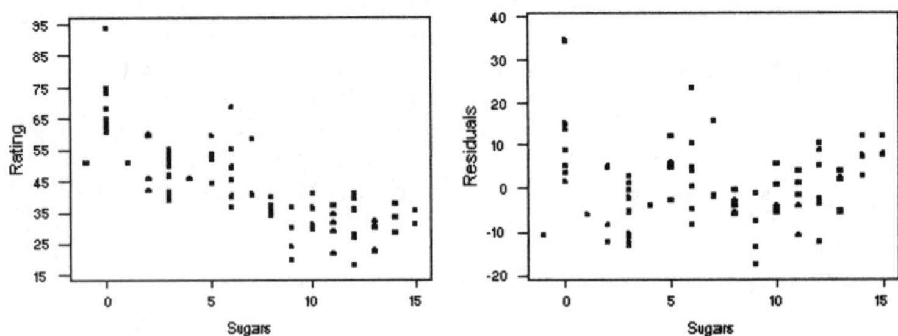

According to the computer output, t=-10.12 and the *P*-Value = 0.000.

Since the *P*-value is less than 5%, we have significant evidence to reject the null hypothesis. It appears the slope of the true regression line may be less than zero.

Section 12.2 Concept 1:

____ *I can use transformations involving powers and roots to find a power model that describes the relationship between two quantitative variables, and use the model to make predictions.*

Plot (diameter$^{0.7}$, length)

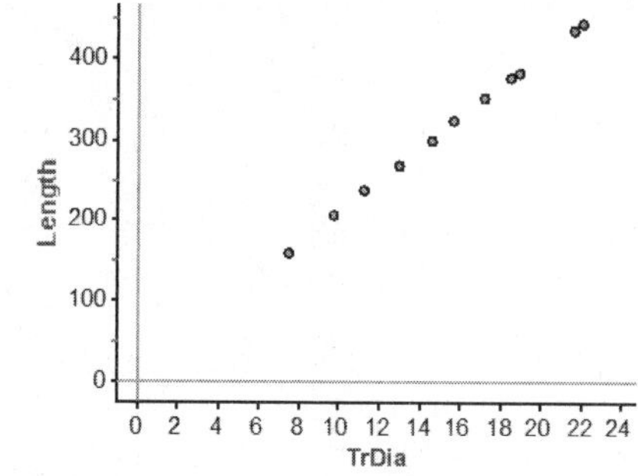

predicted length = 19.5(diameter)$^{0.7}$ +17

If diameter = 47mm,
predicted length = 19.5(47)$^{0.7}$ +17 = 305.74.

Section 12.2 Concept 2:

____ *I can use transformations involving logarithms to find a power model or an exponential model that describes the relationship between two quantitative variables, and use the model to make predictions.*

____ *I can determine which of several transformations does a better job of producing a linear relationship.*

The following data describe the number of police officers (thousands) and the violent crime rate (per 100,000 population) in a sample of states. Use these data to determine a model for predicting violent crime rate based on number of police officers employed. Show all appropriate plots and work.

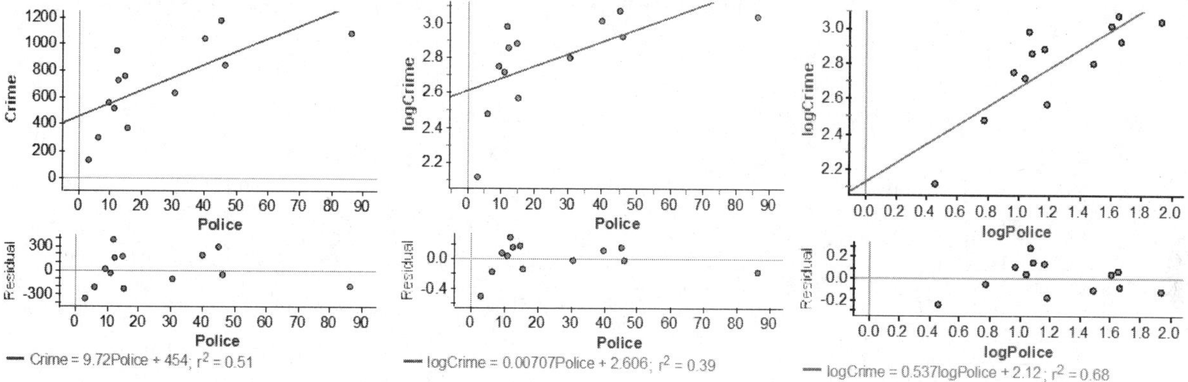

(police, crime) displays a slightly curved pattern.

(log(police), log(crime)) is the most linear of the plots. Therefore, a power model may be the best choice to describe the relationship between police and crime.

$$(\log(crime)\text{-hat}) = 2.12 + 0.537(\log(police))$$

If a state has 25,400 police officers,
log(crime) = 2.12 + 0.537(log(25.4))
log(crime) = 2.874
crime = 748.85

Chapter 12: More About Regression

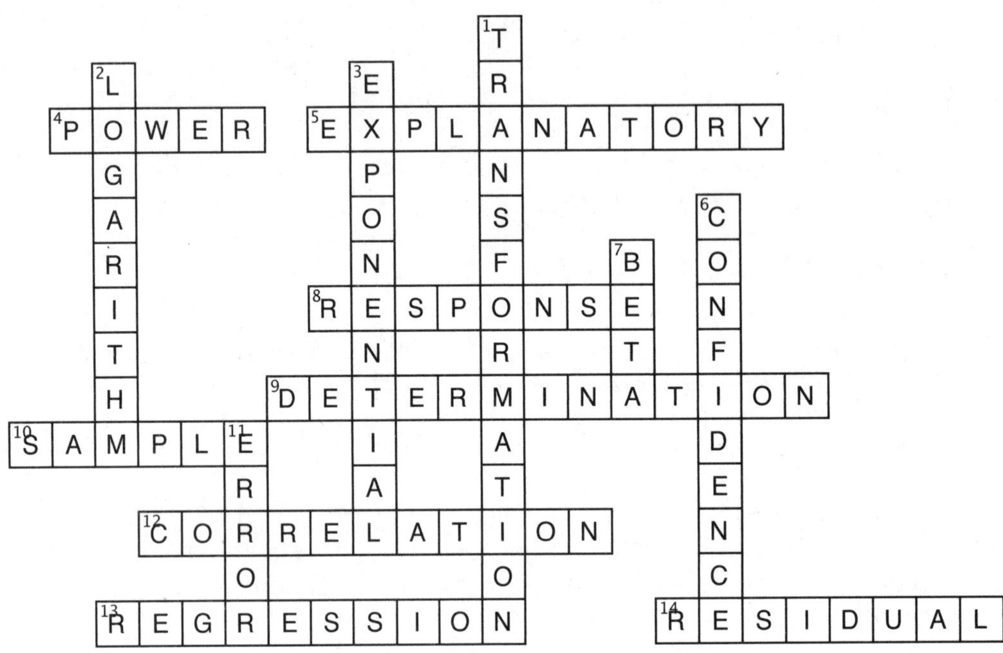

Across

4. if (ln(x), ln(y)) is linear, a/an ____ model may be most appropriate for (x,y) [POWER]
5. another name for the x-variable [EXPLANATORY]
8. another name for the y variable [RESPONSE]
9. the coefficient of ____ indicates the fraction of variability in predicted y values that can be explained by the LSRL of y on x [DETERMINATION]
10. when we construct a LSRL for a set of observed data, we construct a ____ regression line [SAMPLE]
12. the _____ coefficient is a measure of the strength of the linear relationship between two quantitative variables [CORRELATION]
13. the line that describes the relationship between an explanatory and response variable [REGRESSION]
14. observed y - predicted y [RESIDUAL]

Down

1. when data displays a curved relationship, we can perform a ____ to "straighten" it out [TRANSFORMATION]
2. the common mathematical transformation used in this chapter to achieve linearity for a relationship between two quantitative variables [LOGARITHM]
3. if (x, ln(y)) is linear, a/an ____ model may be most appropriate for (x,y) [EXPONENTIAL]
6. to estimate the population (true) slope, we can construct a ____ interval [CONFIDENCE]
7. the Greek letter we use to represent the slope of the population (true) regression line [BETA]
11. to calculate a confidence interval, we must use the standard ____ of the slope [ERROR]

Preparing for the AP® Statistics Examination

After studying and working on statistics all year, you will have a chance to take the AP® Statistics Exam in May. Not only will you be able to apply all that you have learned, you will also have an opportunity to earn college credit for your efforts! If you earn a "passing score" on the exam, you may be eligible to receive AP® credit at a college or university. This means you will have demonstrated a level of knowledge equivalent to that of students completing an introductory statistics course. You may gain credit hours, advanced placement in a course sequence, and possibly a savings in tuition! For these reasons, it is to your advantage to do your best on the exam. Hopefully this guide has helped you understand the concepts in AP® Statistics. Not only is a strong conceptual understanding necessary, but you must also plan and prepare for the exam itself. This section is designed to help you get ready for the exam. Best of luck in your studies!

The number of students taking the AP® Statistics Exam has been increasing in recent years. In 2013, over 169,000 students took the exam and earned scores between 1 and 5.

Score	Qualification	Translation
1	No recommendation	No credit, but you're still better off after having taken the course!
2	Possibly qualified	Credit? Probably not.
3	Qualified	Maybe college credit
4	Well qualified	Most likely college credit
5	Extremely well qualified	Statistical rock star...woohoo!

Of the 217,000 students who took the exam in 2017, 53.8% earned a score of 3 or higher and have a good chance of earning college credit. And just think, the majority of students taking the exam probably didn't prepare as much as you did! After working through this Strive Guide along with your textbook, you are well poised to earn a high score on the exam!

Exam Prep Sections

1. **Sample Schedule**
2. **The AP® Statistics Course Outline**
3. **The AP® Exam Format**
4. **Test Taking Tips**
5. **TPS5e AP® Exam Tips by Chapter**
6. **Planning Your Exam Preparation and Review**
7. **Practice Exams**

Section 1 – Sample Schedule

This section presents a general outline to help you prepare and review for the AP® Statistics Exam. Every student has their own routine and method of studying. Consider these suggestions as you work through the course and create a plan to maximize your chance of earning a 3 or higher on the exam.

1. ***At the start of your AP Statistics course***
   During the first few weeks of the school year, familiarize yourself with the course outline and AP® Statistics Exam format. It has been said that any student can hit a target that is clear and holds still for them. The course outline clearly indicates the topics you will study and the exam format is always the same from year to year. Both the course outline and exam format are presented in the next few sections. Additional information on the exam can be found on the College Board's AP® Central website.

2. ***During your AP Statistics course***
   As you work through each chapter, be sure to read the textbook and do all assigned practice problems. Use this guide to help organize your notes, define key terms, and practice important concepts. Each chapter in this guide includes practice multiple choice and free-response questions. Be sure to try all of them and note any concepts that give you difficulty.

3. ***Six weeks before the exam***
   About six weeks before the exam, begin planning for your review and preparation. If you are currently taking more than one AP® course, be sure you understand when each exam will be given and plan accordingly.

4. ***Four weeks before the exam***
   In early April, you should be wrapping up your studies in the course. This is a good time to attempt a practice test. Your teacher may provide one and/or you may want to take one of the practice exams in this section. Be sure to note any concepts that give you difficulty so you can tailor your preparation and review for the actual exam. There are still several weeks of class left and plenty of time for you to review, practice, and solidify your understanding of the key concepts in the course.

5. ***The week before the exam***
   The week before the exam, you should be done with your studies and should practice, practice, practice! Take this time to refresh your memory on the key concepts and review the topics that gave you the most trouble during the year. Use another practice test to help you get used to the exam format. Be sure to allow 90 minutes for the multiple-choice section and 90 minutes for the free-response section. Use only the formulas and tables provided on the actual exam.

6. ***The day of the exam***
   You have prepared and reviewed as much as possible and are ready for the exam! Make sure you get a good night's sleep, eat a good breakfast the morning of the exam, and check that your calculator is working. Bring several pencils, a clean eraser, and arrive at your exam site early. Good luck!

Section 2 – The AP® Statistics Course Outline

The course outline for AP® Statistics is provided below. This outline lists the topics covered in the course and the percentage of the exam devoted to that material. Additional information on the course outline can be found at the College Board's AP® Central website.

AP® Statistics Course Content Overview
The topics for AP® Statistics are divided into four major themes: exploratory analysis (20-30% of the exam), planning and conducting a study (10-15% of the exam), probability (20-30% of the exam), and statistical inference (30-40% of the exam).

Topic Outline

I. Exploring Data: Describing patterns and departures from patterns (20%–30%)

A. Constructing and interpreting graphical displays of distributions of univariate data (dotplot, stemplot, histogram, cumulative frequency plot)
   1. Center and variability
   2. Clusters and gaps
   3. Outliers and other unusual features
   4. Shape

B. Summarizing distributions of univariate data
   1. Measuring center: median, mean
   2. Measuring spread: range, interquartile range, standard deviation
   3. Measuring position: quartiles, percentiles, standardized scores (z-scores)
   4. Using boxplots
   5. The effect of changing units on summary measures

C. Comparing distributions of univariate data (dotplots, back-to-back stemplots, parallel boxplots)
   1. Comparing center and spread: within group, between group variation
   2. Comparing clusters and gaps
   3. Comparing outliers and other unusual features
   4. Comparing shapes

D. Exploring bivariate data
   1. Analyzing patterns in scatterplots
   2. Correlation and linearity
   3. Least-squares regression line
   4. Residual plots, outliers and influential points
   5. Transformations to achieve linearity: logarithmic and power transformations

E . Exploring categorical data
   1. Frequency tables and bar charts
   2. Marginal and joint frequencies for two-way tables
   3. Conditional relative frequencies and association
   4. Comparing distributions using bar charts

II. Sampling and Experimentation: Planning and conducting a study (10%–15%)

A. Overview of methods of data collection
   1. Census
   2. Sample survey
   3. Experiment
   4. Observational study

B. Planning and conducting surveys
   1. Characteristics of a well-designed and well-conducted survey
   2. Populations, samples and random selection
   3. Sources of bias in sampling and surveys
   4. Sampling methods, including simple random sampling, stratified random sampling and cluster sampling

C. Planning and conducting experiments
   1. Characteristics of a well-designed and well-conducted experiment
   2. Treatments, control groups, experimental units, random assignments and replication
   3. Sources of bias and confounding, including placebo effect and blinding
   4. Completely randomized design
   5. Randomized block design, including matched pairs design

D. Generalizability of results and types of conclusions that can be drawn from observational studies, experiments and surveys

III . Anticipating Patterns: Exploring random phenomena using probability and simulation (20%–30%)

A. Probability
   1. Interpreting probability, including long-run relative frequency interpretation
   2. The Law of Large Numbers
   3. Addition rule, multiplication rule, conditional probability and independence
   4. Discrete random variables and their probability distributions, including binomial and geometric random variables
   5. Simulation of random behavior and probability distributions
   6. Mean (expected value) and standard deviation of a random variable, and linear transformation of a random variable

B. Combining independent random variables
   1. Notion of independence versus dependence
   2. Mean and standard deviation for sums and differences of independent random variables

C. The Normal distribution
   1. Properties of the Normal distribution
   2. Using the Normal distribution table
   3. The Normal distribution as a model for measurements

D. Sampling distributions
   1. Sampling distribution of a sample proportion
   2. Sampling distribution of a sample mean
   3. The Central Limit Theorem
   4. Sampling distribution of a difference between two independent sample proportions
   5. Sampling distribution of a difference between two independent sample means
   6. Simulation of sampling distributions
   7. The *t*-distribution
   8. The chi-square distribution

IV. Statistical Inference: Estimating population parameters and testing hypotheses (30%–40%)

A. Estimation (point estimators and confidence intervals)

1. Estimating population parameters and margins of error
2. Properties of point estimators, including unbiasedness and variability
3. Logic of confidence intervals, meaning of confidence level and confidence intervals, and properties of confidence intervals
4. Large sample confidence interval for a proportion
5. Large sample confidence interval for a difference between two proportions
6. Confidence interval for a mean
7. Confidence interval for a difference between two means (independent and paired samples)
8. Confidence interval for the slope of a least-squares regression line

B. Tests of significance

1. Logic of significance testing, null and alternative hypotheses; p-values; one- and two-sided tests
2. Concepts of Type I and Type II errors; concept of power
3. Large sample test for a proportion
4. Large sample test for a difference between two proportions
5. Test for a mean
6. Test for a difference between two means (independent and paired samples)
7. Chi-square test for goodness of fit, homogeneity of proportions, and independence (one- and two-way tables)
8. Test for the slope of a least-squares regression line

Section 3 – The AP® Statistics Exam Format

The AP® Statistics Exam is divided into two sections. The first section consists of 40 multiple-choice questions and the second section consists of 6 free-response questions. Each section counts for 50% of the total exam score. The number of questions you will be asked from each section of the topic outline corresponds to the percentages provided in the AP® Statistics Course outline. On the free-response section, you will most likely be asked at least one question about exploratory data analysis, at least one about probability, at least one about sampling or experimental design, and at least one about inference. Question 6 on the free-response section is considered an "investigative task" that will stretch your skills beyond what you have learned in the course and is worth almost twice that of each of the other 5 questions. Even though this question most commonly deals with an unfamiliar topic, you should be able to provide a reasonable answer based on your preparation.

The Multiple-Choice Section

You will have 90 minutes to complete the 40 question multiple-choice section of the exam. Each question has five answer choices (A-E), only one of which is correct. Each correct answer earns you one point, while each question answered incorrectly (or left blank) earns you no points. It is in your best interest to answer every question, even if you need to make an educated guess! The questions will cover all of the topics from the course and may require some calculation. Some may require interpreting a graph or reflecting on a given situation. Keep in mind you have just over 2 minutes per question. Don't feel it is necessary to work the questions in order. A good strategy is to work all of the ones you feel are easy and then go back to the ones that might take a little more time. Be sure to keep an eye on the clock!

The Free-Response Section

The second section of the exam is made up of 6 free-response questions including one "investigative task." You will have 90 minutes to complete this section of the exam. You should allow about 25 minutes for the investigative task, leaving just over 12 minutes for each of the other questions. These questions are designed to measure your statistical reasoning and communication skills and are graded on the following 0-4 scale.

4 = Complete Response {NO statistical errors and clear communication}
3 = Substantial Response {Minor statistical error/omission or fuzzy communication}
2 = Developing Response {Important statistical error/omission or lousy communication}
1 = Minimal Response {A "glimmer" of statistical knowledge related to the problem}
0 = Inadequate Response {Statistically dangerous to self and others}

Each problem is graded holistically by the AP® Statistics Readers, meaning your entire response to the problem and all its parts is considered before a score is assigned. Be sure to keep an eye on the clock so you can provide at least a basic response to each question!

Section 4 – Test Taking Tips

Once you have mastered all of the concepts and have built up your statistical communication skills, you are ready to begin reviewing for the actual exam. The following tips were written by your textbook's author, Daren Starnes, and a former AP® Statistics teacher, Sanderson Smith, and are used with permission.

General Advice

Relax, and take time to think! Remember that everyone else taking the exam is in a situation identical to yours. Realize that the problems will probably look considerably more complicated than those you have encountered in other math courses. That's because a statistics course is, necessarily, a "wordy" course.

Read each question carefully before you begin working. This is especially important for problems with multiple parts or lengthy introductions. Underline key words, phrases, and information as you read the questions.

Look at graphs and displays carefully. For graphs, note carefully what is represented on the axes, and be aware of number scale. Some questions that provide tables of numbers and graphs relating to the numbers can be answered simply by "reading" the graphs.

About graphing calculator use: As noted throughout this guide, your graphing calculator is meant to be a tool and is to be used sparingly on some exam questions. Your brain is meant to be your primary tool. Don't waste time punching numbers into your calculator unless you're sure it is necessary. Entering lists of numbers into a calculator can be time-consuming, and certainly doesn't represent a display of statistical intelligence. Do not write directions for calculator button-pushing on the exam and avoid calculator syntax, such as *normalcdf* or *1-PropZTest.*

Multiple-choice questions:

- Examine the question carefully. What statistical topic is being tested? What is the purpose of the question?
- Read carefully. After deciding on an answer, make sure you haven't made a careless mistake or an incorrect assumption.
- If an answer choice seems "too obvious," think about it. If it's so obvious to you, it's probably obvious to others, and chances are good that it is not the correct response.
- Since there is no penalty for a wrong answer or skipped problem, it is to your advantage to attempt every question or make an educated guess, if necessary.

Free-response questions:

- Do not feel it necessary to work through these problems in order. Question 1 is meant to be straightforward, so you may want to start with it. Then move to another problem that you feel confident about. Whatever you do, don't run out of time before you get to Question 6. This Investigative Task counts almost twice as much as any other question.
- Read each question carefully, sentence by sentence, and underline key words or phrases.
- Decide what statistical concept/idea is being tested. This will help you choose a proper approach to solving the problem.
- You don't have to answer a free-response question in paragraph form. Sometimes an organized set

of bullet points or an algebraic process is preferable. NEVER leave "bald answers" or "just numbers" though!

- ALWAYS answer each question in context.
- The amount of space provided on the free-response questions does not necessarily indicate how much you should write.
- If you cannot get an answer to part of a question, make up a plausible answer to use in the remaining parts of the problem.

On problems where you have to produce a graph:

- Label and scale your axes! Do not copy a calculator screen verbatim onto the exam.
- Don't refer to a graph on your calculator that you haven't drawn. Transfer it to the exam paper. Remember, the person grading your exam can't see your calculator!

Communicate your thinking clearly.

- Organize your thoughts before you write, just as you would for an English paper.
- Write neatly. The AP® Readers cannot score your solution if they can't read your writing!
- Write efficiently. Say what needs to be said, and move on. Don't ramble.
- The burden of communication is on you. Don't leave it to the reader to make inferences.
- When you finish writing your answer, look back. Does the answer make sense? Did you address the context of the problem?

Follow directions. If a problem asks you to "explain" or "justify," then be sure to do so.

- Don't "cast a wide net" by writing down everything you know, because you will be graded on everything you write. If part of your answer is wrong, you will be penalized.
- Don't give parallel solutions. Decide on the best path for your answer, and follow it through to the logical conclusion. Providing multiple solutions to a single question is generally not to your advantage. You will be graded on the lesser of the two solutions. Put another way, if one of your solutions is correct and another is incorrect, your response will be scored "incorrect."

Remember that your exam preparation begins on the first day of your AP® Statistics class. Keep in mind the following advice throughout the year.

- READ your statistics book. Most AP® Statistics Exam questions start with a paragraph that describes the context of the problem. You need to be able to pick out important statistical cues. The only way you will learn to do that is through hands-on experience.
- PRACTICE writing about your statistical thinking. Your success on the AP® Statistics Exam depends not only on how well you can "do" the statistics, but also on how well you explain your reasoning.
- WORK as many problems as you can in the weeks leading up to the exam.

Section 5 – TPS6e AP® Exam Tips by Chapter

These AP® Statistics Exam tips will look familiar to you from reading "The Practice of Statistics, 6e" text. We have included them here as a valuable review as you prepare for the AP® Statistics Exam

Chapter 1: Exploring Data

- If you learn to distinguish categorical from quantitative variables now, it will pay big rewards later. You will be expected to analyze categorical and quantitative variables correctly on the AP® exam.
- When comparing distributions of quantitative data, it's not enough just to list values for the center and spread of each distribution. You have to explicitly *compare* these values, using words like "greater than," "less than," or "about the same as."
- If you're asked to make a graph on a free-response question, be sure to label and scale you axes. Unless your calculator shows labels and scaling, don't just transfer a calculator screen shot to your paper.
- You may be asked to determine whether a quantitative data set has any outliers. Be prepared to state and use the rule for identifying outliers.
- Use statistical terms carefully and correctly of the AP® exam. Don't say "mean" if you really mean "median." Range is a single number; so are Q_1, Q_3, and *IQR*. Avoid the use of language like "the outlier *skews* the mean." Skewed is a shape. If you misuse a term, expect to lose some credit.

Chapter 2: Modeling Distributions of Data

- Avoid using calculator speak to show your work on Normal calculations. To get full credit, take the time to draw a normal distribution, note the boundary value(s), and shade the area of interest. Show your work when performing calculations!
- Normal probability plots are not included on the AP® Statistics topic outline. However, these graphs are very useful for assessing Normality. You may use them of the AP® exam if you wish – just be sure that you know what you're looking for (a linear pattern).

Chapter 3: Describing Relationships

- If you are asked to make a scatterplot on a free-response question, be sure to label and scale both axes. Don't copy an unlabeled calculator graph directly onto your paper.
- IF you're asked to interpret a correlation, start by looking at a scatterplot of the data. Then be sure to address direction, form, strength, and outliers (sound familiar?) and put your answer in context.
- When displaying the equation of a least-squares regression line, the calculator will report the slope and intercept with much more precision than we need. However, there is no firm rule for how many decimal places to show for answers on the AP® exam. Our advice: Decide how much to round based on the context of the problem you are working on.
- Students often have a hard time interpreting the value of r^2 on AP® exam questions. They frequently leave out key words in the definition. Our advice: Treat this as a fill-in-the-blank exercise. Write "________% of the variation in [response variable name] is accounted for by the linear model relating [response variable name] to [explanatory variable name]."

Chapter 4: Designing Studies

- If you're asked to describe how the design of a study leads to bias, you're expected to identify the *direction* of the bias. Suppose you were asked, "Explain how using a convenience sample of students in your statistics class to estimate the proportion of all high school students who own a graphing calculator could result in bias." You might respond, "This sample would probably include a much higher proportion of students with a graphing calculator than in the population at large. That is, this method would probably lead to an overestimate of the actual population proportion."
- If you are asked to identify a possible confounding variable in a given setting, you are expected to explain how the variable you choose (1) is associated with the explanatory variable and (2) affects the response variable.
- If you are asked to describe the design of an experiment on the AP® exam, you won't get full credit for a diagram. You are expected to describe how the treatments are assigned to the experimental units and to clearly state what will be measured or compared. Some students prefer to start with a diagram and then add a few sentences. Others choose to skip the diagram and put their entire response in narrative form.
- Don't mix the language of experiments and the language of sample surveys or other observational studies. You will lose credit for saying things like "use a randomized block design to select the sample for this survey" or "this experiment suffers from nonresponse since some subjects dropped out during the study."

Chapter 5: Probability – What Are the Chances?

- On the AP® exam, you may be asked to describe how you will perform a simulation using rows of random digits. If so, provide a clear enough description of your simulation process for the reader to get the same results you did from *only* your written explanation.
- Many probability problems involve simple computations that you can do on your calculator. It may be tempting to just write down your final answer without showing the supporting work. Don't do it! A "naked answer", even if it's correct, will usually earn you no credit on a free response question.

Chapter 6: Random Variables

- If the mean of a random variable should have a non-integer value, but you report it as an integer, your answer will be marked as incorrect.
- When showing your work on a free response question, you must include more than a calculator command. Writing normalcdf (68,70,64,2.7) will *not* earn you full credit for a Normal calculation. At a minimum you must indicate what each of those calculator inputs represents. Better yet, sketch and label a Normal curve to show what you're finding.
- Don't rely on "calculator speak" when showing your work on free-response questions. Writing binompdf(5,0.25,3)= 0.08789 will *not* earn you full credit for a binomial probability calculation. At the very least, you must indicate what each of those calculator inputs represents. For example, "I used binompdf(trials 5,p:0.25, *x* value:3)."

Chapter 7: Sampling Distributions

- Terminology matters. Don't say "sample distribution" when you mean sampling distribution. You will lose credit on free-response questions for misusing statistical terms.
- Notation matters. The symbols $\hat{p}$, $\bar{x}$, p, μ, σ, $\mu_{\hat{p}}$, $\sigma_{\hat{p}}$, $\mu_{\bar{x}}$, and $\sigma_{\bar{x}}$ all have specific and different meanings. Either use notation correctly—or don't use it at all. You can expect to lose credit if you use incorrect notation.

Chapter 8: Estimating with Confidence

- On a given problem, you may be asked to interpret the confidence interval, the confidence level, or both. Be sure you understand the difference: the confidence level describes the long-run capture rate of the method and the confidence interval gives a set of plausible values for the parameter and the confidence level describes the long-run capture rate of the method.
- If a free-response question asks you to construct and interpret a confidence interval, you are expected to do the entire four-step process. That includes clearly defining the parameter and checking conditions.
- You may use your calculator to compute a confidence interval on the AP® exam. But there's a risk involved. If you just give the calculator answer with no work, you'll get either full credit for the "Do" step (if the interval is correct) or no credit (if it's wrong). We recommend showing the calculation with the appropriate formula and then checking with your calculator. If you opt for the calculator-only method, be sure to name the procedure (e.g., one-proportion *z* interval) and to give the interval (e.g., 0.514 to 0.607).
- If a question of the AP® exam asks you to calculate a confidence interval, all the conditions should be met. However, you are still required to state the conditions and show evidence that they are met.
- It is not enough just to make a graph of the data on your calculator when assessing Normality. You must *sketch* the graph on your paper to receive credit. You don't have to draw multiple graphs – any appropriate graph will do.

Chapter 9: Testing a Claim

- The conclusion to a significance test should always include three components: (1) an explicit comparison of the *P*-value to a stated significance level (2) a decision about the null hypothesis: reject or fail to reject H_o, and (3) a statement in the context of the problem about whether or not there is convincing evidence for H_a.
- When a significance test leads to a fail to reject H_o decision, be sure to interpret the results as "we don't have enough evidence to conclude H_a." Saying anything that sounds like you believe H_o is (or might be) true will lead to a loss of credit. And don't write text-message-type responses, like "FTR the H_o."
- You can use your calculator to carry out the mechanics of a significance test on the AP® exam. But there's a risk involved. If you just give the calculator answer with no work, and one or more of your values is incorrect, you will probably get no credit for the "Do" step. We recommend doing the calculation with the appropriate formula and then checking with your calculator. If you opt for the calculator-only method, be sure to name the procedure (one-proportion *z* test) and to report the test statistic ($z = 1.15$) and *P*-value (0.1243).
- It is not enough just to make a graph of the data on your calculator when assessing Normality. You must *sketch* the graph on your paper to receive credit. You don't have to draw multiple graphs – any appropriate graph will do.
- Remember: if you just give calculator results with no work, and one or more values are wrong, you probably won't get any credit for the "Do" step. We recommend doing the calculation with

the appropriate formula and then checking with your calculator. If you opt for the calculator-only method, name the procedure (*t* test) and report the test statistic ($t = -0.94$), degrees of freedom (df = 14), and *P*-value (0.1809).

Chapter 10: Comparing Two Populations or Groups

- The formula for the two-sample z interval for p_1-p_2 often leads to calculation errors by students. As a result, we recommend using the calculator's 2-PropZInt feature to compute the confidence interval of the AP® Exam. Be sure to name the procedure (two-proportion *z* interval) and to give the interval (0.076, 0.143) as part of the "Do" step.
- The formula for the two-sample *z* statistic for a test about p_1-p_2 often leads to calculation errors by students. As a result, we recommend using the calculator's 2-PropZTest feature to compute the confidence interval of the AP® Exam. Be sure to name the procedure (two-proportion *z* test) and to report the test statistic ($z = 1.17$) and *P*-value (0.2427) as part of the "Do" step.
- The formula for the two-sample t interval for $\mu_1 - \mu_2$ often leads to calculation errors by students. As a result, we recommend using the calculator's 2-SampTInt feature to compute the confidence interval of the AP® Exam. Be sure to name the procedure (two-sample t interval) and to give the interval (3.9362, 17.724) and df (55.728) as part of the "Do" step.
- When checking the Normal condition on an AP® exam question involving inference about means, be sure to include a graph. Don't expect to receive credit for describing a graph that you made on your calculator but didn't put on paper.
- The formula for the two-sample t statistic for $\mu_1 - \mu_2$ often leads to calculation errors by students. As a result, we recommend using the calculator's 2-SampTTest feature to compute the confidence interval of the AP® Exam. Be sure to name the procedure (two-sample t test) and to report the test statistic (t=1.600, P-value (0.0644), and df (15.59) as part of the "Do" step.

Chapter 11: Inference for Distributions of Categorical Data

- You can use your calculator to carry out the mechanics of a significance test on the AP® exam. But there's a risk involved. If you just give the calculator answer with no work, and one or more of your values are incorrect, you will probably get no credit for the "Do" step. We recommend writing out the first few terms of the chi-square calculation followed by "...". This approach might help you earn partial credit if you enter a number incorrectly. Be sure to name the procedure (χ^2GOF-Test) and to report the test statistic (χ^2=11.2), degrees of freedom (df= 3), and *P*-value (0.011).
- In the "Do" step, you aren't required to show every term in the chi-square statistic. Writing the first few terms of the sum and the last term, separated by ellipsis, is considered as "showing work." We suggest that you do this and then let your calculator tackle the computations.
- You can use your calculator to carry out the mechanics of a significance test on the AP® exam. But there's a risk involved. If you just give the calculator answer with no work, and one or more of your values is incorrect, you will probably get no credit for the "Do" step. We recommend writing out the first few terms of the chi-square calculation followed by "...". This approach might help you earn partial credit if you enter a number incorrectly. Be sure to name the procedure (x^2-Test for homogeneity) and to report the test statistic (χ^2=18.279), degrees of freedom (df= 4), and *P*-value (0.011).
- If you have trouble distinguishing the two types of chi-square tests for two-way tables, you're better off just saying "chi-square test" than choosing the wrong type. Better yet, learn to tell the difference!

Chapter 12: More About Regression

- The AP® exam formula sheet gives $\hat{y} = b_0 + b_1 x$ for the equation of the sample (estimated) regression line. Many calculators use simpler notation, $\hat{y} = a + bx$. Just remember: the coefficient of *x* is always the slope, no matter what symbol is used.
- The AP® exam formula sheet gives the formula for the standard error of the slope as

$$s_{b_1} = \frac{\sqrt{\frac{\Sigma(y_i - \hat{y}_i)^2}{n-2}}}{\sqrt{\Sigma(x_i - \bar{x})^2}}$$

- The numerator is just a fancy way of writing the standard deviation of the residuals *s*. Can you show that the denominator of this formula is the same as ours?
- The formula for the *t* interval for the slope of a population (true) regression line often leads to calculation errors by students. As a result, we recommend using the calculator's LinRegTInt feature to compute the confidence interval of the AP® Exam. Be sure to name the procedure (*t* interval for slope) and to give the interval (-0.217, -0.108) and df(14) as part of the "Do" step.
- When you see a list of data values on an exam question, don't just start typing the data into your calculator. Read the question first. Often, additional information is provided that makes it unnecessary for you to enter the data at all. This can save you valuable time on the AP® exam.

Section 6 – Planning Your Exam Preparation and Review

This book has been designed to help you identify your statistical areas of strength and areas in which you need improvement. If you have been working through the checks-for-understanding, multiple-choice questions, FRAPPYs, and vocabulary puzzles for each chapter, you should have an idea which topics are in need of additional study or review.

The practice tests that are included in the section of the book are designed to help you get familiar with the format of the exam as well as check your understanding of the key concepts in the course. Plan on taking both of these tests as part of your preparation and review.

Allow yourself 90 minutes for each section of the test and be sure to check your answers in the provided keys. For each question you missed, determine whether it was a simple mistake or whether you need to go back and study that topic again. After you complete the tests, continue practicing problems before the exam date.

The best preparation for the exam (other than having a solid understanding of statistics) is to practice as many multiple-choice and free-response questions as possible. Ask your teacher or refer to the College Board's AP Central website for additional resources to help you with this!

Section 7 – Practice Exams

Use the following two practice exams to help you prepare for the AP® Statistics Exam. Allow yourself 90 minutes for the multiple-choice questions, take a break, and allow yourself 90 minutes for the free-response section. When you finish, correct your exam and note your areas in need of improvement. Answers for each exam are included in the answer key.

Practice Test 1

Multiple-Choice Section

1. A large high school offers AP Statistics and AP Calculus. Among the seniors in this school, 65% take AP Statistics, 45% take AP Calculus, and 30% take both. If a senior class student is randomly selected, what is the probability they are in AP Statistics or AP Calculus, but not both?
   (A) 10%
   (B) 35%
   (C) 50%
   (D) 70%
   (E) 75%

2. An agricultural station is testing the yields for six different varieties of seed corn. The station has four large fields available, which are located in four distinctly different parts of the county. The agricultural researchers consider the climatic and soil conditions in the four parts of the county as being unequal but are reasonably confident that the land in each field is fairly uniform. The researchers divide each field into six sections and then randomly assign one variety of corn seed to each section in that field. This procedure is done for each field. At the end of the growing season, the corn will be harvested, and the yield measured in tons per acre will be compared. Which one of the following statements about the design is correct?
   (A) This is an observational study since the researchers are watching the corn grow.
   (B) This a block design with fields as blocks and seed types as treatments.
   (C) This is a block design with seed types as blocks and fields as treatments.
   (D) This is a completely randomized design since the six seed types were randomly assigned to the four fields.
   (E) This is a completely randomized design with 24 treatments—6 seed types and 4 fields.

3. A large company is testing a new marketing strategy for increasing sales at its stores. Currently store sales average \$250,000 per month. Thus the company wishes to test the following hypotheses: $H_o : \mu = \$250,000$ versus $H_a : \mu > \$250,000$, where μ = true mean monthly sales per store. Which one of the following sample sizes and significance levels would lead to the highest power for this test?
   (A) $n = 10$ *and* $\alpha = 0.01$
   (B) $n = 20$ *and* $\alpha = 0.05$
   (C) $n = 40$ *and* $\alpha = 0.05$
   (D) $n = 20$ *and* $\alpha = 0.10$
   (E) $n = 40$ *and* $\alpha = 0.10$

4. In 1970, the U.S. Armed Services instituted a lottery as a random selection process when 19-year-old males were drafted into service. Each of the 366 possible birth dates was placed in a separate capsule and each capsule was placed in a large plastic drum. Those with the lowest lottery numbers were drafted before others. For example, the May 9th number was 176. This meant that there would be 175 groups of males born on other days drafted before those who had been born on May 9th. The scatter plot below shows the <u>median</u> draft eligible number for each month of the year. The line that connects the points is sometimes called a median trace line. Which one of the following statements is TRUE?

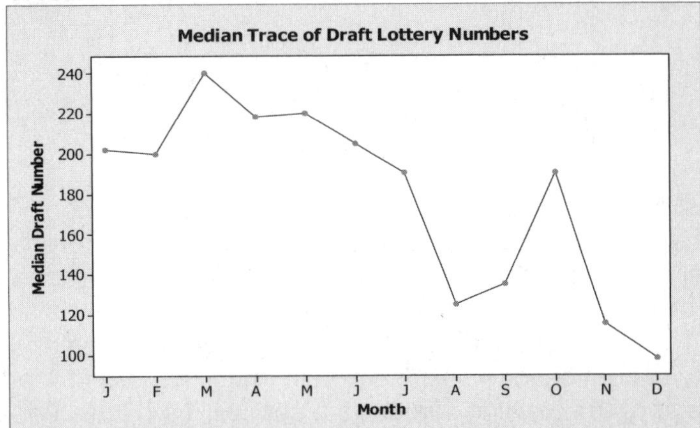

(A) There seems to be no relationship between birth month and median draft lottery number.
(B) There is a very weak negative correlation between birth month and median draft lottery number.
(C) Those people born near the end of the year tended to have lower median draft lottery numbers than those born at the beginning of the year.
(D) Since the spread for each month was about the same (30 days), the median draft lottery number for each month is about the same.
(E) Since each birth date is independent, the scatterplot indicates that you were just as likely to be drafted early in the process as late in the process.

5. Traffic engineers studied the traffic patterns of two busy intersections on opposite sides of town at rush hour. At the first intersection, the average number of cars waiting to turn left was 17 with a standard deviation of 4 cars. At the second intersection, the average number of cars waiting to turn left was 25 cars with a standard deviation of 7 cars. The report combined the mean number of cars at both intersections. Assuming that the number of cars waiting to turn left at each intersection is independent, what is the standard deviation of the total of the number of cars waiting to make a left turn at the two intersections?

(A) 11
(B) $\sqrt{65}$
(C) 8
(D) $\sqrt{33}$
(E) $\sqrt{11}$

6. Two drugs are available to treat the flu. Adiflu is effective in curing 80% of flu cases, but results in unpleasant side effects for 45% of individuals who take it. Beneflu cures 55% of flu cases but results in unpleasant side effects for 25% of individuals who take it. Suppose 70 people take Adiflu and 90 people take Beneflu. What is the expected number of people who will experience unpleasant side effects?
(A) 37.6
(B) 54
(C) 58
(D) 105.5
(E) 110.5

7. A pharmaceutical company has developed a new medication to lower a person's cholesterol. The advertisement for the new drug cites the results of an experiment comparing the drug with a placebo stating that the cholesterol level had been lowered by an average of 13.2 mg/l with a *P*-value < 0.01. Which one of the following best explains the meaning of *P*-value?

(A) There was a less than 1% difference in the mean cholesterol levels between those taking the placebo and those taking the new medication using the same experimental method.
(B) Less than 1% of the people who were given the new cholesterol-reducing drug experienced any serious side effects.
(C) If the mean cholesterol levels for the placebo and medication groups were really equal, we would have obtained a difference such as this or greater, in less than 1% of samples by random chance.
(D) All but 1% of the people who were given the experimental drug experienced a drop of 13.2 mg/l when compared to the placebo group.
(E) The difference in the mean cholesterol levels between those taking the placebo and those taking the medication was not significant at the 1% level.

8. A simple random sample of 100 batteries is selected from a process that produces batteries with a mean lifetime of 32 hours and a standard deviation of 3 hours. Thus, the standard deviation of the sampling distribution of the mean, $\sigma_{\bar{x}}$, is equal to 0.3. If the sample size had been 400, how would the value of $\sigma_{\bar{x}}$ change?
(A) It is one-fourth as large as when $n = 100$.
(B) It is one-half as large as when $n = 100$.
(C) It is twice as large as when $n = 100$.
(D) It is four times as large as when $n = 100$.
(E) The value of $\sigma_{\bar{x}}$ does not change.

9. A poll of 83 AP Statistics students asked, "How many hours a week do you spend studying statistics?" The results are summarized in the computer output below.

Variable	N	Mean	Median	TrMean		StDev
Hours	83	8.5	7.2	7.38		4.9
	Minimum		Maximum	Q1	Q3	
Hours	0.00		25.00	5	12	

If a boxplot of this data was constructed, what data points would be marked as outliers?

(A) Any points greater than 10.5
(B) Any points greater than 17
(C) Any points greater than 17.7
(D) Any points greater than 22.5
(E) Any points greater than 25

10. Prospective salespeople for a high-tech company are now being offered a sales training program before they start working for the company. Previous data indicates that the mean number of sales per month for those who did not participate in the program is 28. To determine whether the sales-training program is effective, a random sample of 32 new sales personnel is given the training. The company is interested in assessing whether the sales program significantly increases the mean monthly sales as there will be a considerable investment in time and money to conduct the training program. The null and alternative hypotheses would be $H_o : \mu = 28 \textit{ and } H_a : \mu > 28$.Which of the following describes a Type II error?

(A) The company finds evidence that sales have increased when, in fact, they have. The company implements the sales program and recovers the cost of the training program.
(B) The company finds evidence that sales have increased when, in fact, they have Not. The company implements the sales program and doesn't recover its costs for the training.
(C) The company finds no evidence that sales have increased when, in fact, they have not. The company does not implement the sales program and doesn't waste the money on training.
(D) The company finds no evidence that sales have increased when, in fact, they have. The company does not implement the sales program and misses out on a chance to increases sales.
(E) The company takes more samples to ensure that they make the correct decision.

11. A child psychologist claims the mean attention span for a 1-year old is 12 seconds. An advocacy group claims exposure to a popular cartoon results in a decrease in attention span. To investigate this claim, which of the following hypotheses would be appropriate?
    (A) H_0: the mean attention span for children who watch the cartoon is less than 12 seconds.
    (B) H_0: the mean attention span for children who watch the cartoon is greater than 12 seconds.
    (C) H_a: the mean attention span for children who watch the cartoon is less than 12 seconds.
    (D) H_a: the mean attention span for children who watch the cartoon is greater than 12 seconds.
    (E) H_a: the mean attention span for children who watch the cartoon is less than or equal to 12 seconds.

12. Researchers wanted to know if playing soft music improved students' performance on tests. Twelve students worked a complicated paper-and-pencil maze while listening to music and again while not listening to music. The order of music and non-music was randomly decided, and the mazes were completed one week apart. The time to complete each maze (in seconds) was recorded and the results are shown in the following table.

Subject	1	2	3	4	5	6	7	8	9	10	11	12
Time with music	83	79	73	76	75	79	74	67	80	69	71	63
Time without music	89	82	85	82	81	90	75	69	74	82	78	68

Which one of the following test procedures would be the most appropriate in order determine if there is a significant difference in time to complete the maze?

(A) A chi-square test for independence.
(B) A two-proportion z test.
(C) A t test on the slope of the regression line.
(D) A two-sample t test for difference in means.
(E) A matched pair t test for mean difference.

13. A cereal manufacturer is packaging collectible cards from a popular movie in each box of cereal. One card, a holographic card of a memorable scene, is more rare than the others and occurs in only 3% of boxes. Which of the following represents the probability that the fifth box you purchase will contain your first holographic card?

(A) $({}_5C_1)(.97)^4(.03)$
(B) $(.97)^4(.03)$
(C) $(.97)(.03)^4$
(D) $1-(.97)^5$
(E) $1-(.03)^5$

14. A local bridge club has 120 members. A cumulative relative frequency graph of their ages is shown in the figure below. Approximately how many of the bridge club members are more than 80 years of age?

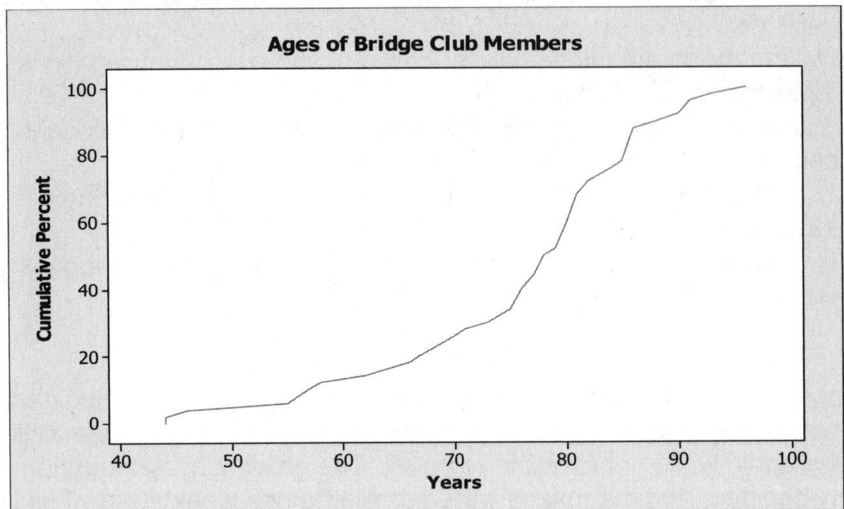

(A) 42
(B) 50
(C) 70
(D) 85
(E) 92

15. A study tested a recently advertised claim that a newly developed "gas pill" boosts mileage when added to a tank of gas. Ten randomly selected new cars were used in the study and would be run over an identical course twice, once with the additive and once without the additive. The order of the two treatments (pill and no pill) was randomized. Each car's tank was filled with a standard gasoline and one of the treatments applied. The car was then driven until the tank was empty. The car's tank was then refilled with the same standard gasoline and the process repeated for the other treatment. Professional drivers drove over the same course at identical speeds and did not know which treatment was used. The miles per gallon for each car were calculated under both treatments. Let μ_1 = mean mileage for all cars that would be driven without the gas pill and μ_2 = mean mileage for all cars that would be driven with the gas pill. The 95% confidence interval for the mean difference is given by (-0.13, 1.73).

Based on the information, which one of the following statements is TRUE?

(A) Ninety-five percent of the time, the confidence interval will contain 0.
(B) A two-sample *t* test should have been used instead of a paired *t* test.
(C) The true mean mileage is higher for cars without the additive than those with the additive since the mean difference is positive.
(D) The sample size is too small to draw any conclusions.
(E) Since the interval contains 0, we cannot conclude that using the gas pill made a difference in mileage.

16. A veterinarian was interested in comparing the effects of two drugs, glucosamine and chondroitin, on arthritis in beagles. She believes glucosamine is more effective in increasing mobility for dogs that suffer from arthritis in their hips. A sample of 150 beagles showed no difference in mobility between dogs treated with each drug. When the sample size was increased to 475 beagles, the results showed that glucosamine was significantly more effective in increasing mobility. This is an example of
(A) larger sample sizes leading to higher power.
(B) using a test for means instead of a test for proportions.
(C) reducing the probability of a Type I error.
(D) using blocking to reduce variation.
(E) always using large samples.

17. Eight people who suffer from hay fever volunteer to test a new medication that will relieve the symptoms. The last names of the volunteers are

1. Rodriguez
2. Liu
3. Brown
4. Kim
5. Harris
6. Munoz
7. Klein
8. Scott

Four of the volunteers will receive the new medication, while the other four will receive a placebo as part of a double-blind experiment. Starting at the left of the list of random numbers below and reading from left to right, assign the four people to be given the medication.

07119 97336 71048 08178 77233 13916 47564 81056 97025 85977 29372

The four people assigned to the treatment group are

(A) Rodriguez, Liu, Brown, Klein
(B) Liu, Harris, Klein, Scott
(C) Rodriguez, Brown, Munoz, Klein
(D) Rodriguez, Brown, Klein, Scott
(E) Rodriguez, Kim, Klein, Scott

18. A farmer randomly sampled 46 sheep from a very large herd to construct a 95% confidence interval to estimate the mean weight of wool (in pounds) from a sheep in her herd. The interval obtained was (23.7, 33.6). If the farmer had used a 90% confidence interval instead, the confidence interval would have been

(A) narrower and the resulting estimate would have been more precise
(B) narrower and the resulting estimate would have been less precise
(C) narrower but the precision cannot be determined.
(D) wider and the resulting estimate would have been more precise.
(E) wider and the resulting estimate would have been less precise.

19. Company A has 500 employees and Company B has 5000 employees. Union negotiators want to compare the salary distributions for the two companies. Which one of the following would be the most useful for accomplishing this comparison?
(A) Dotplots for A and B drawn on the same scale.
(B) Back-to-back stemplots for A and B.
(C) Two frequency histograms for A and B drawn on the same scale.
(D) Two relative-frequency histograms for A and B drawn on the same scale.
(E) A scatterplot of A versus B

20. Most flights at airports leave at their posted departure time. The frequency table below summarizes the times in the last month at a major airport where flights have been delayed past their posted departure time. The delays occurred for a variety of reasons—mechanical problems with the plane, congestion on the tarmac, the previous flight at a gate was late in leaving, weather, etc.

Delay Time	Frequency
Less than 10 minutes	11
At least 10 but less than 20 minutes	37
At least 20 but less than 30 minutes	33
At least 30 but less than 40 minutes	30
At least 40 but less than 50 minutes	26
At least 50 but less than 60 minutes	14
At least 60 but less than 70 minutes	6
At least 70 minutes	3

Which one of the following represents possible values for the median and mean delay times for flights from this major airport?

(A) median = 27 minutes and mean = 24 minutes
(B) median = 29 minutes and mean = 29 minutes
(C) median = 29 minutes and mean = 32 minutes
(D) median = 31 minutes and mean = 35 minutes
(E) median = 35 minutes and mean = 39 minutes

21. A national tabloid newspaper wants to estimate the true proportion of all people who believe that life exists elsewhere in the universe. What is the least number of people that should be sampled in order to estimate the true proportion of those who believe that life exists elsewhere within 4% of the real answer with 95% confidence? Past data indicate that 30% of the general population holds this belief.
(A) 11
(B) 356
(C) 505
(D) 610
(E) 711

22. Consumers frequently complain that there is a large variation in prices charged by different pharmacies and drug stores for the same medication. A survey of a large sample of the pharmacies in a major city revealed that the prices charged for one bottle containing 50 tablets of a popular pain reliever were approximately normally distributed. The charge of $10.48 for this bottle was at the 90th percentile and 20% of the bottles cost less than $8.61. What are, respectively, the approximate mean and standard deviation of the price distribution for bottles of this pain reliever?

(A) $9.35 and $0.88
(B) $9.41 and $2.67
(C) $9.55 and $0.88
(D) $9.55 and $1.70
(E) $9.73 and $1.70

23. A study of the effect of a fertilizer on the yield of 10 plots of tomato plants resulted in the following computer output:
The regression equation is
Yield = 10.1 + 1.15 Fertilize

Predictor	Coef	SE Coef	T	P
Constant	10.100	0.7973	12.67	0.000
Fertilize	1.150	0.1879	6.12	0.000

S = 0.8404 R-Sq = 82.4% R-Sq(adj) = 80.2%

Which of the following would represent a 99% confidence interval to estimate the true slope of the regression line relating amount of fertilizer to yield? Assume all conditions for inference are met.

(A) $1.150 \pm 3.355(0.8404)$

(B) $1.150 \pm 3.169\left(\frac{0.8404}{\sqrt{10}}\right)$

(C) $1.150 \pm 6.12(0.1879)$

(D) $1.150 \pm 3.355(0.1879)$

(E) $1.150 \pm 3.169(0.1879)$

24. Flossing helps prevent tooth decay and gum disease. Yet reports suggest millions of Americans never floss. You are planning a sample survey to determine the proportion of the students at your high school who never floss. Which of the following will result in the largest margin of error?

(A) A sample size of 200 and a confidence level of 95%
(B) A sample size of 200 and a confidence level of 96%
(C) A sample size of 200 and a confidence level of 99%
(D) A sample size of 300 and a confidence level of 96%
(E) A sample size of 300 and a confidence level of 99%

25. A statistics class took a random sample of students at the school to find the proportion of those who claimed to be vegetarians. This class found 12 out of the 150 students questioned were vegetarians. Another statistics class in another school took a similar random sample of the students at its school and found that 9 out of 90 claimed to be vegetarians. Which one of the following represents the approximate 90% confidence interval for the difference between the proportions of students of the two schools that would claim to be vegetarians?

(A) $0.02 \pm 1.645\left(\frac{(0.08)(0.92)+(0.10)(0.9)}{\sqrt{150+90}}\right)$

(B) $0.02 \pm 1.645\left(\frac{(0.08)(0.92)}{\sqrt{150}}+\frac{(0.1)(0.9)}{\sqrt{90}}\right)$

(C) $0.02 \pm 1.645\left(\sqrt{\frac{(0.08)(0.92)}{150}+\frac{(0.1)(0.9)}{90}}\right)$

(D) $0.02 \pm 1.645\left(\sqrt{\frac{(0.08)(0.92)+(0.1)(0.9)}{150+90}}\right)$

(E) $0.02 \pm 1.645\left(\sqrt{\frac{(0.08)}{150}+\frac{(0.1)}{90}}\right)$

26. A study performed by a psychologist determined that a person's sense of humor is linearly related to their IQ. The equation of the least squares regression line is $\widehat{humor} = -49 + 1.8(IQ)$. What is the residual for an individual with an IQ score of 110 and a humor score of 140?

(A) -30
(B) -9
(C) 9
(D) 30
(E) Cannot be determined since we don't know the original data points.

27. A large sales company recruits many graduating students from universities for its workforce. Thirty percent of those hired for management positions come from private universities and colleges and the rest from public colleges and universities. It is very expensive and time-consuming to train new managers, so the company is examining its retention rate (those still working for the company after six years) of these hires. Over the past six years, 35% of those managers who were hired from private schools had left for other jobs, while 20% of those from public schools had done so. What is the probability that a randomly selected person, who left the company within the past six years, was hired from a private university or college?

(A) 0.105
(B) 0.4286
(C) 0.2308
(D) 0.35
(E) 0.2450

28. In a recent social research poll, 4800 randomly selected adults aged 18 to 26 were asked to record the number of hours they spent watching television and the number of non-work-related hours they spent surfing the Web during a typical week. Parallel boxplots of the data are given below.

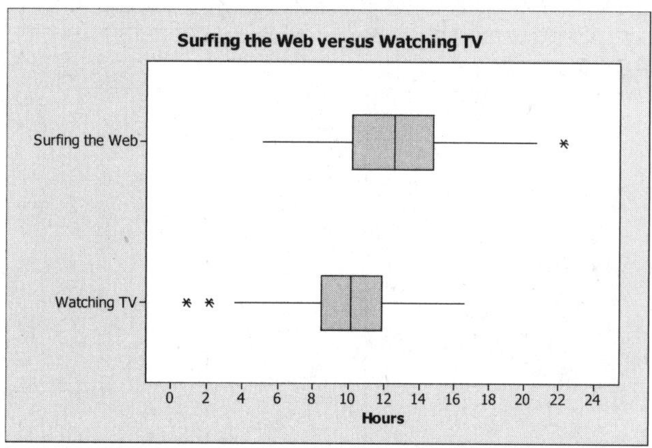

Based on the plots, which one of the following statements is **FALSE**?

(A) The range of the distribution of hours spent surfing the Web is higher than the range for the distribution of hours spent watching television.

(B) On average, adults aged 18 to 26 spent more time surfing the Web than watching television.

(C) The median number of hours spent surfing the Web is higher than the median number of hours spent watching television.

(D) The *IQR* for hours spent surfing the Web is smaller than the IQR for hours spent watching TV.

(E) Both distributions of hours are a slightly skewed.

29. Three distributions are shown below. One represents the distribution of original values, one represents the sampling distribution of means for samples of size $n = 5$, and one is the distribution of sample means for samples of size 20. If the distributions are ordered by original, means for $n = 5$, and means for $n = 20$, which of the following gives the correct order?

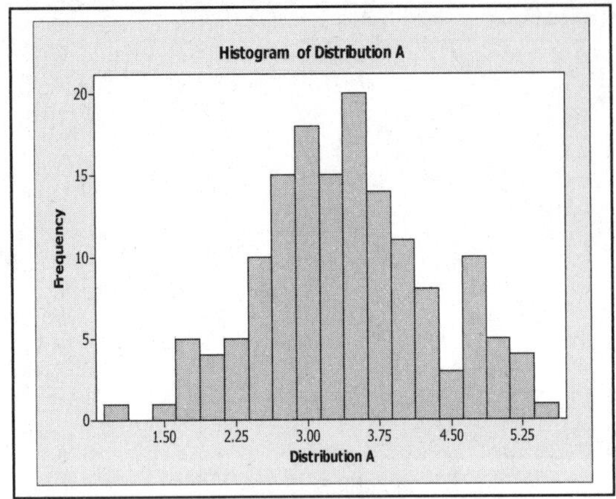

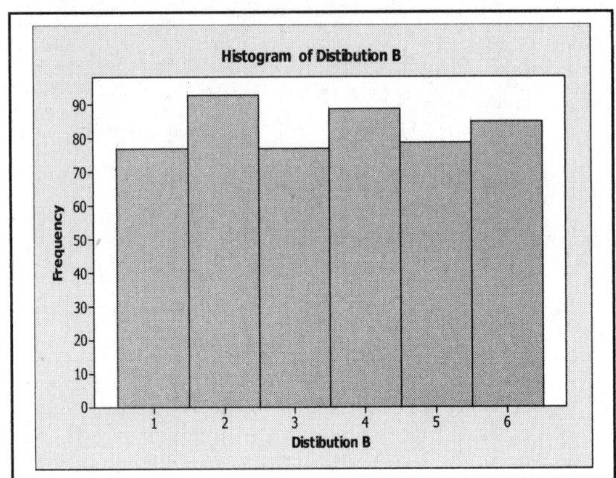

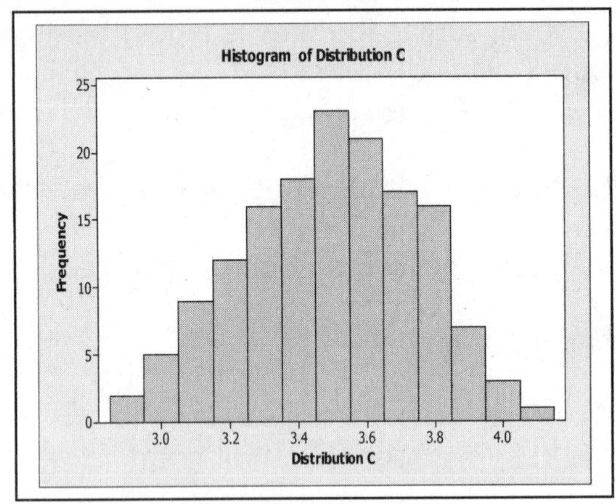

(A) A, B, C

(B) B, A, C

(C) B, C, A

(D) C, A, B

(E) C, B, A

30. A dice game pays a player \$5 for rolling a 3 or a 5 with a single die. The player has to pay \$2 if any other number is rolled. If a person plays the game 30 times, what is the approximate probability that the person will win at least \$15?

(A) 0.0030
(B) 0.0643
(C) 0.2767
(D) 0.3085
(E) 0.3910

31. A study is conducted to determine the effectiveness of mental exercises on short-term memory. A randomized comparative experiment is conducted, and the number of items recalled on a short-term memory test is recorded for 25 individuals who practiced mental exercises and 35 individuals who did not. The individuals who practiced mental exercises recalled 16.0 items on average with a variance of 6.25. The individuals that did not do any mental exercises recalled 12.5 items on average with a variance of 7.29. Assuming the population of items recalled is Normal for each group, what is the standard deviation of the sampling distribution of the difference in mean items recalled for the two groups?

(A) $\sqrt{6.25+7.29}$

(B) $\sqrt{\frac{6.25}{25}+\frac{7.29}{35}}$

(C) $\sqrt{6.25-7.29}$

(D) $\sqrt{\frac{6.25}{25}-\frac{7.29}{35}}$

(E) $\sqrt{\frac{6.25}{60}+\frac{7.29}{60}}$

32. In a random sample of older patients at a large medical practice, the age of a patient and a measure of that patient's hearing loss were recorded. The correlation between age and hearing loss of the patients in the sample was found to be 0.7. Which one of the following would be a correct statement if the age of a patient were used to predict the amount of hearing loss for a patient?

(A) Forty-nine percent of the time, the least-squares regression line accurately predicts hearing loss.
(B) Forty-nine percent of the variation in hearing loss can be explained by the least-squares regression line relating hearing loss and age.
(C) About 70% of a person's hearing loss can be explained by age, according to the regression line relating hearing loss and age.
(D) About 70% of the time, age will correctly predict the amount of hearing loss.
(E) The least-squares regression line relating hearing loss to age will have a slope of approximately 0.7.

33. There are an ever-increasing number of sources for getting the daily news—traditional newspapers, online, radio, nightly television newscasts, comedy shows, cell phones, podcasts, etc. In a recent telephone survey, 3204 randomly selected adults were asked to cite their primary source of daily news. Four in 10 adults said that they read a newspaper, either in print or online, almost every day. A 98% confidence interval to estimate the true proportion of adults who would read either a print newspaper or its online equivalent for their daily news is given by (0.38, 0.42). Which of the following is a correct interpretation of the confidence level?
   (A) Ninety-eight percent of all samples of this size would yield a confidence interval of (0.38, 0.42).
   (B) There is a 98% chance that the true proportion of readers who would read either a print newspaper or its online equivalent for their daily news is in the interval (0.38, 0.42).
   (C) The procedure used to generate this interval will capture the true proportion of readers who would read either a print newspaper or its online equivalent for their daily news in 98% of samples.
   (D) Ninety-eight percent of all of the samples of size 3204 lie in the confidence interval (0.38, 0.42).
   (E) There is a 98% chance that a randomly selected reader is one of the 40% who would read a print newspaper or its online equivalent for their daily news.

34. A school counselor believes that students who enroll in a fine arts class (like band, orchestra, or choir) are more likely to participate in a fine arts activity (such as the school musical, play, or speech). The table below shows data that were collected from a random sample of high school students in a large city.

		Class			
		Band	Orchestra	Choir	Total
Activity	One-Act Play	42	21	5	68
	Musical	120	59	10	189
	Speech	20	10	15	45
	Total	182	90	30	302

What is the expected count for the cell for the musical and orchestra?

(A) 59

(B) $\frac{(59)(90)}{302}$

(C) $\frac{(59)(189)}{302}$

(D) $\frac{(90)(189)}{302}$

(E) $\frac{(59)(302)}{189}$

35. A tax assessor has developed a model to predict the value (in thousands of dollars) of a home based on the square footage of the home. The accompanying computer printout summarizes the findings.

Parameter Estimates

Variable	DF	Parameter Estimate	Standard Error	T for H0: Parameter = 0	Prob > |T|
INTERCEP	1	35.70032	6.9533214	13.049	0.0237
SQFEET	1	0.14875	0.0030725	4.014	0.0017

Based on the information, which of the following is the best interpretation for the slope of the least squares regression line?

(A) Each additional square-foot increase in the size of the house will increase the home value by approximately $35.70.
(B) Each additional square-foot increase in the size of the house will increase the home value by approximately $14.88.
(C) Each additional square-foot increase in the size of the house will increase the home value by approximately $148.75.
(D) Each additional 1000 square-foot increase in the size of the house will increase the home value by approximately $35,700.32.
(E) Each additional square-foot increase in the size of the house will increase the home value by approximately $35.70.

36. In a recent 2017 national survey of 920 randomly selected U.S. adults, only 36% could name the three branches of the United Sates government. Which of the following represents a 98% confidence interval to estimate the proportion of all U.S. adults who could name the three branches of the United Sates government.

(A) $36 \pm 2.326 \bullet \sqrt{\frac{(36)(64)}{920}}$

(B) $36 \pm 2.576\left(\frac{(36)(64)}{\sqrt{920}}\right)$

(C) $0.36 \pm 2.326\left(\frac{(0.36)(0.64)}{\sqrt{920}}\right)$

(D) $0.36 \pm 2.576 \bullet \sqrt{\frac{(0.36)(0.64)}{920}}$

(E) $0.36 \pm 2.326 \bullet \sqrt{\frac{(0.36)(0.64)}{920}}$

37. A certain candy has different color wrappers for various holidays. During holiday #1 the candy wrappers are 50% red and 50% green. During holiday #2 the wrapper colors are 30% red, 30% silver, and 40% pink. Fifty pieces of candy are randomly selected from the holiday #1 distribution and 50 pieces of candy are randomly selected from the holiday #2 distribution. Respectively, what is the expected number and standard deviation of the total number of red wrappers from both holiday distributions?
(A) 40, 23
(B) 40, 4.796
(C) 40, 6.776
(D) 80, 6.776
(E) Cannot be determined from the given information.

38. There is intense competition among Internet service providers to get their home customers to switch from dial-up service to high-speed broadband service. A survey of 1758 randomly- selected U.S. households conducted in 2006 asked whether or not they had Internet service. If the response was yes, then they were asked about household income level and whether the Internet service was dial-up or high-speed broadband. The results in 2006 were compared to a similar poll of 1494 randomly-selected households in 2005. The table below summarizes the data for household income and the number of households with high-speed broadband service for the two years of the poll.

Household Income	**2006**	**2005**
Under $30,000	196	184
$30,000 – under $50,000	414	316
$50,000 – under $75,000	488	382
At least $75,000	660	612

A chi-square test was performed, which resulted in a *P*-value of 0.0760. Which of the following conclusions is correct?

(A) At the 5% level of significance, we have strong evidence that the rate of broadband use across income levels has not changed between 2005 and 2006.
(B) At the 5% level of significance, we can conclude that the rate of broadband use across various income levels has changed between 2005 and 2006.
(C) At the 5% level of significance, there is sufficient evidence to conclude that an association exists between rate of broadband use across income levels and year.
(D) At the 10% level of significance, we can conclude that the rate of broadband use across income groups has not changed between 2005 and 2006.
(E) At the 10% level of significance, we can conclude that the rate of broadband use across income levels has changed between 2005 and 2006.

39. A pharmaceutical company wants to test two new acid reflux medications, A and B, to see which is most effective in relieving the reflux in the shortest amount of time. The researchers plan to give the two medications in three dosages: 50 mg, 100 mg, and 150 mg. They have 120 volunteers available for the study. For this part of the study, researchers think that the weight of the patient might play a role in the incidence of reflux disease. Volunteers are separated into 20 groups—the six heaviest, the next six heaviest, and so on, down to the six lightest. Within each group of six, three are randomly assigned to medication A and three to medication B. Within each medication group of three, one is randomly assigned to each dosage. Which one of the following statements best describes this experiment?

(A) There are two treatments, participants were blocked on weight, and the response variable is time until pain relief is achieved.

(B) There are two treatments, participants were blocked on dosage, and the response variable is time until pain relief is achieved.

(C) There are three treatments, participants were blocked on medication A and B, and the response variable is proportion of participants who achieved pain relief.

(D) There are six treatments, participants were blocked on weight, and the response variable is time until pain relief is achieved.

(E) There are six treatments, participants were blocked on weight, and the response variable is proportion of participants who achieved pain relief.

40. A new variety of turf grass developed for use on golf courses is claimed to have a germination rate of 85%. To evaluate this claim, 100 seeds are randomly selected and planted in a greenhouse so that they are exposed to identical conditions. If the germination rate is 85% as claimed, which one of the following represents the approximate probability that 92 seeds or more will germinate?

(A) $\binom{100}{92}(0.85)^{92}(0.15)^{8}$

(B) $\binom{100}{92}(0.15)^{92}(0.85)^{8}$

(C) $P\left(z > \dfrac{0.92-0.85}{\sqrt{\dfrac{(0.85)(0.15)}{100}}}\right)$

(D) $P\left(z > \dfrac{0.85-0.92}{\sqrt{\dfrac{(0.92)(0.08)}{100}}}\right)$

(E) $P\left(z > \dfrac{0.92-0.85}{\dfrac{(0.85)(0.15)}{\sqrt{100}}}\right)$

Practice Test 1

Free-Response Section

1. In a recent survey of 1500 randomly selected U.S. adults, 68% of the respondents agreed with the statement, "I should exercise more than I do."
   (a) Construct and interpret a 96% confidence interval to estimate the proportion of the U.S. adult population that would agree with this statement.

   (b) For this study, state one source of potential bias and how it would affect the estimate of the proportion of adults who would agree with the statement, "I should exercise more than I do."

2. A study in Sweden compared former elite soccer players with people of the same age who had played soccer recreationally but not at the elite level. Of the 71 former elite soccer players surveyed, 10 had developed arthritis of the hip or knee by their mid-50s, compared to only 9 of the 215 recreational soccer players. Consider these groups of soccer players to be representative of soccer players at both the elite and recreational levels, respectively.
   (a) Was there a significant difference between the proportion of elite soccer players who develop arthritis of the hip or knee versus the proportion of recreational soccer players who develop arthritis of the knee or hip. Support your answer with appropriate statistical evidence.

   (b) Can we conclude that playing elite soccer leads to more knee or hip arthritis? Justify your answer.

3. A candy manufacturer is marketing a gift box containing four cream-filled nuggets and five pieces of fudge. The manufacturing process for each candy is designed so that the mean weight of a cream-filled nugget is 3 ounces with a standard deviation of 0.2 ounces and the mean weight of a piece of fudge is 4 ounces with a standard deviation of 0.3 ounces. The boxes have a mean weight of 3 ounces with a standard deviation of 0.1 ounce.

(a) Assuming that the weights of the boxes, nuggets, and fudge are independent, what is the mean and standard deviation of the weight of the box of candy?

(b) Assuming that the weights of the boxes, the nuggets, and the fudge are approximately normally distributed, what is the probability that a randomly selected box of candy will weigh less than 34 ounces?

(c) What is the probability that at least one of three randomly selected boxes of candy will weigh less than 34 ounces?

(d) Determine the probability that a random sample of three boxes of candy will have a mean weight of less than 34 ounces.

4 A certain large city wants to set up a citywide program to recycle newspapers, glass, and plastics as a way to reduce the amount of landfill, preserve the environment, and ultimately save the city some money. An analyst for one of the companies interested in bidding on the recycling contract has computed that they would be able to make a profit if the mean weekly household contribution to recycling exceeds 20 pounds. If this company can make a profit, then it is financially feasible for them to spend the money necessary to build a recycling plant, set up a collection system, and hire the necessary workers.

(a) Identify appropriate hypotheses for this situation.

(b) Describe a Type I error and a Type II error in this setting.

(c) Describe the economic consequences to the company for each of these types of errors.

5. A simple random sample of 592 students was selected. The eye color and natural hair color for each student was recorded. The resulting data are summarized in the following table.

	Hair Color				
Eye Color	Black	Brown	Red	Blonde	Total
Brown	68	119	26	7	220
Blue	20	84	17	94	215
Hazel	15	54	14	10	93
Green	5	29	14	16	64
Total	108	286	71	127	592

(a) What is the probability that a person chosen at random from this sample will have brown hair?

(b) What is the probability that a person chosen at random from this sample who has blue eyes will also have brown hair?

(c) Based on your answers to (a) and (b), are hair color and eye color independent?

6. Suppose that in soccer, in games where the teams are not evenly matched, the better team has a 72% chance of scoring the first goal in overtime of any game. If a team has six overtime games, and is considered the better team in all of them, what is the probability that this team will score the first overtime goal in at least five out of these six games? Answer the question using simulation.

   (a) Describe the design of a simulation using the table of random digits below to estimate the desired probability.

   (b) Using the random number table below, run your simulation 12 times. Make your procedure clear by marking on or above the table so that someone can follow what you did. Make a table of your results. From your simulation, determine the estimated probability of scoring the first overtime goal in at least five of the six games.

84177 06757 17613 15582 51506 81435 41050 92031 06449 05059

59884 31180 53115 84469 94868 57967 05811 84514 75011 13006

63395 55041 15866 06589 13119 71020 85940 91932 09488 74987

54355 52704 90359 02649 47496 71567 94268 08844 26294 64759

08989 57024 97284 00637 89283 03514 59195 07635 03309 72605

29357 23737 67881 03668 33876 35841 52869 23114 15864 38942

(c) Below are two simulations of 200 runs each for the playing of six overtime games. One represents a team that is fairly evenly matched with its opponent, and the other represents a team that is better than its opponent. Which distribution represents which team? Explain your reasoning.

Distribution A

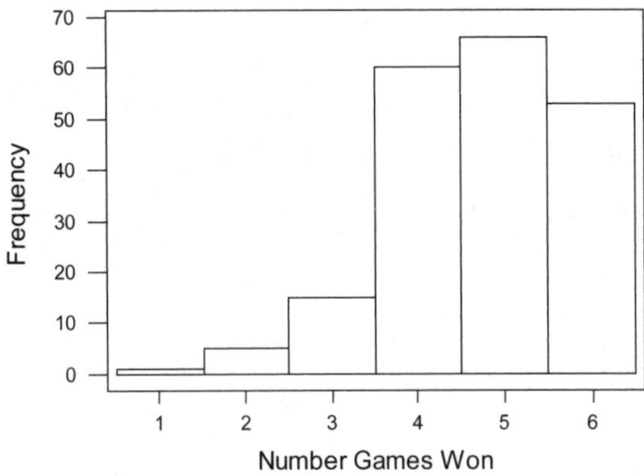

Distribution B

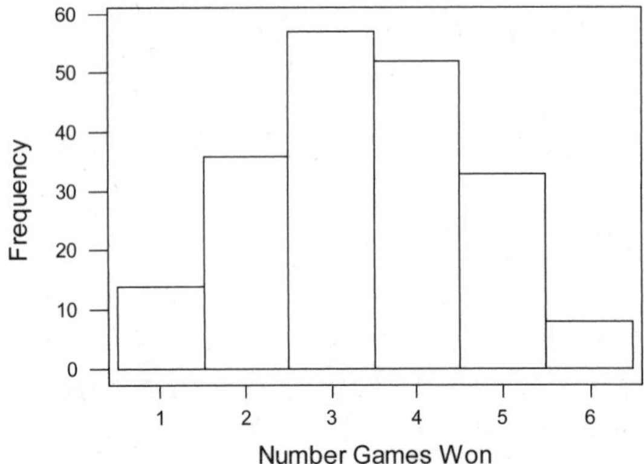

Answer Key for Practice Test 1

1.

	S	S'
C	.30	.15
C'	.35	.20

P(Calc and not Stat OR Stat and not Calc) = .35 + .15 = .50

ANS: C

2 Since the four fields are different and we want to account for these differences we will block by field and plant all six varieties of seed in each field. The varieties of seed corn are the treatments that are being applied to the fields.

ANS: B

3. To increase the power of a test you can increase the sample size *n* and increase α.

ANS: E

4. The monthly median draft number is trending lower as the year progresses. People born the near end of the year tended to have lower lottery numbers than those born near the first of the year. No regression can be computed using only the medians.

ANS: C

5. Since the number of cars at the two intersections is independent, we add variances.
$\sigma = \sqrt{Var_1 + Var_2} = \sqrt{4^2 + 7^2} = \sqrt{65}$

ANS: B

6. Expected (individuals with side effects from Adiflu + individuals with side effects from Beneflu) = $(0.8)(0.45)(70) + (0.55)(0.25)(90) = 37.575$.

ANS: A

7. The *P*-value is the probability of obtaining a sample statistic at least as extreme as that found in the experiment (a difference of 13.2), given that there was no difference between the mean cholesterol levels for those taking the placebo and for those taking the new medication.

ANS: C

8. Since $\sigma_{\bar{x}} = \frac{\sigma}{\sqrt{n}}$, multiplying the sample size by 4 would divide the standard deviation of the sampling distribution for the mean by $\sqrt{4n} = 2\sqrt{n}$, i.e., in half.

ANS: B

9. A data point is an outlier if it lies above $Q_3 + 1.5(IQR)$ or below $Q_1 - 1.5(IQR)$. The lower bound is $Q_1 - 1.5(IQR) = 5 - 1.5(12-5) = -5.5$. Since negative data points are impossible in this context, there cannot be any low outliers. The upper bound is $Q_3 + 1.5(IQR) = 12 + 1.5(12-5) = 22.5$. Any point above this value is an outlier.

ANS: D

10. A Type II error occurs when we fail to reject a null hypothesis that is not true. In this case, that would indicate that we fail to reject $H_0: \mu = 28$ (sales have not increased), but it is, in fact, false. Thus the company concludes that sales have not increased when they actually have. The company loses the chance to make more money.

ANS: D

11. The null hypothesis always includes an equal sign. The one-sided alternative hypothesis in this case is less than 12 seconds.

ANS: C

12. There are two measurements on each subject (time with music versus time without music). Each subject serves as his or her own control. This is a matched pairs experiment.

ANS: E

13. This is a geometric distribution with *P*(success) = 0.03. Since the fifth customer is the "first success", we have four failures and a success or $P(X = 5) = (0.97)^4(0.03)^1$

ANS: B

14. Mark a vertical line at 80 years on the *x*-axis. At the point where this line intersects the cumulative relative frequency graph, make a horizontal line. This is about the 65% mark. This means that 65% of the all members are less than 80 years of age. Therefore 35% are at least 80 years of age and 0.35(120) = 42 members.

ANS: A

15. (A) False. The confidence either contains 0 or it does not.
    (B) False. The data are clearly paired.
    (C) False. Even though the mean difference is positive, it is not positive enough to be significant.
    (D) False. A sample of size 10 can be sufficient to make a conclusion.
    (E) True. Since the confidence interval contains zero, there is no statistically significant difference between the additive and the non-additive. We cannot conclude that the pill is effective in improving gas mileage.

ANS: E

16. The power of a significance test is its ability to detect a difference. The higher the power of a test the better is the test's ability to detect a significant result. One way to increase power is to increase the sample size.

ANS: A

17. The subjects are labeled as 1, 2, 3, 4, 5, 6, 7, 8. Ignore repeats since subjects were not used more than once. The random numbers chosen are 0 7 1 1 9 9 7 3 3 6 resulting in the selection of 7, 1, 3, 6. This corresponds to Klein, Rodriguez, Brown, and Munoz.

ANS: C

18. Decreasing the confidence level would make the confidence interval narrower since the critical z-value would be smaller. A narrower interval leads to a more precise estimate.

ANS: A

19. The number of data points for the two companies would not lend itself to easily using dotplots or stemplots. Since you have only one variable, a scatterplot is inappropriate. Since the companies are quite different in size, a relative-frequency histogram would let you display the percent of workers in each salary group for comparison purposes.

ANS: D

20. There are 160 observations in the data set. The median would be the value between the 80th and 81st data points. This clearly falls in the upper end of the "at least 20 but less than 30 minutes" interval. This narrows the choices to (A), (B), or (C). In addition, the distribution of waiting times is skewed toward the higher numbers, causing the mean to be higher than the median. This eliminates choices (A) and (B).

ANS: C

21. ANS: C

$$n=\left[\frac{z^*_{0.95}\,\sigma}{ME}\right]^2=\left[\frac{1.96\sqrt{(0.3)(0.7)}}{0.04}\right]^2=(22.45)^2=504.21\approx 505$$

22. From the z-table or by using a calculator, InvNormal(0.20) = −0.84 and InvNormal (0.90) = 1.28

$$z=\frac{x-\mu}{\sigma}\Rightarrow -0.84=\frac{8.61-\mu}{\sigma}\text{ and }1.28=\frac{10.48-\mu}{\sigma}$$

Solving the system: $\begin{cases}-0.84\sigma=8.61-\mu\\ 1.28\sigma=10.48-\mu\end{cases}$, $\mu=\$9.35$ and $\sigma=\$0.88$ ANS: A

23. For n = 10, there are n–2 degrees of freedom in regression on slope. Therefore, with a 99% confidence level df = 8 and $t^*=3.355$. Weeks of experience is the predictor, so the slope coefficient is 0.1879

ANS: D

24. The margin of error is defined by $z^*\sqrt{\frac{\hat{p}(1-\hat{p})}{n}}$. Assuming that $\hat{p}$ remains the same, a higher level of confidence yields a large margin of error. A larger sample size yields a smaller margin of error. So a higher level of confidence and a small sample size yields the largest margin of error.

ANS: C

25. p_1 = true proportion who claim to be vegetarians at school #1. $\hat{p}_1=\frac{9}{90}=0.10$

p_2 = true proportion who claim to be vegetarians at school #2. $\hat{p}_2=\frac{12}{150}=0.08$

$$C.I.=(0.10-0.08)\pm 1.645\left(\sqrt{\frac{(0.10)(0.90)}{90}+\frac{(0.08)(0.92)}{150}}\right)$$

ANS: C

26. $\widehat{humor}=-49+1.8(110)=149$. Residual $=y-\hat{y}=140-149=-9$

ANS: B

27. **New solution**

$$P(\text{private}\mid\text{left})=\frac{P(\text{private}\cap\text{left the company})}{P(\text{left the company})}=\frac{(0.3)(0.35)}{(0.3)(0.35)+(0.7)(0.2)}=0.4286$$

ANS: B

28. The distribution of hours spent surfing the Web exhibits more variation (has a larger *IQR*) than does the distribution of hours spent watching TV.

ANS: D

29. From the Central Limit Theorem, we know the larger that the sample size the more normal-like is the sampling distribution and the lower the spread of scores. Histogram B is the original population, Histogram A is sample size 5, and Histogram C is the most normal-like of the three and has the smallest spread of scores.

ANS: B

30. The mean or the expectation on each roll is $E(\text{winnings}) = (\$5)\left(\frac{1}{3}\right) + (-\$2)\left(\frac{2}{3}\right) = \$\frac{1}{3}$

with a standard deviation of $\sigma(\text{winnings}) = \sqrt{\left(5-\frac{1}{3}\right)^2\left(\frac{1}{3}\right) + \left(-2-\frac{1}{3}\right)^2\left(\frac{2}{3}\right)} = \$3.30$

For 30 turns, the expected winnings are $E(total) = 30\left(\frac{1}{3}\right) = \$10$

The standard deviation of the 30 turns is $\sigma(\text{total}) = \sqrt{30(3.30)^2} = \$18.07$

$$P(\text{total} > \$15) = P\left(z > \frac{\$15 - \$10}{\$18.07}\right) = P(z > 0.277) \approx 0.3910$$

ANS: E

31. Since the 60 individuals were randomly divided, we can consider the two groups as being independent. The variances are added even though we are computing the differences. In this problem you are given the variances rather than the standard deviations. The standard error of the sampling distribution for the difference in two independent means is given by

$$\sqrt{\frac{Var_1}{n_1} + \frac{Var_2}{n_2}} = \sqrt{\frac{6.25}{25} + \frac{7.29}{35}}$$

ANS: B

32. You are asked to interpret the coefficient of determination, given by r^2. Since $r = 0.7$, then

$r^2 = 0.49$. The coefficient of determination, r^2, is the percent of variation in the *y*-variable that can be explained by the regression line, using the *x*-variable as the predictor. In this case, the *y*-variable is the amount of hearing loss and the *x*-variable is the age of the patient.

ANS: B

33. This question is asking for an interpretation of the confidence level not the interval. The 98% level indicates how often, on average, the procedure will produce an interval that will capture the true population parameter of interest.

ANS: C

34. The expected counts for each cell are determined by $\frac{\text{(row total)(column total)}}{\text{grand total}}$

ANS: D

35. The equation is HOMEVALU = 35.700 + 0.14875*SQFEET.
The slope is 0.14875. Every increase of one square foot in house size leads to an increase in the house value of 0.14875($1000) = $148.75.

ANS: C

36. To construct a 98% confidence interval, we use $\hat{p} \pm z^* \sqrt{\frac{\hat{p}(1-\hat{p})}{n}}$ where $\hat{p} = 0.36$.

With a 98% confidence level, z*= 2.326.

ANS: E

37. This is a binomial situation and we have independent random variables, therefore
E(red) = $n_1 p_1 + n_2 p_2$ = (50)(.5) + (50)(.3) = 40
Var (red) = *Var* (holiday 1) + *Var* (holiday 2) = $n_1 p_1 (1 - p_1) + n_2 p_2 (1 - p_2)$
SD(red) = $\sqrt{50(0.5)(0.5) + 50(0.3)(0.7)} = 4.796$

ANS: B

38. We would reject the null hypothesis if the *P*-value < the α level of significance. With a *P*-value of 0.0760, we would fail to reject H_0 at the 5% level, eliminating choices A, B, and C. The *P*-value would be sufficient to reject at the 10% level. Choice (D) indicates that there is little difference across income groups between the two years (i.e., no change), which is equivalent to failing to reject H_0 at 10%. Choice (E) rejects H_0 and concludes that a difference exists.

ANS: E

39. The treatments are the two types of medication, each at three different dosages, for a total of six treatments. Subjects are blocked into groups of six according to their weight. Because the company is testing which medication is most effective in the least amount of time, time until pain relief is the response variable.

ANS: D

40. $p = 0.85 \text{ and } \hat{p} = 0.92.\ P(\hat{p} > 0.92) = P\left(z < \dfrac{\hat{p} - p}{\sqrt{\dfrac{p(1-p)}{n}}} \right)$

ANS: C

Textbook Correlation Multiple-Choice Practice Exam 1

Question	Correct Answer	Textbook Section	Question	Correct Answer	Textbook Section
1	C	5.2	21	C	8.2
2	B	4.2	22	A	2.2
3	E	9.1	23	D	12.1
4	C	3.1	24	C	8.1 & 8.3
5	B	6.2	25	C	10.1
6	A	6.1	26	B	8.3
7	C	9.1	27	B	5.3
8	B	7.3	28	D	1.3
9	D	1.3	29	B	7.1 & 7.3
10	D	9.1 & 9.3	30	E	6.2
11	C	9.1 & 9.3	31	B	10.2
12	E	9.3	32	B	3.2
13	B	6.3	33	C	8.1
14	A	2.1	34	D	11.2
15	E	10.2	35	C	3.2
16	A	4.1	36	E	8.2
17	C	4.1	37	B	6.3
18	A	8.1	38	E	11.2
19	D	1.2	39	D	4.2
20	C	1.2 & 1.3	40	C	7.2

Part II Free-Response Section Solutions

1. (a) Let p = true proportion of adults who would agree with the statement, "I should exercise more than I do."

A one-sample confidence interval for proportions will be constructed

$$\hat{p} \pm z^* \sqrt{\frac{\hat{p}(1-\hat{p})}{n}}$$

- Random: We are told that a random sample of adults was taken.
- Large Counts: $n \cdot \hat{p} = 1500(0.68) = 1020$ and $n(1-\hat{p}) = 1500(0.32) = 480$ Since both are greater than 10, the sample size is large enough to use z procedures.

$$0.68 \pm 2.054\sqrt{\frac{(0.68)(0.32)}{1500}} \Rightarrow (0.655, 0.705)$$

We are 96% confident that the true proportion of adults who would agree with the statement, "I should exercise more than I do," lies between 0.655 and 0.705.

(b) There is a good possibility of response bias. Given all of the media attention concerning the issue of overweight Americans, some of the respondents in the survey may feel compelled to agree with the statement in order to make themselves appear more health-conscious than they really are. This would lead to an overestimation of the proportion of people who would agree with the above statement.

2. (a) p_1 = true proportion of elite soccer players who develop arthritis of the hip or knee

p_2 = true proportion of recreational soccer players who develop arthritis of the hip or knee

$$H_o : p_1 = p_2$$
$$H_a : p_1 \neq p_2$$

1. Random:. We are told that these two groups are representative of all soccer players at these levels.
2. Independent: There are at least 710 elite soccer players and 2100 recreational soccer players in the respective populations.

3. Large sample size. $n_1\hat{p}_1 = 10 \; and \; n_1(1-\hat{p}_1) = 61$
$n_2\hat{p}_2 = 9 \; and \; n_2(1-\hat{p}_2) = 206$

All are at least 5, so we have a large sample.

All conditions are met. We will use a two-proportion *z* test.

$$z = \frac{\hat{p}_1 - \hat{p}_2}{\sqrt{\hat{p}(1-\hat{p})\left(\frac{1}{n_1}+\frac{1}{n_2}\right)}} \quad \text{where } \hat{p} = \frac{10+9}{71+210} = 0.0676$$

$$z = \frac{0.1408 - 0.0429}{\sqrt{(0.0676)(0.9324)\left(\frac{1}{71}+\frac{1}{210}\right)}} = 2.84$$ *P*-value = 2 *P*(*z* > 2.84) = .0046

Let $\alpha = 0.05$. Since 0.0046 < 0.05, H_o is rejected. There is sufficient evidence to conclude that the proportion of elite soccer players who develop arthritis of the hip or knee is
significantly different than for recreational soccer players.

(b) No. Since this was an observational study rather than an experiment, no cause-effect conclusions are possible.

3. (a) $\mu_w = 4(3) + 5(4) + 1(3) = 35$ ounces

$\sigma_w = \sqrt{4(0.2)^2 + 5(0.3)^2 + 1(0.1)^2} = 0.79$ ounces

(b) $P(w < 34) = P\left(z < \frac{34-35}{0.79}\right) = P(z < -1.27) = 0.1020$

The probability that a randomly selected box of candy will weigh less than 34 ounces is 0.102.

(c) *P*(a box does <u>not</u> weigh < 34) $= 1 - P(w < 34) = 1 - 0.102 = 0.898$

Let *x* = number of boxes that weigh less than 34 ounces

$P(x \geq 1) = 1 - P(x = 0) = 1 - (0.898)^3 = 0.2758$

(d) $P(\overline{w} < 34) = P\left(z < \frac{34-35}{\frac{0.79}{\sqrt{3}}}\right) = P(z < -2.19) = 0.0143$

If three boxes of candy were randomly selected, only about 1.43% of the time would their average weight be less than 34 ounces.

4. (a) μ = true mean weekly household contribution to recycling.

$H_o : \mu = 20$

$H_a : \mu > 20$

(b) A Type I error is committed when a true null hypothesis is rejected. The company concludes that there is a sufficient amount of recycling (more than 20 pounds per household) when, in fact, there is not a sufficient amount to be recycled.

A Type II error is committed when a false null hypothesis is not rejected. The company concludes that there is an insufficient amount of recycling (no more than 20 pounds per household) when, in fact, there is more than that to be recycled.

(c) With a Type I error, the company builds the recycling plant, hires workers, and sets up the collection system only to find that not enough recyclables are available. The company loses money.

With a Type II error, the company decides not to build the collection system even though a sufficient amount of recyclables was available. The company loses an opportunity to make money.

5. (a) P(brown hair) = 286/592 = 0.4831

(b) P(brown hair | blue eyes) = 84/215 = 0.3907

(c) If hair color and eye color are independent, then the two probabilities in (a) and (b) should be approximately the same. That is, P(brown hair) should equal P(brown hair | blue eyes). However, the two probabilities are not equal. Thus, hair color and eye color are not independent.

6.

(a) Let *x* = the number of games in which the team scores the first overtime goal.
P(the team scores first) = 0.72

Let the digits 00 through 71 represent scoring the first goal in overtime. Let 72 through 99 represent not scoring the first goal in overtime. Two digits will represent one game and it will take 12 digits to represent the six games played in overtime. Starting at the left, taking two digits at a time, count the number of games in which the team scores first until you have counted six games. For each trial, record the number of games in which the team scored the first goal. Mark the end of each trial with a vertical line. Above each trial write the number of games in which the team scored the first overtime goal.

(b)

---

5 5 6 5

8 4 1 7 7 0 6 7 5 7 1 7 |6 1 3 1 5 5 8 2 5 1 5 0| 6 8 1 4 3 5 4 1 0 5 0 9| 2 0 3 1 0 6 4 4 9 0 5 0| 5 9

4 5 2 5

5 9 8 8 4 3 1 1 8 0| 5 3 1 1 5 8 4 4 6 9 9 4| 8 6 8 5 7 9 6 7 0 5 8 1| 1 8 4 5 1 4 7 5 0 1 1 1| 3 0 0 6

6 4 4 5

6 3 3 9 5 5 5 0| 4 1 1 5 8 6 6 0 6 5 8 9| 1 3 1 1 9 7 1 0 2 0 8 5| 9 4 0 9 1 9 3 2 0 9 4 8| 8 7 4 9 8 7

# Games	Frequency
0	0
1	0
2	1
3	0
4	3
5	6
6	2

Based on the above simulation,

$$P(\text{win at least 5 games}) = P(x \geq 5) = \frac{8}{12}$$

(c) If the teams were fairly evenly matched, then each team would be expected to win approximately 3 out of 6. This situation would best be represented by the histogram in Distribution B. If one team was clearly better than the other, it would be expected to win more games than its opponent and this is represented by Distribution A.

Practice Test 2

Multiple-Choice Section

1. Which of the following situations would be difficult to explore using a census?
   (A) You wish to know the proportion of teachers in a school district that have a Master's degree.
   (B) You want to know the average amount of time spent on homework by students in your high school.
   (C) You are interested in the difference in performance on the AP Statistics exam for students taught by two different teachers.
   (D) You want to know the proportion of homes in a suburban community that have wireless internet access.
   (E) You want to know the proportion of trees in a large state forest that are infected with Dutch Elm disease.

2. It is known that 15% of the seniors in a large high school enter military service upon graduation. If a group of 20 seniors are randomly selected, what is the probability of observing at most one senior who will be entering military service?

   (A) $20(0.15)^1(0.85)^{19}$
   (B) $1-20(0.15)^1(0.85)^{19}$
   (C) $(0.85)^{20}+20(0.15)^1(0.85)^{19}$
   (D) $(0.85)^{20}$
   (E) $1-(0.85)^{20}$

3. A study of two popular sleep-aids measured the mean number of hours slept for 200 individuals administered the drugs. In the group of 200 volunteer subjects, 100 were randomly chosen to be administered Drug 1 and the other 100 were administered Drug 2. A two-sample *t* test was performed on the difference in the mean sleep times experienced by the subjects. The *p*-value was 0.12. Which of the following is a correct interpretation of the *p*-value?
   (A) Approximately 12% of the subjects did not sleep at all.
   (B) There was a 12% increase in the mean sleep time by subjects in the study.
   (C) There was a 12% difference between the mean sleep times of the two groups.
   (D) Assuming both drugs are equally effective, we would expect to see a difference in the mean sleep times at least as extreme as the observed difference for 200 subjects 12% of the time.
   (E) Approximately 12% of the subjects in the study experienced the same sleep time.

4. Pediatricians were interested in determining if a relationship exists between preschoolers who snore or who had nocturnal coughing, and the prevalence of asthma among those children. The results of a survey of 1048 randomly selected children aged 2 to 5 are given below.

	Snored four or more times per week	**Snored three or less times per week**	**Total**
Had nocturnal coughing	59	255	314
Had asthma	22	222	244
Did not have asthma or nocturnal coughing	115	375	490
Total	196	852	1048

The hypotheses that the pediatricians used were given as:

H_o: There is no association between frequency of snoring and the prevalence of asthma symptoms and nocturnal coughing.

H_a: An association does exist between frequency of snoring and the prevalence of asthma symptoms and nocturnal coughing.

If the test statistic for this procedure is $\chi^2 = 22.38$, which one of the following is a correct conclusion?

(A) Since the *P*-value is large, an association does exist between frequency of snoring and the prevalence of asthma symptoms and nocturnal coughing.
(B) The results are not significant at the 5% level of significance.
(C) The results are not significant at $\alpha = 0.01$ but are significant at $\alpha = 0.001$.
(D) Since the *P*-value is < 0.001, there is an association between frequency of snoring and the prevalence of asthma symptoms and nocturnal coughing.
(E) Since the test statistic is so large, no association exists between frequency of snoring and the prevalence of asthma symptoms and nocturnal coughing.

5. In order to evaluate the yield per acre of three different varieties of corn, 12 one-acre plots of land will be used. One variety of corn will be randomly assigned to 4 of the plots, the second variety to 4 other plots, and the third variety to the 4 remaining plots of land. At the end of the growing season, the yield per acre will be evaluated. Which one of the following is a true statement?

(A) The type of land is being used as a blocking factor.
(B) The treatments in this experiment are the plots of land.
(C) The experimental units in this situation are the varieties of corn.
(D) Our randomization process will reduce the variation that exists due to the plots of land.
(E) Replication of corn variety on four plots of land allows us to measure variability across plots of land.

6. Biologists have collected data relating the age (in years) of one variety of oak tree to its height (in feet). A scatterplot for 20 of these trees is given below.

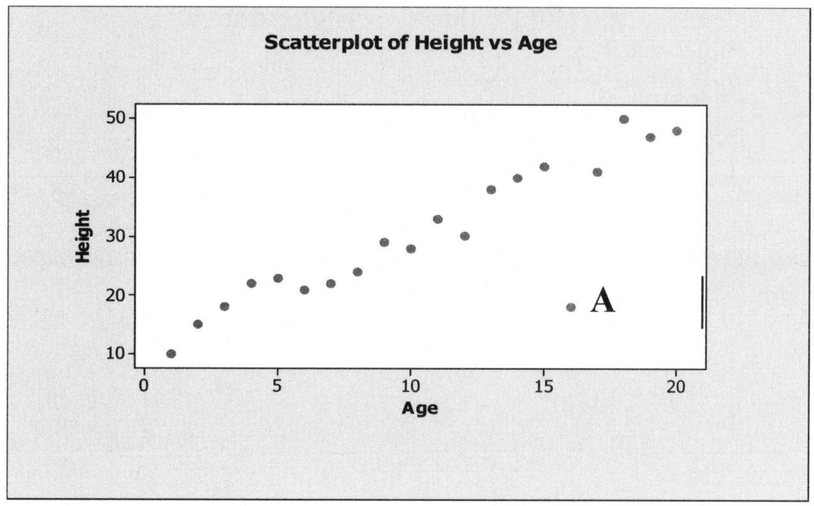

If the point labeled A in the above scatterplot were changed to have a height of 40 feet, which one of the following statements would be TRUE?

(A) The slope of the least-squares regression line would decrease, and the correlation would increase.
(B) The slope of the least-squares regression line would decrease, and the correlation would decrease.
(C) The slope of the least-squares regression line would increase, and the correlation would increase.
(D) The slope of the least-squares regression line would increase, and the correlation would decrease.
(E) The slope of the least-squares regression line, and the correlation would remain the same.

7. A school principal is interested in calculating a 96% confidence interval for the true mean number of days students are absent during the school year. The attendance office lists the number of days absent for each student during the month of December. Which of the following is the best reason the principal cannot construct the confidence interval?
   (A) A confidence interval cannot be constructed unless all data for the population are known.
   (B) The critical value for 96% is not listed in the table, therefore it can't be calculated.
   (C) The attendance office should report the average for the month, not the number for each individual student.
   (D) Because most students wouldn't miss any days, the average will be too small to construct a confidence interval.
   (E) The number of absences in December may not be representative of the rest of the months during the year. Data should be collected from all months.

8. A study of drug addicts in Amsterdam recorded how often each addict had recently injected drugs and whether or not the addict was infected with HIV, the virus that causes AIDS. Here is a two-way table of the numbers of addicts in each condition:

		HIV Positive	**HIV Negative**
Injected Drugs?	Daily	32	45
	Less than Daily	20	18
	Never	18	23

If an addict is chosen at random, what is the probability that the addict is infected with HIV, given that he/she injects daily?

(A) 0.42
(B) 0.45
(C) 0.49
(D) 0.52
(E) 0.55

9. A sports physiologist wants to compare the effects of two different exercise machines, A and B, on the flexibility of gymnasts. There are 20 gymnasts available. All gymnasts have been given a flexibility rating before the experiment starts. A similar ratings test will be given at the end of the experiment. Which one of the following methods of assigning gymnasts to machines would be optimal in order to assess the difference in flexibility due to the machines?

(A) Randomly assign a number (01 through 20) to each gymnast. Match the gymnasts by consecutive number (i.e., #1 with #2, #3 with #4, etc.). Then assign the odd-numbered gymnast in each pair to machine A and the even-numbered gymnast to machine B.

(B) Have all the gymnasts use both machines. A coin flip will determine which machine is used first.

(C) Flip a coin to assign the gymnasts into two groups of 10. Flip the coin again to assign one group to use machine A and the other group to use machine B.

(D) Match each gymnast with a gymnast who has the same years of experience, training level, and gymnastic ability. Then flip a coin to assign each pair of gymnasts to one of the two machines.

(E) Match each gymnast with a gymnast who has the same years of experience, training level, and gymnastic ability. Flip a coin to assign one member of the pair to machine A and the other to machine B.

10. According to the Insurance Institute for Highway Safety in 2017, 58% of 18-year-olds admitted to texting while they are driving. In a large city in the Midwest, a random sample of 250 18-year-old drivers found that 160 of them had admitted to texting while driving. If the texting while driving rate of 18-year-old drivers in this Midwestern city is the same as the national rate, which of the following represents the probability of getting a random sample of 250 18-year-old drivers whose texting rate is 64% or greater?

(A) $P\left(z > \dfrac{0.64 - 0.58}{\sqrt{\dfrac{(0.58)(0.42)}{250}}}\right)$

(B) $P\left(z > \dfrac{0.64 - 0.58}{\sqrt{\dfrac{(0.64)(0.36)}{250}}}\right)$

(C) $P\left(z > \dfrac{0.58 - 0.64}{\sqrt{\dfrac{(0.64)(0.36)}{250}}}\right)$

(D) $\binom{160}{250}(0.58)^{160}(0.42)^{90}$

(E) $\binom{160}{250}(0.64)^{160}(0.36)^{90}$

11. A poll was conducted to determine the level of support within a large suburban school district for the construction of a new football stadium. Of the 655 people who were surveyed, 57% said they would support spending the money to construct the stadium. The poll had a margin of error of 4%. Which is the correct interpretation for this margin of error?

(A) About 4% of the school district residents polled refused to respond to the question.
(B) There is most likely less than a 4% difference between the proportion of support in the sample and the true proportion of support in the district.
(C) There is most likely more than a 4% difference between the proportion of support in the sample and the true proportion of support in the district.
(D) If we conducted repeated samples from this population, the results would vary by no more than 4% from the current sample result of 57%.
(E) If the poll had sampled more people, the proportion of district residents who would have said they supported the stadium would have been about 4% higher.

12. In order to study the adage "You get what you pay for," the following data were obtained with regard to the price of refrigerators and their lifetime (in years).

		Lifetime (in years)			
		<5	5 to 10	>10	Total
	Low	45	60	30	135
Cost	Medium	40	65	40	145
	High	35	70	70	175
	Total	120	195	140	455

What is the expected count for the cell denoting medium cost and a lifetime of more than 10 years?

(A) 40

(B) $\frac{(40)(140)}{455}$

(C) $\frac{(40)(145)}{455}$

(D) $\frac{(140)(145)}{455}$

(E) $\frac{(40)(455)}{145}$

13. As of August 2017, there were more than 116,000 men, women and children in the United States were awaiting transplants of a variety of organs such as livers, hearts, and kidneys. A national organ donor organization is trying to estimate the proportion of all people who would be willing to donate their organs after their death to help transplant recipients. Which one of the following would be the most appropriate sample size required to ensure a margin of error of at most 3 percent for a 98% confidence interval estimate of the proportion of all people in the U.S. who would be willing to donate their organs?

(A) 175
(B) 191
(C) 1510
(D) 1740
(E) 1845

14. The number of taps per minute a person can do was measured before and after a person consumed a popular energy drink. The researcher is interested in the effect of energy drinks on a person's physical actions. The results are below:

Subject	1	2	3	4	5	6	7	8	9	10	$\bar{x}$	s
Before Drink	105	99	93	96	95	99	94	87	100	89	95.7	5.36
After Drink	103	102	105	102	101	110	95	89	102	98	100.8	5.75
Difference	2	-3	-12	-6	-6	-11	-1	-2	-2	-9	-4.64	4.48

Which one of the following represents the test statistic for the mean difference in the number of taps before the drink and after the drink? There appears to be no evidence of non-normality in the data.

(A) $t = \dfrac{95.7 - 100.8}{\sqrt{\dfrac{5.36}{10} + \dfrac{5.75}{10}}}$

(B) $t = \dfrac{95.7 - 100.8}{\sqrt{\dfrac{5.36^2}{10} + \dfrac{5.75^2}{10}}}$

(C) $t = \dfrac{95.7 - 100.8}{\sqrt{\dfrac{5.36^2}{9} + \dfrac{5.75^2}{9}}}$

(D) $t = \dfrac{-4.64}{\dfrac{4.48}{\sqrt{10}}}$

(E) $t = \dfrac{-4.64}{\dfrac{4.48}{\sqrt{9}}}$

15. The number of goals scored per game in a full season of games for a professional soccer league is strongly skewed to the right with a mean of 2.3. An SRS of size $n = 15$ is taken from the population and the sample mean is computed. This process is repeated a total of 375 times. Which one of the following best describes the shape of the resulting distribution of sample means?

    (A) Skewed to the right with a mean of 2.3 goals.

    (B) Skewed to the right with a mean of $\frac{2.3}{\sqrt{15}}$ goals.

    (C) Approximately normally distributed with a mean of 2.3 goals.

    (D) Approximately normally distributed with a mean of $\frac{2.3}{\sqrt{15}}$ goals.

    (E) Binomially distributed with a mean of 2.3 goals.

16. The enrollment rate of graduating seniors who have been accepted into the freshman class of a random selection of public and private colleges and universities was compared. For example, at a highly selective academic institution, only 12% of the applicants to the school may be accepted to the incoming class but 75% of those accepted may choose to enroll at that school. Below is the side-by-side boxplots of the percent of students accepted who enrolled at the randomly selected schools.

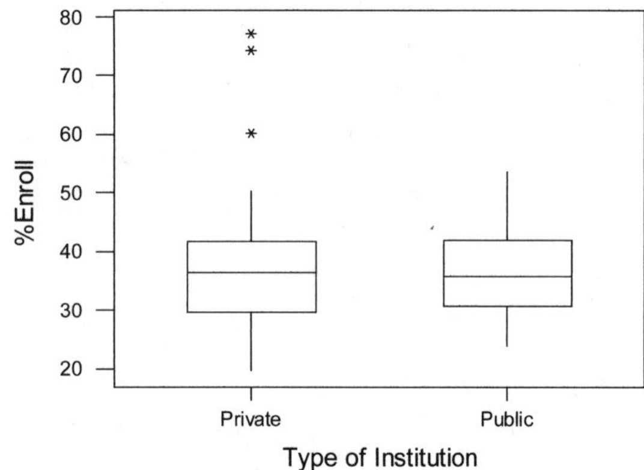

Which one of the following statements is most TRUE about the enrollment rate distributions?

(A) The *IQR* of the private enrollment rate distribution is smaller than the *IQR* of the public enrollment rate distribution.

(B) The median of the private enrollment rate distribution is smaller than the median of the public enrollment rate distribution.

(C) The mean of the private enrollment rate distribution is larger than the mean of the public enrollment rate distribution.

(D) The private enrollment rate distribution is skewed to the left.

(E) The outliers of the private enrollment rate distribution have no effect on the statistics.

17. In an attempt to reduce repeat offenders, the Department of Traffic Safety wants to randomly select 5 drivers, without replacement, from a population of 50 drivers convicted of speeding in order to try a new intensive program. If the drivers are labeled 01, 02, 03, ... , 50 and the following line is from a random number table

22368 46573 25595 85393 30995 89198 27982 53401 93965 34095 52666 19174

Which one of the following represents the sample of 5, starting from the left end of the table?

(A) 22, 36, 8, 46, 32
(B) 22, 46, 25, 30, 27
(C) 22, 36, 25, 30, 27
(D) 22, 36, 46, 32, 39
(E) 22, 23, 36, 46, 32

18. The following frequency table summarizes the final exam scores for 150 students in a statistics class at a large university.

Interval	Frequency
90–99	16
80–89	27
70–79	46
60–69	37
50–59	19
40–49	5

Which of the following is closest to the percentile rank of a score of 80 in this distribution?

(A) 25
(B) 30
(C) 50
(D) 70
(E) 75

19. A local dealer has two video stores in a town, one located on Foothill Drive and the other one on Grand Avenue. The Foothill Drive store does 70% of the dealer's business in the town, and the Grand Avenue store does the rest. In the Foothill Drive store, 40% of all rentals are DVDs. At the Grand Avenue store, 30% of all rentals are DVDs. If a customer is selected at random, what is the approximate probability that the customer rented a DVD?

(A) 0.175
(B) 0.33
(C) 0.35
(D) 0.37
(E) 0.70

20. A car salesperson receives a commission of $300 for each car she sells. The number of cars she sells each day follows the probability distribution given below.

Cars Sold	0	1	2	3	4
Probability	0.15	0.25	0.35	0.20	0.05

Based on this distribution, what are the mean and standard deviation of the daily commission that the salesperson should receive?

(A) Mean = $525 and standard deviation = $87.46
(B) Mean = $525 and standard deviation = $326.92
(C) Mean = $525 and standard deviation = $335.41
(D) Mean = $600 and standard deviation = $335.41
(E) Mean = $600 and standard deviation = $474.34

21. An insurance company provides insurance policy coverage for automobile owners at a fixed yearly premium. If customers are a poor risk, the company is likely to pay out more in damages than it collects in premiums. However, if the customers are good risks, the company stands to make a profit, since the company will collect the premium but will likely not have to pay out in damages. The company would like to have this customer's business. Each time a customer applies for insurance the company is faced with a decision based on the following hypotheses.

H_o: The customer is a good risk.
H_a: The customer is a bad risk.

Which of the following represents a Type II error and its consequence for the company?

(A) The company decides that the customer is a bad risk, but he was, in fact, a good risk. The company misses an opportunity to make a profit.
(B) The company decides that the customer is a bad risk, but he was, in fact, a good risk. The company loses money.
(C) The company decides that the customer is a good risk, but he was, in fact, a bad risk. The company loses money.
(D) The company decides that the customer is a good risk, but he was, in fact, a bad risk. The company makes a profit.
(E) The company decides that the customer is a bad risk and he is a bad risk and the company avoids losing money.

22. A family-run convenience store/gas station in a resort town experiences a lot of variation in sales, depending whether it is during the summer vacation season or the regular year. Sales during the off-season do not vary that much, but the vacation season can be very unpredictable. The family is interested in predicting the total daily sales from the number of customers who buy gas and various other items in the store each day during the summer vacation season. A random sample of 20 days from the past two summer seasons is taken. Some of the summary statistics about the number of customers per day and daily sales (in $1000) are given below.

mean number of customers = 721.47standard deviation of customers = 182.0
mean daily sales ($1000) = 38.778 standard deviation of daily sales ($1000) = 7.442
$r^2 = 0.743$

Which one of the following would be <u>closest</u> to the predicted daily sales if there were 800 customers that day?

(A) $10,820
(B) $14,555
(C) $16,770
(D) $41,135
(E) $41,545

23. A spring-loaded launcher is used to propel a ping pong ball for a game at a school carnival. If the ball lands in a bucket, the participant wins a prize. In testing the launcher to determine where the bucket should be placed, 20 test shots are taken. The mean distance traveled is 64.5 inches with a standard deviation of 2.1 inches. A stemplot reveals no evidence of non-normality. Which one of the following is the approximate 98% confidence interval estimate for μ, the true mean distance traveled by the ping pong ball?

(A) $64.5 \pm 2.539\left(\frac{2.1}{\sqrt{19}}\right)$

(B) $64.5 \pm 2.539\left(\frac{2.1}{\sqrt{20}}\right)$

(C) $64.5 \pm 2.528\left(\sqrt{\frac{2.1}{20}}\right)$

(D) $64.5 \pm 2.528\left(\frac{2.1}{\sqrt{9}}\right)$

(E) $64.5 \pm 2.528\left(\frac{2.1}{\sqrt{20}}\right)$

24. Automobile engineers wanted to check the effect of a new automatic transmission on mileage. They selected 50 cars of the same make and model. Twenty-five of the cars used the old transmission, while the other 25 cars used the new model of transmission. A hypothesis test on the difference between the means was conducted. Which one of the following is a condition necessary to conduct this test?

    (A) The standard deviation of each group should be different.
    (B) The sampling distribution of the difference between the sample means is bimodal.
    (C) Since the data were random, no linear regression was possible.
    (D) Twenty-five cars are randomly selected with the new transmission and 25 with the old transmission.
    (E) The difference between the means was within one standard deviation.

25. The boxplots below show the reaction times (in milliseconds) for two groups of subjects who were exposed to either threatening or non-threatening stimuli.

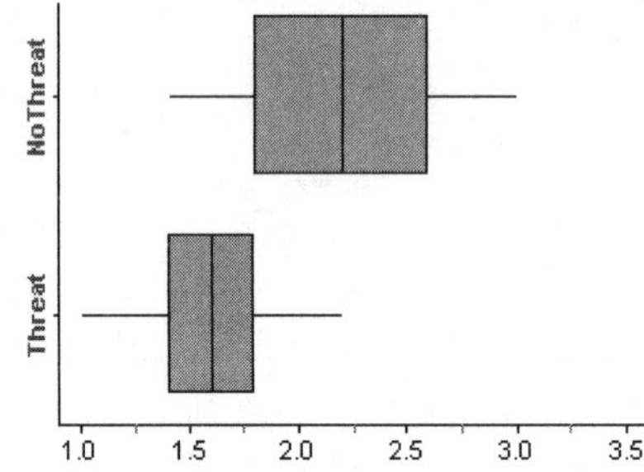

    Which of the following statements is TRUE?
    (A) The median reaction time of the group exposed to non-threatening stimuli is less than that of the group exposed to threatening stimuli.
    (B) The range of the group exposed to non-threatening stimuli is less than that of the group exposed to threatening stimuli.
    (C) The first quartile of the group exposed to non-threatening stimuli is greater than all reaction times for the group exposed to threatening stimuli.
    (D) Half of the reaction times for the group exposed to non-threatening stimuli are greater than all of the reaction times for the group exposed to threatening stimuli.
    (E) There were more subjects exposed to non-threatening stimuli than threatening stimuli.

26. Which one of the following would be a correct interpretation if you have a *z*-score of -2.0 on an exam?

    (A) It means that you missed two questions on the exam.
    (B) It means that you got twice as many questions wrong as the average student.
    (C) It means that your grade was two points lower than the mean grade on this exam.
    (D) It means that your grade was in the bottom 2% of all grades on this exam.
    (E) It means that your grade is two standard deviations below the mean for this exam.

27. At a local car dealership, a particular salesperson sells an average of 3.1 cars per week with a standard deviation of 1.1 cars. The dealership pays her $300 a week plus a commission of $250 for each car that she sells. What are the mean and standard deviation, respectively, of her total weekly pay?

(A) $1705.00, $605.00
(B) $1705.00, $275.00
(C) $1075.00, $775.00
(D) $1075.00, $302.50
(E) $1075.00, $275.00

28. A large insurance company has 95 agents within a mid-size state. The histogram below shows the amount of insurance sold (in $100,000) for the period October through December in a recent year. The highest bar, for example, indicates that 23 agents sold between $7.25 million and $7.75 million of insurance during the three-month period.

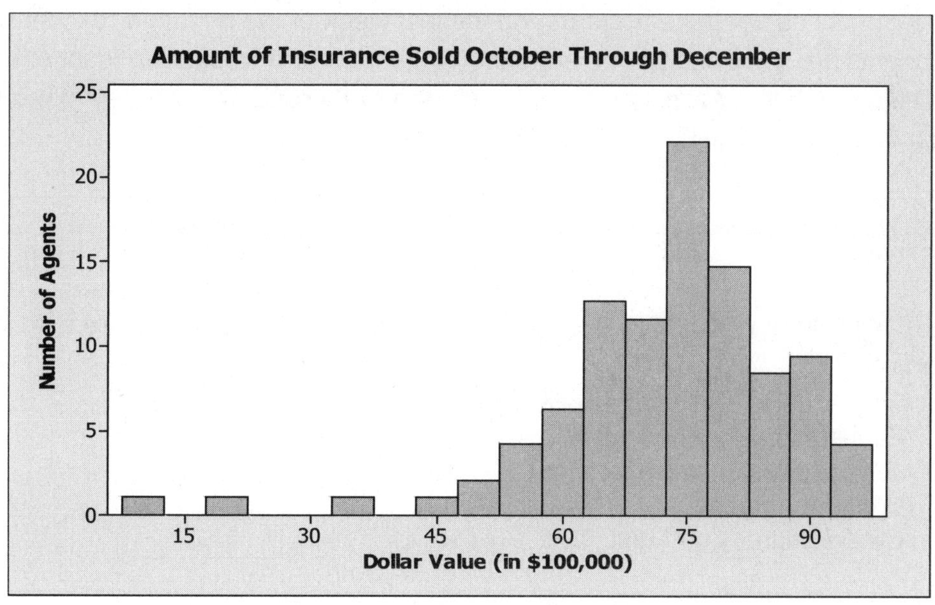

Which one of the following represents a boxplot of the same data?

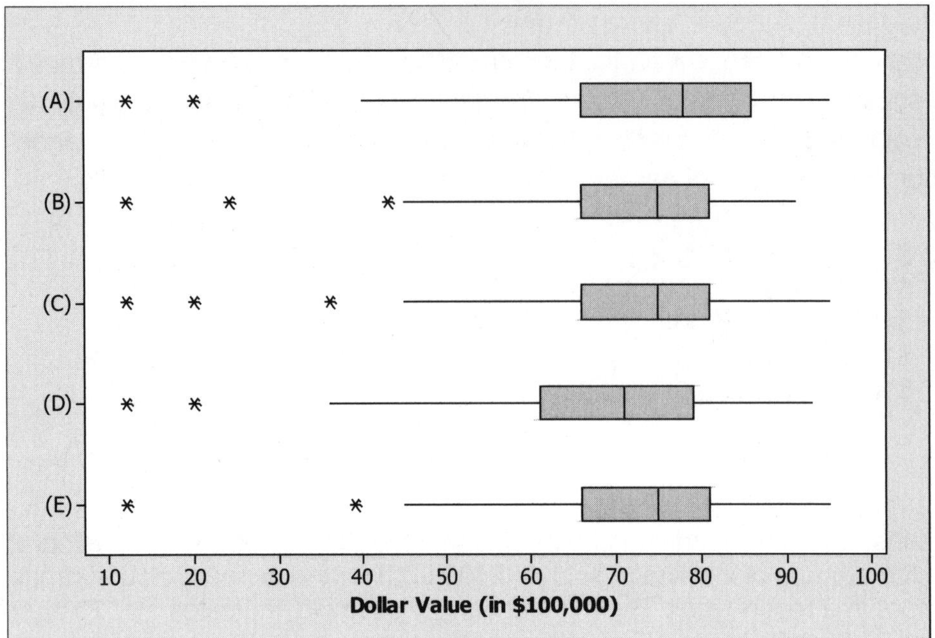

29. Because most chapter tests occur during the last week of the month, a teacher believes that students are more likely to be absent during that fourth week. The teacher collected data on the number of absences during each week of a random sampling of months over the past 5 years. The results are displayed below:

Week	One	Two	Three	Four
Absences	125	110	150	185

Which of the following would be the most appropriate inference procedure to test the teacher's claim?

(A) A chi-square test for association.
(B) A chi-square goodness-of-fit test.
(C) Multiple *z*-tests for proportions.
(D) A *t*-test for means.
(E) A linear regression *t*-test.

30. Students in an elementary school class ran a 100-yard race and timed themselves. They then practiced sprinting for the next four weeks. At the end of that month, the students again timed themselves over the same distance. The dot plots below record the differences in the times for each student before and after the practice sprinting (differences = after – before). Which of the dot plots below would indicate an improvement in time for the class?

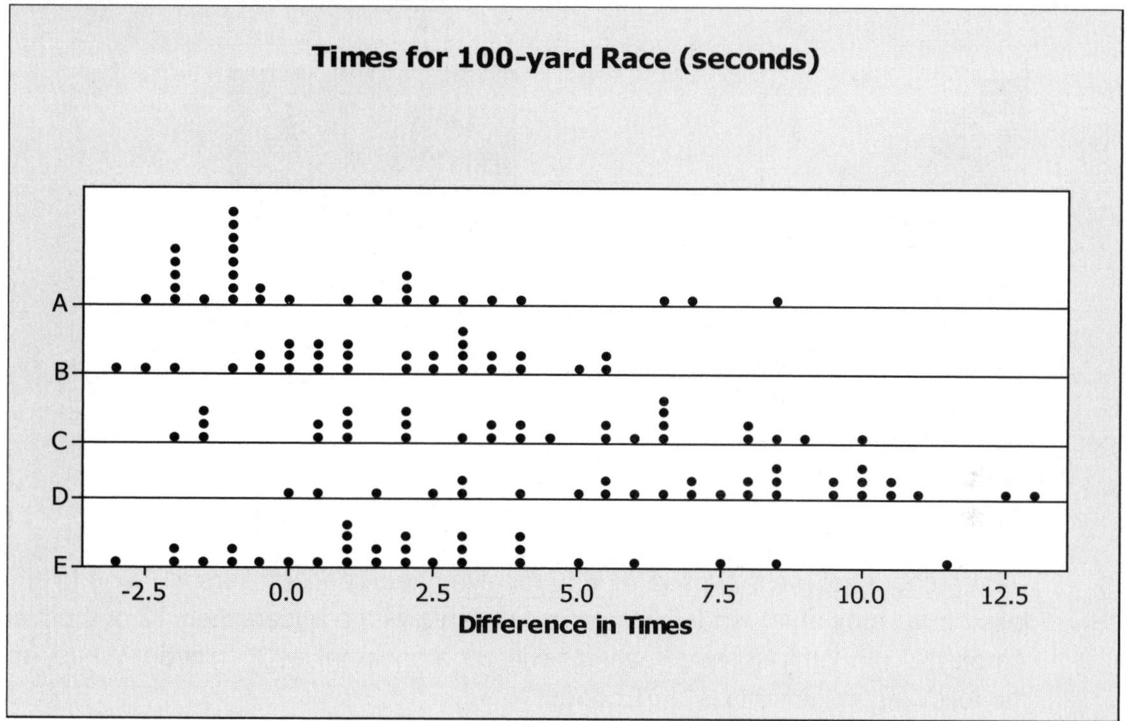

31. In which of the following distributions is the mean most likely greater than the median?

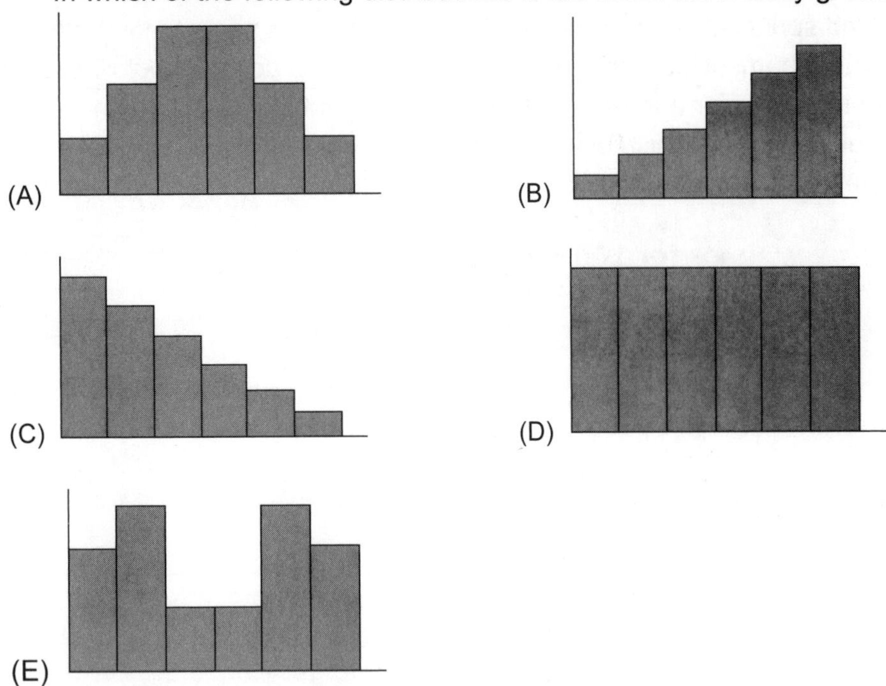

32. Researchers believe that patients who received heart pacemakers will tend to snore less. In a study of 40 randomly selected patients with a pacemaker, 12 of them snored. Among 60 randomly selected patients without a pacemaker, 25 snored. Which one of the following statements is NOT correct?

Let p_1 = proportion of patients with pacemakers who snore

Let p_2 = proportion of patients without pacemakers who snore

(A) Appropriate hypotheses for this problem would be $H_o: p_1 = p_2$ and $H_a: p_1 < p_2$.

(B) The 95% confidence interval for the difference between the proportions of those who snore in the two groups is $(-0.31, 0.07)$.

(C) Since the 95% confidence interval contains 0, there is no significant difference between the two proportions.

(D) At the 10% level, we would reject the null hypothesis and conclude that patients who received pacemakers snored less than those who did not.

(E) If a 98% confidence interval were constructed based on this data, it would also contain 0.

33. To determine the effectiveness of speed-boosting a computer by "overclocking" its processor, researchers randomly selected 16 computers of various makes and models and paired each of them with a second identical make and model. For each pair, it was randomly decided which of the two would be "overclocked" and which would be unmodified. The 95% confidence interval for the mean difference in speeds (not modified – "overclocked") was (1.3, 17.3). Assuming the conditions for inference are reasonably met, what can we conclude?

(A) The difference in mean speeds will be greater than 1.3 MHz approximately 95% of the time.
(B) The unmodified computers are anywhere from 1.3 to 17.3 MHz faster than those that were overclocked 95% of the time.
(C) Since the interval does not contain 0, there is a significant difference between the mean speeds for the two procedures (not modified – "overclocked").
(D) 95% of the differences between speeds will be between 1.3 and 17.3 MHz.
(E) A two-sample t test should have been used instead of an interval.

34. Suppose the mean rate of return on the common stocks in a large diversified investment portfolio was 12% last year. If the rates of return of the stocks within the portfolio are approximately normally distributed and a rate of return of 15% represents the 80th percentile, what is the approximate standard deviation of the rates of return in the portfolio?

(A) 1.79%
(B) 2.8%
(C) 3.57%
(D) 3.98%
(E) 4.1%

35. A newspaper wants to determine the level of support in a large town regarding the establishment of a city-wide residential recycling program. Which one of the following would represent a method of obtaining a stratified sample?

(A) Randomly select eight residential blocks in the town and ask everyone who lives on those blocks if they would be in favor of a city-wide residential recycling program.
(B) Select every fourth person who enters City Hall until the desired number of people is selected.
(C) Take an SRS of people from the town phone directory.
(D) Take a random sample of residents from each of the northwest, northeast, southwest, and southeast quadrants of the city.
(E) Number the residents of the town using the latest census data. Use a random number generator to pick the sample.

36. A biologist in the Northeast has gathered data on a random sample of 22 brown bears in New England and performed a regression analysis on the weight of the bears and their length. Some of the output is given below.

Regression Analysis: Weight versus Length

The regression equation is
Weight = - 441 + 10.3 Length

Predictor	Coef	SE Coef	T	P
Constant	-441.39	29.91	-14.76	0.000
Length	10.3382	0.4825	21.42	0.000

S = 53.7777 R-Sq = 76.5% R-Sq(adj) = 76.3%

Which of the following would represent a 95% confidence interval to estimate the true slope of the regression line relating brown bear weight to length? The biologist believes that he has a representative sample of bears and that the conditions for regression are reasonably met.

(A) $10.3382 \pm 2.086(53.7777)$

(B) $10.3382 \pm 2.080\left(\frac{53.777}{\sqrt{22}}\right)$

(C) $10.3382 \pm 21.42(0.4825)$

(D) $10.3382 \pm 2.086(0.4825)$

(E) $10.3382 \pm 2.080\,(53.7777)$

37. A simple random sample of 50 adults is surveyed to determine the true proportion of adults who visit the dentist at least once per year and a confidence interval for the proportion is constructed. Suppose the researcher had surveyed a random sample of 450 adults instead and got the same sample proportion. How would the width of the confidence interval for 450 adults compare to that of the interval for 50 adults?

(A) The width would be about one-ninth the width of the original interval.
(B) The width would be about one-third the width of the original interval.
(C) The width would be the same as the width of the original interval.
(D) The width would be about three times the width of the original interval.
(E) The width would be about nine times the width of the original interval.

38. A biologist has gathered data on a population of bears in the forests of the northeast. The distribution of the weights of the sample of bears and their sex is given below. Based on the plot, which statement below is TRUE?

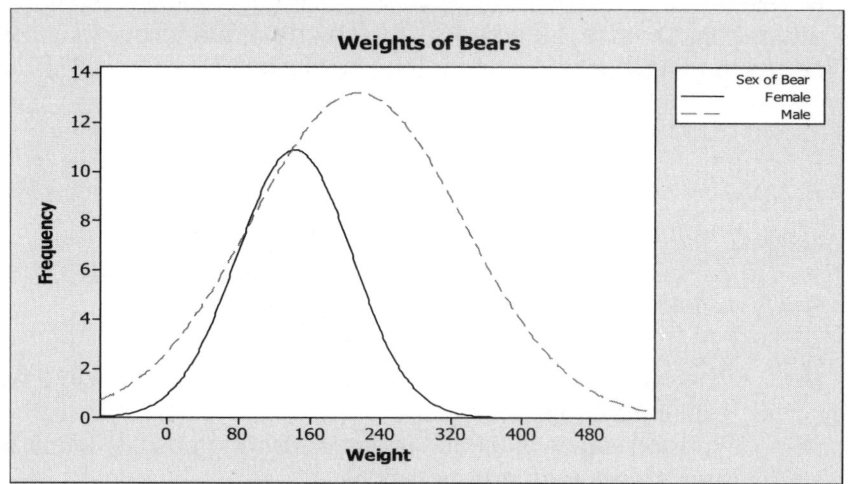

(A) Since the distributions overlap, there is not much difference between male and female bears.

(B) The female bears have a higher mean weight than the male bears and also exhibit more variability in those weights.

(C) The female bears have a higher mean weight than the male bears and also exhibit less variability in those weights.

(D) The male bears have a higher mean weight than the female bears and also exhibit more variability in those weights.

(E) The male bears have a higher mean weight than the female bears and also exhibit less variability in those weights.

39. As part of a bear population study, data were gathered on a sample of black bears in the western United States to examine the relationship between the bear's neck girth (distance around the neck) and the weight of the bear. Below is some of the output from a least-squares regression analysis examining the linear relationship between neck girth and weight of each of the bears. Which one of the following is the correct value and corresponding interpretation for the correlation?

Predictor	Coef	SE Coef	T	P
Constant	-293.53	19.27	-15.23	0.000
Neck.G	22.6447	0.8574	26.41	0.000

S = 30.1994 R-Sq = 93.6% R-Sq(adj) = 93.4%

(A) The correlation is 0.936, and 93.6% of the variation in a bear's weight can be explained by its neck girth.
(B) The correlation is 0.936. There is a strong positive linear relationship between a bear's neck girth and its weight.
(C) The correlation is 0.967, and 96.7% of the variation in a bear's weight can be explained by its neck girth.
(D) The correlation is –0.967. There is a strong negative linear relationship between a bear's neck girth and its weight.
(E) The correlation is 0.967. There is a strong positive linear relationship between a bear's neck girth and its weight.

40. Echinacea is widely used as a herbal remedy for the common cold but does it work? In a double-blind experiment, healthy volunteers agreed to be exposed to common-cold-causing rhinovirus type 39 and have their symptoms monitored. The volunteers were randomly assigned to take either a placebo or an echinacea supplement daily for 5 days following the viral exposure. Among the 103 taking a placebo, 88 developed a cold, whereas 75 of 116 subjects taking echinacea developed a cold. Which of the following represents the 95% confidence interval for the difference in the proportion of individuals developing a cold after viral exposure between the echinacea treatment and the placebo?

(A) $(0.647-0.854)\pm 1.645\sqrt{(0.744)(0.256)\left(\frac{1}{116}+\frac{1}{103}\right)}$

(B) $(0.647-0.854)\pm 1.645\sqrt{\frac{(0.647)(0.353)}{116}+\frac{(0.854)(0.146)}{103}}$

(C) $(0.647-0.854)\pm 1.960\sqrt{\left(\frac{75}{116}\right)\left(\frac{88}{103}\right)\left(\frac{1}{116}+\frac{1}{103}\right)}$

(D) $(0.647-0.854)\pm 1.960\sqrt{\frac{(0.647)(0.353)}{116}+\frac{(0.854)(0.146)}{103}}$

(E) $(0.647-0.854)\pm 1.960\sqrt{(0.744)(0.256)\left(\frac{1}{116}+\frac{1}{103}\right)}$

Practice Test 2

Free-Response Section

1. Today's baseball players are bigger, faster, and stronger than baseball players from the early days of the game. However, there is some question as to whether or not they are better hitters. The following summary statistics show the number of home runs hit in a single season by 15 randomly selected All-Stars from the early 1900s and 12 randomly selected all stars from the 2000s.

Time	Minimum	Q1	Median	Q3	Maximum	Mean	St Dev	N
1900s	22	35	46	54	60	43.93	11.25	15
2000s	9	27	39	50.5	70	37.83	18.48	12

(a) Use this information to construct side-by-side boxplots that compare the number of home runs hit during each time period.

(b) Compare the distribution of home runs for each time period.

(c) If you wanted to determine if there was a significant difference in the number of home runs hit during the two different time periods, what test of significance would you use? Explain your reasoning. Do not carry out the test.

(d) Given your choice in (c), are the conditions for the test met? Explain. Do not carry out the test.

2. Recently a group of adults who swim regularly for exercise were evaluated for depression. It turned out that these swimmers were less likely to be depressed than the general population. The researchers said the difference was statistically significant

a) Is this an experiment or an observational study? Explain.

b) News reports claimed this study proved that swimming can prevent depression. Explain why this conclusion is not justified by the study. Include an example of a possible extraneous variable that might confound the results of the study. Explain why this variable is confounding.

c) But perhaps it is true that exercise wards off depression? We wonder whether anaerobic exercise (such as weightlifting) is as effective as aerobic exercise (swimming, for example). We have 120 volunteers who are not currently engaged in a regular program of exercise. Explain how these volunteers would be randomly assigned to the two treatments: aerobic and anaerobic exercise.

d) If you were to design an experiment, explain why a control group might be useful.

3. The Rio Grande Railroad knows that adding more cars to a train increases fuel consumption but is uncertain as to how much cost should be assigned to each additional rail car on a particular route between Wyoming and Texas. The company randomly selects 12 trips between the two states and records the number of rail cars in each train and fuel consumption in units per mile. The scatterplot showed a fairly linear relationship. A regression analysis was conducted and some of the resulting computer output is given below.

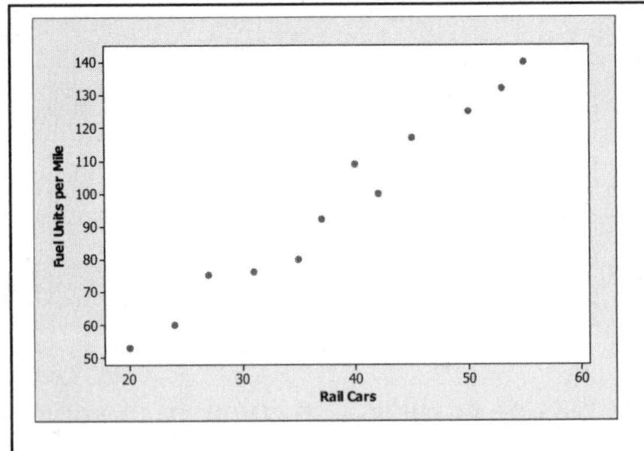

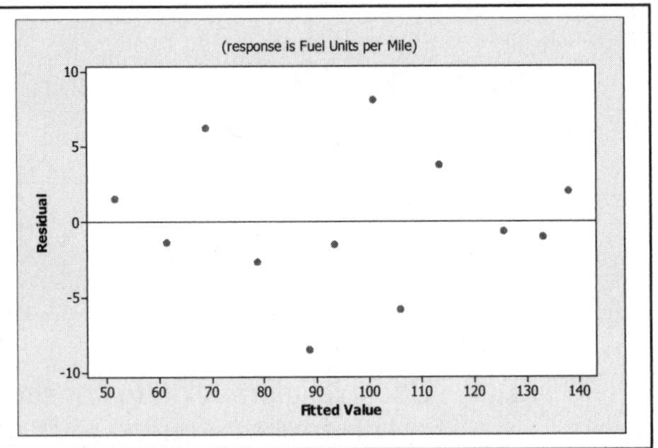

Predictor	Coef	SE Coef	T	P
Constant	2.068	5.208	0.40	0.700
Rail Cars	2.4710	0.1309	18.87	0.000
S = 1.14	R-sq = 87.3%		R-sq (adj) = 82.2%	

(a) Describe the relationship between the number of rail cars and the fuel units per mile.

(b) Is it reasonable to use a linear model to describe this data? Clearly explain your reasoning.

(c) Write the equation of the least-squares regression line relating the number of rail cars and the fuel units per mile.

(d) Identify and interpret the slope of the least-squares regression line.

(e) Interpret the value of *s* in the context of the problem.

4. The National Council on Economic Education asked more than 2200 randomly-selected teenagers and adults to take a 24-question quiz on economics. The data on gender and grades are presented in the following two-way table. Source: *What American Teens and Adults Know about Economics*, National Council on Economic Education; 8 November 2005.

Grade	Male	Female
A	39	23
B	60	71
C	146	207
D	137	177
F	548	827

Can we conclude that the grade received is dependent on gender? Support your answer with appropriate statistical evidence.

5. Radon is a radioactive gas formed by the natural radioactive decay of uranium in rock, soil and water. It occurs in all 50 sates but is more prevalent in some than in others. Problems can occur when the gas accumulates in houses, especially basements, and any exposure carries some health risk particularly lung cancer. The EPA recommends that radon mitigation be undertaken if the radon level exceeds 4pCi/L (picocuries per liter of air). A builder is looking to construct 150 new homes in a first phase of a large development but is concerned whether or not radon measures are within acceptable levels. If they are not, she will have to install radon resistant features in each new home as part of construction. She takes 22 readings at randomly selected spots within the development boundaries. These resulted in a sample mean radon level of 4.3 pCi/L and a standard deviation of 0.7. The distribution of radiation levels from the 22 samples did not exhibit much skewness and there are no outliers.

a) Can the developer conclude that the radon in this area has a mean radiation level that exceeds the EPA standard? Support your answer with appropriate statistical evidence.

b) Explain the concept of power in the context of this situation.

6. Researchers are interested in measuring the role that home environment plays in academic achievement. Since genetic differences might influence academic achievement, identical twins were used in the study. The twin sets, who were adopted at a very early age, were identified based on the fact that at the time of adoption, one of the children had been placed in a home where academics were emphasized and the other had been placed in a home where academics was not emphasized. The 12 twin sets used in the study were randomly selected from among those sets that shared this characteristic.

An academic achievement test was given to each child when they were 14 years old. The data is given below.

Set of Twins	Academic	Nonacademic	Diff = Academic - Nonacademic
1	80	73	7
2	75	70	5
3	86	88	-2
4	92	85	7
5	65	70	-5
6	94	92	2
7	65	57	8
8	83	78	5
9	98	93	5
10	52	55	-3
11	77	73	4
12	54	51	3
	$\bar{x} = 76.75$ $s = 15.15$	$\bar{x} = 73.75$ $s = 14.24$	$\bar{x} = 3$ $s = 4.22$

(a) Construct a 95% confidence interval to estimate the mean difference in academic achievement between twin pairs where one was raised in a home where academics were emphasized and the other was raised in a home where academics were not emphasized. Make sure to interpret your interval.

(b) Based on your interval, can you conclude that a significant difference exists in academic achievement exists between twin pairs where one was raised in a home where academics were emphasized and the other was raised in a home where academics were not emphasized? Justify your answer.

(c) Two data points were omitted from the original analysis as they were thought to be in error. After careful examination, it was found that the measurements were correct and should be included in the analysis. The data from the two sets of twins are given below.

Set of Twins	Academic	Nonacademic
13	78	98
14	72	84

With the addition of these data, which condition for the inference procedure you used in part (a) is no longer met? Justify your answer.

(d) Given that one of the necessary inference conditions was violated with the addition of the new data, a different test of significance will be conducted.

1. Calculate the differences for each of the 14 pairs of twins.
2. Take the absolute value of each difference.
3. List the absolute values of the differences in increasing order.

2, 2, 3, 3, 4, 5, 5, 5, 5, 7, 7, 8, 12, 20

Assign ranks (lowest ranked number = 1, next lowest ranked number = 2, etc.). Average if there are ranks that are tied. For example, there are two values of 2 in the list above. Since they hold the first and second positions (ranks in order) the average of 1 and 2 is 1.5. If three identical numbers held the 7th, 8th, and 9th positions, for example, then they would each be assigned a rank of 8—the average of 7, 8, and 9.

4. If the difference had a negative in the 4th column, assign a negative in the rank-with-sign column. For example, one of the values of 2 was originally negative. Assign a negative sign to its rank of 1.5.

Set of Twins	Academic	Nonacademic	Difference	Absolute value of differences	Rank	Rank with Sign
1	80	73	7	7		
2	75	70	5	5	7.5	7.5
3	86	88	-2	2	1.5	−1.5
4	92	85	7	7		
5	65	70	-5	5	7.5	−5
6	94	92	2	2	1.5	1.5
7	65	57	8	8		
8	83	78	5	5	7.5	7.5
9	98	93	5	5	7.5	7.5
10	52	55	-3	3	3.5	−3.5
11	77	73	4	4	5	5
12	54	51	3	3	3.5	3.5
13	78	98	-20	20		
14	72	84	-12	12		

5. Complete the two rankings columns that have been started for you.

6. Let T_{-} = sum of all the ranks which have negative signs in the last column.

   Let T_{+} = sum of all the ranks which have positive signs in the last column.

   Define the test statistic as T = the <u>smaller</u> of $|T_{-}|$ or T_{+}

   - Find the value of *T*.

(e) How unusual is your value of T? Two hundred values of T were generated, assuming a null hypothesis of no difference in the achievement scores between twin pairs where one was raised in a home where academics were emphasized and the other was raised in a home where academics were not emphasized. The results are shown in the dotplot below. [Values are graphed in bins 5 units wide, i.e., 20 represents values from 17.5 to 22.49]

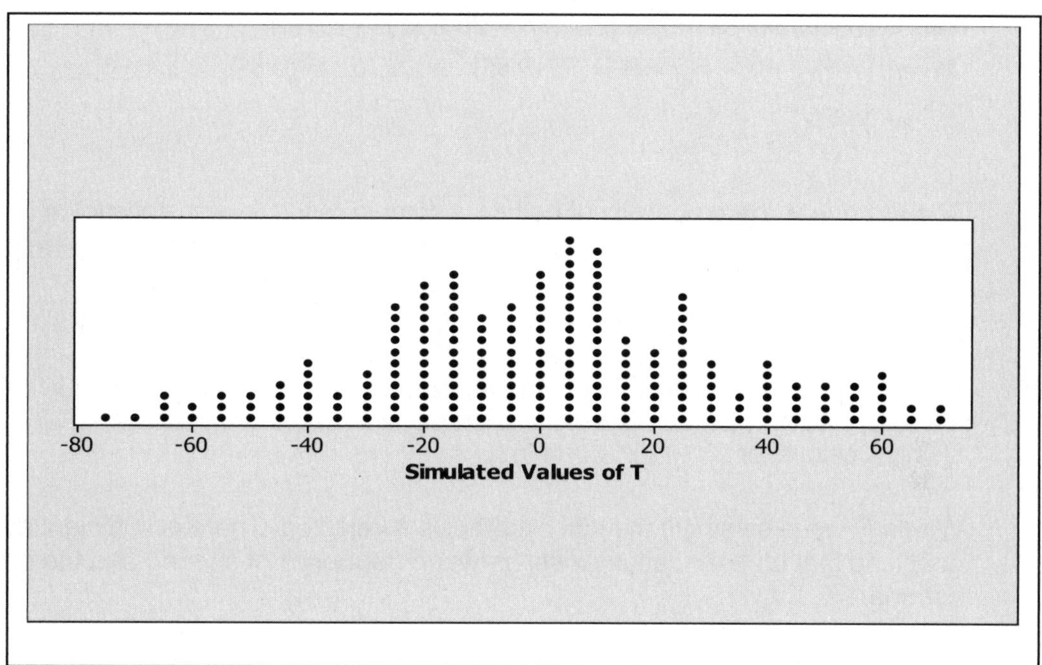

Use the value of T you calculated and the simulated values of the statistic above to determine if the observed data provide evidence of a significant difference in academic achievement between twin pairs where one was raised in a home where academics were emphasized and the other was raised in a home where academics were not emphasized. Explain your reasoning.

Answer Key for Practice Test 2

1. A census involves measuring every element in the population. It would be difficult to check every tree in a large state forest.

ANS: <u>E</u>

2. This is a binomial distribution with $n = 20$ and $p = 0.15$.

$$P(x \le 1) = P(x = 0) + P(x = 1) = {}_{20}C_0(0.85)^{20} + {}_{20}C_1(0.15)^1(0.85)^{19}$$

ANS: <u>C</u>

3. The *P*-value is the probability of getting a sample result (a test statistic) at least as extreme as the observed result you have, given that the null hypothesis is true.

ANS: <u>D</u>

4. With a chi-square value of 22.38, the *P*-value is $P(\chi^2 > 22.38) = 0.0000138$

Using a calculator $\chi^2 cdf(22.38, 1000, 2)$

With a *P*-value so small, the null hypothesis is rejected. There is sufficient evidence to conclude that an association exists between frequency of snoring and the prevalence of asthma.

ANS: <u>D</u>

5. This is a completely randomized experiment with three treatments—varieties of corn. The experimental units are the plots of land. Randomization is used to create approximately equivalent groups. It does not eliminate any variability that exists in the plots of land. Having four plots of land for each treatment allows the researcher to measure the variability of crop yield due to being planted in different types of soil.

ANS: <u>E</u>

6. Moving point A would increase the slope since point A acts as an influential point by drawing the regression line toward the point. In addition, the correlation would increase since all the points would now be closer to the regression line.

ANS: <u>C</u>

7. One of the conditions is that the sample must be a random sample from the population of interest or at least is representative of the population.
ANS: <u>E</u>

8. $$P(\text{infected} \mid \text{injects daily}) = \frac{P(\text{infected} \cap \text{injects daily})}{P(\text{injects daily})} = \frac{32/156}{77/156} = \frac{32}{77} = 0.416$$

ANS: <u>A</u>

9. You are checking for the improvement in flexibility for each gymnast. There is likely a lot of variability between gymnasts so you would want some way to reduce this variability. Therefore, a matched-pairs design is best, which means either choice (B) (where each gymnast uses both machines) or choice (E) (where each gymnast uses only one machine). Having a gymnast use both machines is not ideal, because the residual benefits of one machine might carry over to using the other machine. The researchers wouldn't know which of the two machines was the real contributor to improved flexibility. Therefore, choice (E) is the best approach.

ANS: E

10. $\hat{p} = \frac{160}{250} = 0.64$ $\qquad P\left(z > \frac{\hat{p} - p}{\sqrt{\frac{p(1-p)}{n}}}\right) = P\left(z > \frac{0.64 - 0.58}{\sqrt{\frac{(0.58)(0.42)}{250}}}\right)$

ANS: A

11. The margin of error is the extent to which the sample statistic you calculated and the unknown population parameter would most likely differ. A 4% margin of error would mean that the difference between the sample proportion and the population proportion is most likely less than 4%.

ANS: B

12. $\text{Expected(medium cost and more than 10 years)} = \frac{\text{(RowTotal)(Column Total)}}{\text{GrandTotal}} = \frac{(140)(145)}{455}$

ANS: D

13. To find the sample size for a proportion use Bound $= z^* \sqrt{\frac{p(1-p)}{n}}$

$0.03 = 2.326\sqrt{\frac{(0.5)(0.5)}{n}}, \quad n = \left(\frac{2.326\sqrt{(.5)(.5)}}{0.03}\right)^2$ = 1502.85 ANS: C

14. The question asks for the difference in the number of taps before and after the drink. This is a matched-pair situation, since there are two measurements on each subject. The appropriate test is a one-sample *t* test on paired data with μ_d = true mean difference in dexterity (*d* = before - after)

$$t = \frac{\bar{x}_d - \mu}{\frac{s_d}{\sqrt{n}}} = \frac{-4.64 - 0}{\frac{4.48}{\sqrt{10}}}$$

ANS: D

15. The mean of a sampling distribution is equal to the mean of the population. When the sample size is small ($n < 30$), the shape of a sampling distribution will still be somewhat skewed to the right but would become more normal-like as n increases.

ANS: A

16. The medians for both types of institutions are approximately the same. There are no outliers for the public institutions, but this type has a slight skew toward the larger numbers. Therefore, the mean for the public institutions will be just a bit above its median. Since the distribution of enrollment rates for the private institutions is much more skewed toward the higher numbers this will raise its mean substantially above its median, making the private mean higher than the public mean.

ANS: C

17. 22 368 46573 25595 85393 30995 89198 27982 53401
Numbers selected are 22, 36, 25, 30, and 27.
ANS: C

18. A score of 80 lies at $\frac{5+19+37+46}{150} = \frac{107}{150} = 0.713$ or about the 71st percentile.

ANS: D

19. $P(\text{DVD}) = (0.70)(0.40) + (0.30)(0.30) = 0.28 + 0.09 = 0.37$

ANS: D

20. $E(X) = 0(0.15) + 300(0.25) + 600(0.35) + 900(0.20) + 1200(0.05) = \$525.00$

$$\sigma_x = \sqrt{0(0-525)^2 + 0.25(300-525)^2 + ... + 0.05(1200-525)^2} = \sqrt{106875} = \$326.92$$

ANS: B

21. A Type II error is committed when we fail to reject a null hypothesis that is false. In this case the company fails to reject the null hypothesis (the customer is a good risk) but it is false. The company then enrolls a risky customer who costs them money.

ANS: C

22. x = number of customers and y = daily sales. Start with $\hat{y} = a + bx$.

$$b = r\left(\frac{s_y}{s_x}\right) = 0.862\left(\frac{7.442}{182}\right) = 0.0352 \text{ and } a = \bar{y} - b\bar{x} = 38.778 - (0.0352)(721.47) = 13.382$$

Therefore, the equation is $\hat{y} = 13.382 + 0.0352x$. If $x = 800$, then $\hat{y}$ =41,542.
ANS: E

23. A 98% confidence interval for a mean is given by $\mu \pm t^*\left(\frac{s}{\sqrt{n}}\right)$. For this situation df = 19 and t^* = 2.539(from the table). $C.I. = 64.5 \pm 2.539\left(\frac{2.1}{\sqrt{20}}\right)$

ANS: B

24. A condition of any experiment and the associated hypothesis test is that the data come from a randomized situation, that the sample was drawn from an approximately normal population, and that a random sample was taken.

ANS: D

25. In this problem the median of the non-threatening group is equal to the maximum of the threatening group. Therefore, all reaction times for the threatening group are less than the median for the non-threatening group.

ANS: D

26. A *z*-score is a measure of relative rank and indicates the number of standard deviations above or below the mean of a distribution that a particular numerical value is located.

ANS: E

27. Let X = number of cars sold each week and let Y = salesperson's weekly wages

$Y = 300 + 250X$. $E(Y) = 300 + 250 \cdot E(X) = 300 + 250(3.1) = \$1075$

$Var(Y) = (250)^2 \cdot Var(X) = (250)^2(1.1)^2 = 75625$

$SD(Y) = \sqrt{75625} = \$275.00$ or $SD(Y) = |250|(1.1) = \$275.00$

ANS: E

28. By looking at outliers, it is clear that the two lowest values can be used to eliminate choices (E) and (B). The lowest value is less than 15 and the next value is around 20. The median for choice (D) is too low for the histogram and the median for choice (A) is too high.

ANS: C

29. Are absences equally distributed over the four weeks? This is goodness-of-fit.

ANS: B

30. Since the difference = after – before, the after-training times should mostly be lower as students should be faster. This means that most of the differences should be negative.

ANS: A

31. A distribution which is skewed to the right will most likely have a mean greater than the median.

ANS: C

32. (A) Correct. The alternate hypothesis indicates that those with pacemakers will snore less.
    (B) Correct. Using the calculator and performing a two-proportion *z*- interval at 95%, you get $(-0.31,\ 0.07)$.
    (C) Correct. If the confidence interval contains 0, there is no statistically significant difference between the two proportions (we fail to reject the hypothesis of no difference).
    (D) Incorrect. The *P*-value for this would be 0.118. Since $P > \alpha$, we fail to reject H_o.
    (E) Correct. The 98% confidence interval would be wider and would still contain 0.

ANS: D

33. This is a matched pair interval and we are interested in whether a difference exists or not. Since 0 is not in the interval, it is not a plausible value, so we would reject the null hypothesis of no difference in favor of the alternative.

ANS: C

34. Using your calculator or your table; $z = invnorm(0.80) = 0.84$

$$z = \frac{x-\mu}{\sigma} \Rightarrow 0.84 = \frac{0.15-0.12}{\sigma} \qquad \sigma = \frac{0.03}{0.84} = 0.0357$$

ANS: C

35. A stratified sampling process would involve taking the population of interest and separating it into clearly identifiable subgroups and then drawing a random sample from each subgroup. One way to think of this is "some from all strata". Choice (A) is a cluster sample, choice (B) is a systematic sample, choices (C) and (E) are simple random samples.

ANS: D

36. The confidence interval for a slope estimate is given by $b_1 \pm t^* \times SE(b_1)$ for $\text{df} = 20$
From the computer output and the *t*-table, the 95% confidence interval is given by $10.3382 \pm 2.086(0.4825)$.

ANS: D

37. Since $\sigma_{\bar{x}} = \frac{\sigma}{\sqrt{n}}$, multiplying the sample size by 9 would change the denominator of the standard deviation of the sampling distribution of the mean by $\sqrt{9n} = 3\sqrt{n}$, i.e., divides by 3.

ANS: B

38. The distribution of weights for male bears is wider than that for female bears, which means the male weights are more highly variable. Since both distributions are relatively symmetric, their peaks are at the means. The mean for males is about 240 and is about 160 for females.

ANS: D

39. Since the neck girth of the bear increases as the bear's weight increases and $r = \sqrt{.936} = +0.967,$ there is a strong, positive, linear relationship between a bear's neck girth and its weight.

ANS: E

40. This is a two-proportion z-confidence interval where $\hat{p}_{Echinacea} = \frac{75}{116} = 0.647$ and

$\hat{p}_{Placebo} = \frac{88}{103} = 0.854$

ANS: D

Textbook Correlation Multiple-Choice Practice Exam 2

Question	Correct Answer	Textbook Section	Question	Correct Answer	Textbook Section
1	E	4.1	21	C	9.1
2	C	6.3	22	E	3.2
3	D	9.1	23	B	8.3
4	D	11.2	24	D	10.2
5	E	4.2	25	D	1.2&1.3
6	C	3.1 & 3.2	26	E	2.1
7	E	8.1	27	E	6.2
8	A	5.3	28	C	1.2
9	E	4.2	29	B	11.1
10	A	7.2	30	A	1.2 & 10.3
11	B	8.1 & 8.2	31	C	1.3
12	D	11.2	32	D	10.1
13	C	8.2	33	C	10.2
14	D	9.3	34	C	2.2
15	A	7.1 & 7.2	35	D	4.1
16	C	1.3	36	D	12.1
17	C	4.1	37	B	8.1
18	D	2.1	38	D	2.1 & 2.2
19	D	5.3	39	E	3.1
20	B	6.1	40	D	10.1

Part II Free-Response Section

1. (a)

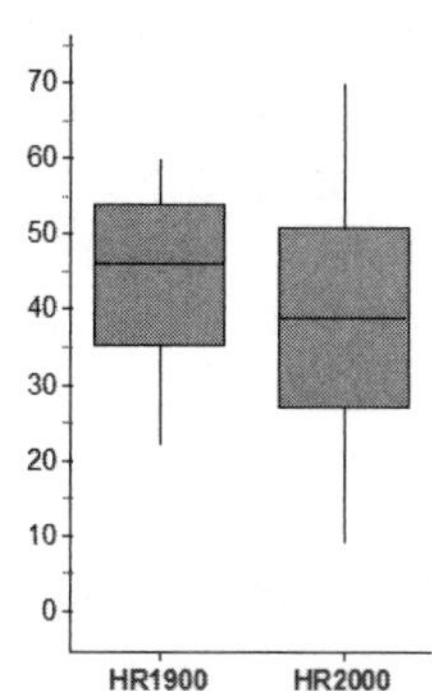

(b) The distribution of home runs hit in the 1900s is slightly skewed left, while the distribution for the 2000s is symmetric. The median number of home runs hit in the 1900s is higher than that of the 2000s, while the maximum number of home runs hit in the 2000s is much higher than that of the 1900s. The distribution of home runs hit in the 2000s has a larger range and standard deviation than the distribution for the 1900s. It is difficult to say if one era had a higher number in general since there is a great deal of overlap between the two distributions.

(c) Since we are comparing counts of home runs, we should compare the average number of home runs hit during each era. This could be accomplished by conducting a two-sample *t*-test for comparing means.

(d) Condition 1: The data come from two independent random samples or from two groups in a randomized experiment. This condition is met since the All Stars were randomly selected from each decade.

Condition 2: For each sample, the corresponding population distribution is Normal or the sample size is large. Both sample sizes are less than 30 so we cannot appeal to the large sample size argument. However, the boxplots in part (a) do not indicate strong skewness or outliers in either distribution, so we are safe to use the *t*-procedures, so condition 2 is met.

2. (a) This is an observational study. Subjects were not randomly assigned to swim or not swim and the level of depression was merely observed.

(b) It is not possible to establish cause-effect from an observational study. Two variables are confounded when their effects on the response variable cannot be distinguished from one another. Perhaps those who swim regularly also eat more healthfully, socialize more, do other exercises as well as swim, or just take better care of themselves in general. If there is less depression among this group, we wouldn't know whether to attribute this lower level of depression to the swimming or the other activity(ies) that swimmers are engaged in.

(c) Assign each subject a number from 1 through 120. Use a random number generator to generate 60 unique numbers between 1 and 120, inclusive. The subjects with these numbers will be assigned to the aerobic exercise group and the remaining 60 will be assigned to the anaerobic group.

(d) A control group would be useful in that we could get a measure of how depression changes depending on external variables not being measured, such as time of year, holidays, etc. For example, suppose we observe subjects from winter into summer. The mere fact that progressively sunnier days might alleviate some of the depressive symptoms, could coincide with both exercise groups showing a reduction in depression but the reduction isn't attributable to the exercise groups, rather, it is due to the sunnier days. A control group can show whether or not either treatment is effective.

3. (a) There is a strong, positive, linear relationship between the number of rail cars and the fuel units per mile.

(b) Yes, it is reasonable to use a linear model for this data. The scatterplot indicates that a linear relationship exists, and the residual plot has no pattern. The computer output also indicates the test of significance for slope $(H_0 : \beta = 0)$ has a *P*-value of 0, meaning that there is a significant linear relationship between number of rail cars and fuel units per mile.

(c) $\text{predicted fuel units per mile} = 2.068 + 2.471(\text{number of rail cars})$

(d) slope = 2.471. For every extra rail car added to the train, the predicted fuel units per mile increases by 2.471.

(e) S = 1.14. On average, the difference between the predicted fuel units per mile and the actual fuel units per mile is 1.14.

4. We will use the chi-square test of independence.

H_o : Grade received on the quiz and sex of test-taker are independent

H_a: Grade received on the quiz and sex of test-taker are not independent

$$\chi^2 = \sum \frac{(\text{obs - exp})^2}{\text{exp}}$$

- Random: Teenagers and adults were randomly selected
- Independent: There are more than 2235(10) = 22,350 American teens and adults.
- Large sample size: All expected counts are greater than 5

$$\begin{bmatrix} 25.8 & 36.2 \\ 54.5 & 76.5 \\ 146.9 & 206.1 \\ 130.7 & 183.3 \\ 572.1 & 802.9 \end{bmatrix}$$

$$\chi^2 = \sum \frac{(\text{obs - exp})^2}{\text{exp}} = \frac{(39-25.8)^2}{25.8} + \frac{(23-36.2)^2}{36.2} + \ldots. + \frac{(827-802.9)^2}{802.9} = 14.798$$

P-value $= 0.0051$ and $df = 4$.

Since 0.0051 < 0.05, we reject the null hypothesis. There is sufficient evidence to conclude that sex of test-taker and grade on the economics quiz are not independent.

5. (a) μ = mean maximum allowable radon level from all sites in the development

$$H_0 : \mu = 4 \quad versus \quad H_a : \mu > 4$$

$$t = \frac{\bar{x} - \mu}{\frac{s}{\sqrt{n}}}$$

i) Data comes from a random sample: Housing sites were randomly selected.
ii) Also, there were likely to be more than 22(10) = 220 house sites
iii) Normal/Large Sample: The problem states that there is not much skewness and no outliers

$$t = \frac{4.3 - 4}{\frac{0.7}{\sqrt{22}}} = 2.01$$

P-value = P ($t > 2.01$) = 0.0287

Let $\alpha = 0.05$. Since 0.0287 < 0.05, we reject the null hypothesis. There is sufficient evidence to conclude that the mean maximum radon level for the building sites in this development significantly exceeds the EPA standard. The developer will have to install radon resistant features in all the new houses she is building.

(b) The power of a test is the probability of rejecting the null hypothesis when the null hypothesis is false. In this situation, the developer finds evidence that the radon readings exceed the EPA standard and they do exceed that standard.

6. (a) Let μ_d = mean difference in achievement scores between twin pairs where one was raised in a home where academics were emphasized and the other was raised in a home where academics were not emphasized (emphasized – not emphasized).

$$\bar{x}_d \pm t^* \frac{s_d}{\sqrt{n}}$$

- Random: We believe that the twin pairs in this study were randomly selected from among all twin sets who met the criteria for being included in the study stated in the problem.
- Independent: We believe that the sets of twins are independent of each other. Also, the number of sets of twins in the population who met the criteria is greater than 10 (12) = 120.

- Normal:

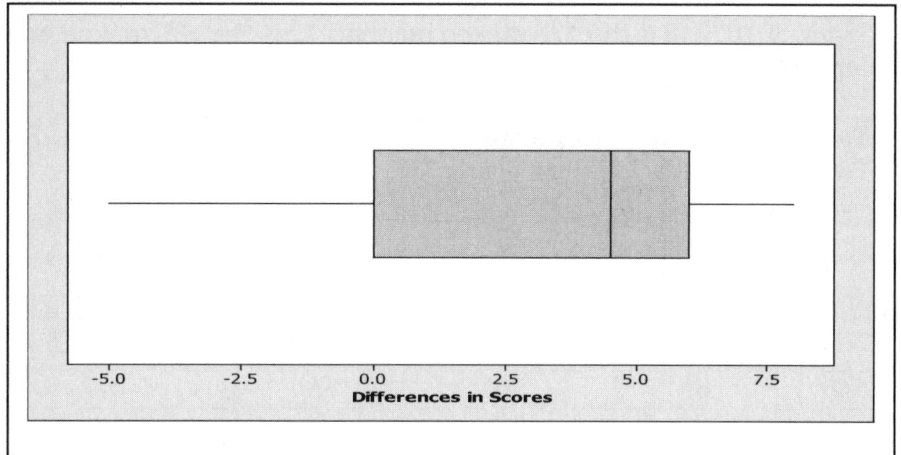

The boxplot shows some skewness but no outliers so we can reasonably believe that the differences come from an approximately Normal distribution.

$$\bar{x}_d \pm t^* \frac{s_d}{\sqrt{n}} = 3 \pm (2.201)\left(\frac{4.22}{\sqrt{12}}\right) = 3 \pm 2.68 \Rightarrow (0.32, 5.68) \text{ with } df = 11.$$

We are 95% confident that the mean difference in achievement scores between twin pairs where one was raised in a home where academics were emphasized and the other was raised in a home where academics were not emphasized lies between 0.32 and 5.68.

(b) Since 0 does not lie in the 95% confidence interval calculated in part (a), there appears to be a significant difference between twin pairs where one was raised in a home where academics were emphasized and the other was raised in a home where academics were not emphasized.

(c) Based on the boxplot below, the condition of Normality appears to be violated as there is an outlier among the differences in scores and the distribution of the differences is highly skewed to the left.

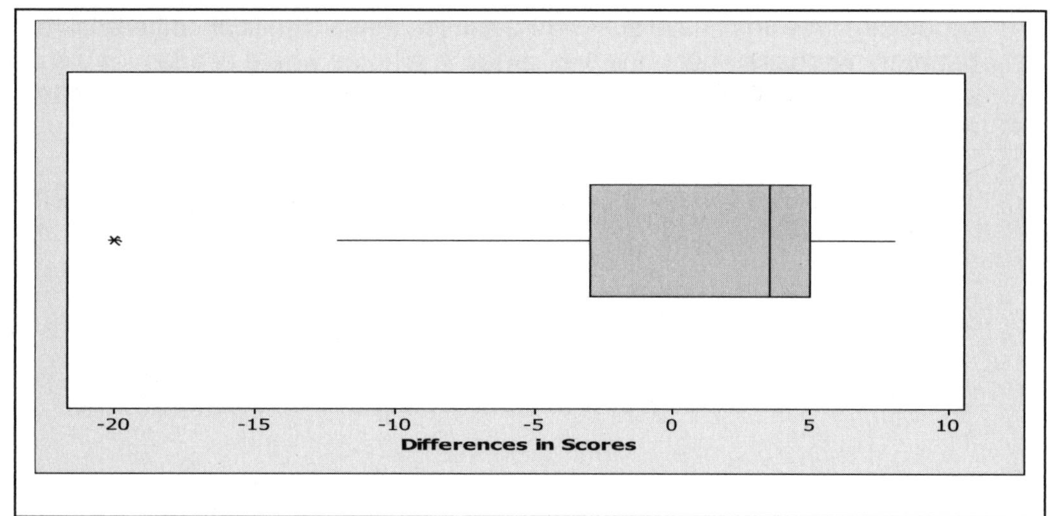

(d)

The two values of seven rank as the 10th and 11th values so get assigned an average rank of 10.5; 8 is the 12th ranked number; 12 is the 13th ranked number; and 20 is the 14th ranked number.

Set of Twins	Academic	Nonacademic	Difference	Absolute value of differences	Rank	Rank with Sign
1	80	73	7	7	10.5	10.5
2	75	70	5	5	7.5	7.5
3	86	88	-2	2	1.5	−1.5
4	92	85	7	7	10.5	10.5
5	65	70	-5	5	7.5	−7.5
6	94	92	2	2	1.5	1.5
7	65	57	8	8	12	12
8	83	78	5	5	7.5	7.5
9	98	93	5	5	7.5	7.5
10	52	55	-3	3	3.5	−3.5
11	77	73	4	4	5	5
12	54	51	3	3	3.5	3.5
13	78	98	-20	20	14	−14
14	72	84	-12	12	13	−13

$$T_- = (-1.5)+(-3.5)+(-7.5)+(-13)+(-14) = -39.5$$
$$T_+ = 1.5+3.5+5+7.5+7.5+7.5+7.5+10.5+10.5+12 = 65.5$$
$$T = 39.5$$

(e) $P(T \geq 39.5) \approx \dfrac{27}{200} = 0.135$. Since the probability of getting a test statistic as large or larger than 39.5 occurs roughly 135 out of every 1000 times, just by chance when the null hypothesis is true, which is greater than 0.05, we are unable to reject the null hypothesis. We don't have sufficient evidence that a significant difference exists between twin pairs where one was raised in a home where academics were emphasized and the other was raised in a home where academics were not emphasized.

The addition of the two extra data points was enough to change our conclusion from part (b).